P9-DGM-669

A BRIEF TOUR OF HUMAN CONSCIOUSNESS

From Impostor Poodles to Purple Numbers

V. S. Ramachandran, M. D.

PI PRESS
NEW YORK

An Imprint of Pearson Education, Inc.
1185 Avenue of the Americas, New York, New York 10011

Pi Press offers discounts for bulk purchases. For information contact U.S. Corporate and Government Sales, 1-800-382-3419, or corpsales@pearsontechgroup.com. For sales outside the U.S.A., please contact: International Sales, 1-317-581-3793, or international@pearsontechgroup.com.

Printed in the United States of America
First Printing

The BBC in association with Profile Books Ltd.
First published in Great Britain in 2003 by
Profile Books Ltd.
58a Hatton Garden
London Ec1N 8lX

Library of Congress Cataloging-in-Publication Data

A CIP catalog record for this book can be obtained from the Library of Congress.
Pi Press books are listed at www.pipress.net

ISBN 0-13-148686-1

Pearson Education Ltd.
Pearson Education Australia Pty., Limited
Pearson Education Singapore, Pte. Ltd.
Pearson Education North Asia Ltd.
Pearson Education Canada, Ltd.
Pearson Educación de Mexico, S.A. de C.V.
Pearson Education — Japan
Pearson Education Malaysia, Pte. Ltd.

For my parents, Vilayanur Subramanian and Vilayanur Meenakshi

For Diane, Mani and Jaya

For Semmangudi Sreenivasa Iyer

For President Abdul Kalam, for launching the youth of our country into the new millennium

For Shiva Dakshinamurthy, Lord of Gnosis, music, knowledge and wisdom

Contents

Preface

My goal in writing this book has been to make neuroscience—the study of the brain—more accessible to a broad audience, to "workingmen," as Thomas Huxley would have said. The overall strategy is to investigate neurological dysfunction caused by a change in a small part of a patient's brain and ask: Why does this patient display these curious symptoms? What do the symptoms tell us about the workings of the normal brain? Can a careful study of these patients help us explain how the activity of a hundred billion nerve cells in the brain gives rise to all the richness of our conscious experience? I have chosen to focus both on areas in which I have worked directly (such as phantom limbs, synesthesia and visual processing) and ones that have a broad interdisciplinary appeal, in order, ultimately, to bridge the gap that now separates C. P. Snow's "two cultures"—the sciences and the humanities.

The book emerged from the annual BBC Reith lectures that I delivered in Great Britain in 2003. It was an honor for me to be invited to give these lectures, the first physician/experimental psychologist to do so since they were begun by Bertrand Russell in 1948. In the last five decades these lectures have enjoyed a distinguished place in the intellectual and cultural life of the Western world. I was delighted to accept the invitation, knowing that

I would be joining a long list of previous lecturers whose works inspired me as a teenager: Peter Medawar, Arnold Toynbee, J. Robert Oppenheimer, John Galbraith and Russell, to mention only a few. I realized that theirs would be a tough act to follow, given their towering stature and the pivotal role that many of them played in defining the intellectual ethos of our age. Even more daunting was the requirement that I would have to make the lectures not only interesting to the specialist but also intelligible to the "common people," thereby fulfilling Lord Reith's original mission for the BBC. Given the enormous amount of research on the brain, the best I could do was to provide an impressionistic survey rather than try to be comprehensive. In doing this I was worried I might have oversimplified many of the issues involved and so run the risk of annoying some of my specialist colleagues. But as Lord Reith himself once said, "There are some people whom it is one's duty to annoy!"

Chapter 3 (based on the third lecture) deals with an especially controversial subject, the neurology of artistic experience, "neuro-aesthetics," that is usually considered off limits by scientists. I take a stab at it just for fun and to indicate how a neuroscientist might approach this problem. I make no apology for the fact that it is speculative. As Peter Medawar said, "all good science begins as an imaginative excursion into what might be true." Speculation is fine, provided it leads to testable predictions and so long as the author makes it clear when he is merely speculating—skating on thin ice—as opposed to when he's on solid ground. I have taken pains to preserve this distinction throughout the book, often adding qualifying remarks in what have become extensive endnotes.

There is also a tension in the field of neurology between the "single case study" approach, the intensive study of just one or two patients with a syndrome, and sifting through a large number of patients and doing a statistical analysis. The criticism is sometimes made that it is easy to be misled by single strange cases, but this is nonsense. Most of the syndromes in neurology that have stood the test of time—for example, the major aphasias (language disturbances), amnesia (explored by Brenda Milner, Elizabeth Warrington, Larry Squire and Larry Weiskrantz), cortical color blindness, neglect, blindsight, "split brain" syndrome (commissurotomy), etc.—were initially discovered by a careful study of single cases, and I don't know of even one that was discovered by averaging results from a large sample. The best strategy, in fact, is to begin by studying individual cases and then to make sure that the observations are reliably repeatable in other patients. This is true for a majority of findings described here, such as phantom limbs, the Capgras (impostor) delusion, synesthesia and neglect. The findings are remarkably consistent across patients and have been confirmed in several laboratories.

Many colleagues and students often ask me when I became interested in the brain and why. It is not easy to trace the lineage of one's interests, but I'll give it a shot. I have been interested in science from about the age of eleven. I remember being a somewhat lonely child and socially awkward—although I did have one very good science playmate in Bangkok: Somthau ("Cookie") Sucharitkul. I always felt companioned by Nature and perhaps science was my "retreat" from the social world with all its arbitrariness and mind-numbing conventions. I spent a lot of time collecting seashells and geological specimens and fossils. I enjoyed dabbling in ancient archaeology, cryptography (the Indus script), comparative anatomy and palaeontology; I found it endlessly fascinating that the tiny bones inside our ears, which we mammals use for amplifying sounds, had originally evolved from the jawbones of reptiles. As a schoolboy I was passionate about chemistry and often mixed chemicals just to see what would happen (a burning piece of magnesium ribbon could be plunged into water—it would continue to burn underwater by extracting the oxygen from H_2O). Another passion was biology. I once tried placing various sugars, fatty acids and individual amino acids inside the "mouths" of Venus flytraps to see what triggered them to stay shut and secrete digestive enzymes. And I did an experiment to see if ants would hoard and consume saccharin—showing the same fondness for it as they do for sugar. Would the saccharin molecule "fool" their taste buds the same way it fools ours?

All these pursuits, Victorian in inspiration, are quite remote from the neurology and psychophysics I specialize in now. Yet those childhood preoccupations must have left a mark that profoundly influences my adult personality and my style of doing science. While engaged in such pursuits, I felt that I was in my own private playground, my own parallel universe inhabited by Darwin and Cuvier and Huxley and Owen and William Jones and Champollion. For me these people were more alive than most "real" people I knew. Perhaps this escape into my own private world made me feel special rather than isolated, "weird" or different. It allowed me to rise above the tedium and monotony—the humdrum existence that most people call a normal life—to a place where, to quote Russell, "one at least of our nobler impulses can escape from the dreary exile of the actual world".

Such an escape is especially encouraged at the University of California, San Diego, an institution that is both venerable and vibrantly modern. Its neuroscience program was recently ranked number one in the country by the National Academy of Sciences. If you include the Salk Institute and Gerry Edelman's Neurosciences Institute, there is a higher concentration of eminent neuroscientists in La Jolla's "neuron valley" than anywhere else in the world. I can't think of a more stimulating environment for anyone interested in the brain.

Science is most fun when it is still in its infancy, when its practitioners are still driven by curiosity and it hasn't become "professionalized" into just another nine-to-five job. Unfortunately this is no longer true for many of the most successful areas of science, such as particle physics or molecular biology. It is now commonplace to see a paper in *Science* or *Nature* cowritten by thirty authors. For me this takes some of the joy out of it (and I imagine it does for the authors too). This is one of two reasons I instinctively gravitate toward old-fashioned Geschwindian neurology, where it is still possible to ask naive questions starting from first principles—the kinds of very simple questions that a schoolboy might ask but are embarrassingly hard for experts to answer. It's a field where it's still possible to do Faraday-style research and come up with surprising answers. Indeed, many of my colleagues and I see it as an opportunity to revive the golden age of neurology, the age of Charcot, Hughling Jackson, Henry Head, Luria and Goldstein.

The second reason I chose neurology is more obvious; it's the same reason you picked up this book. As human beings we are more curious about ourselves than about anything else, and this is a research enterprise that takes you right into the heart of the problem of who we are. I got hooked on neurology after examining my very first patient in medical school. He was a man with a pseudo-bulbar palsy (a kind of stroke), who alternately laughed and wept uncontrollably every few seconds. It struck me as an instant replay of the human condition. Were these just mirthless joy and crocodile tears, I wondered? Or was he actually feeling alternately happy and sad, the same way a manic-depressive might, but on a compressed timescale?

We will be considering many such questions throughout the book: What causes phantom limbs? How do we construct a body image? Are there artistic universals? What is a metaphor? Why do some people see musical notes as colored? What is hysteria? Some of these questions I will answer, but the answers to the others remain tantalizingly elusive, such as the big question: What is consciousness? But whether I answer them or not, if this book at least whets your appetite to learn more about this fascinating field, it will have served its purpose. Endnotes, a glossary, a bibliography, and an index are provided at the end for the benefit of those who wish to probe these topics more deeply. As my colleague Oliver Sacks said of one of his books: "The real book is in the endnotes, Rama."

I would like to dedicate this work to the patients who volunteered to endure hours of testing at our center. I have often learned more from my conversations with them, despite their damaged brains, than from my learned colleagues.

I

A Pain in the Brain

The history of mankind in the last three hundred years has been punctuated by major upheavals in human thought that we call scientific revolutions — upheavals that have profoundly affected the way in which we view ourselves and our place in the cosmos. First there was the Copernican revolution — the notion that far from being the center of the universe, our planet is a mere speck of dust revolving around the sun. Then there was the Darwinian revolution, culminating in the view that we are not angels but merely hairless apes, as Thomas Henry Huxley once pointed out. And, third, there was Freud's discovery of the "unconscious" — the idea that even though we claim to be in charge of our destinies, most of our behavior is governed by a cauldron of motives and emotions of which we are barely conscious. Your conscious life, in short, is nothing but an elaborate post-hoc rationalization of things you really do for other reasons.

But now we are poised for the greatest revolution of all — understanding the human brain. This will surely be a turning point in the history of the human species for, unlike those earlier revolutions in science, this one is not about the outside world, not about cosmology or biology or physics, but about ourselves, about the very organ that made those earlier revolutions possible. And I want to emphasize that these insights into the human brain will have a profound impact not just on scientists but also on the humanities, and indeed they may even help us bridge what C. P. Snow called the two cultures — science on the one hand and arts, philosophy and humanities on the other. Given the enormous amount of research on the brain, all I can do here is to provide a very impressionistic survey rather than try to be comprehensive. The five chapters cover a very wide spectrum of topics, but two recurring themes run through all of them. The first broad theme is that by studying neurological syndromes which have been largely ignored as curiosities or mere anomalies we can sometimes acquire novel insights into the functions of the normal brain — how the normal brain works. The second theme is that many of the functions of the brain are best understood from an evolutionary vantage point.

The human brain, it has been said, is the most complexly organized structure in the universe and to appreciate this you just have to look at some numbers. The brain is made up of one hundred billion nerve cells or "neurons" which form the basic structural and functional units of the nervous system (Figure 1.1). Each neuron makes something like one thousand to ten thousand contacts with other neurons and these points of contact are called synapses. It is here that exchange of information occurs.

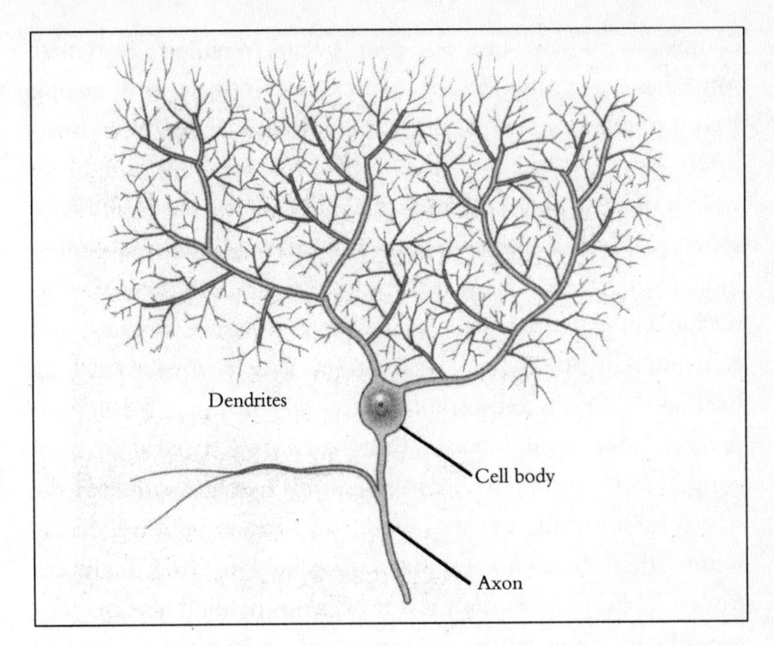

Figure 1.1 *Drawing of a neuron showing dendrites which receive information from other neurons and a single long axon that sends information out to other neurons.*

Based on this information, it has been calculated that the number of possible permutations and combinations of brain activity, in other words the number of brain states, exceeds the number of elementary particles in the known universe. Even though it is common knowledge, it never ceases to amaze me that all the richness of our mental life – all our feelings, our emotions, our thoughts, our ambitions, our love lives, our religious sentiments and even what each of us regards as his or her own intimate private self – is simply the activity of these little specks of jelly in our heads, in our brains. There is nothing else. Given this staggering complexity, where does one even begin? Well, let's start with some basic anatomy. In the twenty-first century most people have a rough idea of what the brain looks like. It has two mirror-image halves, called the cerebral hemispheres, and resembles a walnut sitting on top of a stalk, called the brain stem. Each hemisphere is divided into four lobes: the frontal lobe, the parietal lobe, the occipital lobe and the temporal lobe (Figure 1.2). The occipital lobe in the back is concerned with vision. Damage to it can result in blindness. The temporal lobe is concerned with hearing, emotions and certain aspects of visual perception. The parietal lobes of the brain – at the sides of the head – are concerned with creating a three-dimensional representation of the spatial layout of the external world, and also of your own body within that three-dimensional representation. And lastly the frontal lobes, perhaps the most mysterious of all. They are concerned with some very enigmatic aspects of the human mind and human behavior such as your moral sense, your wisdom, your ambition and other activities of the mind which we know very little about.

There are several ways of studying the brain, but my approach is to look at people who have had some sort of damage or change to a small part of their brain. Interestingly, people who have had a small lesion in a specific part of the brain do not suffer an across-the-board reduction in all their cognitive capacities; no blunting of their mind. Instead there is often a highly selective loss of one specific function while other functions are preserved intact, which is a good indication that the affected part of the brain is somehow involved in mediating the impaired function. I could cite many examples, but here are some of my favorites.

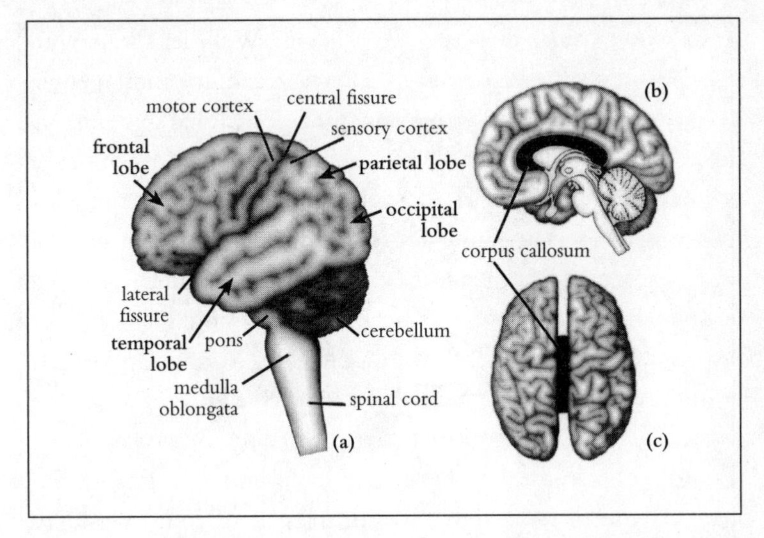

Figure 1.2 *Gross anatomy of the human brain. (a) Shows the left side of the left hemisphere. Notice the four lobes: frontal, parietal, temporal and occipital. The frontal is separated from the parietal by the central or rolandic sulcus (furrow or fissure), and the temporal from the parietal by the lateral or sylvian fissure. (b) Shows the inner surface of the left hemisphere. Notice the conspicuous corpus callosum (black) and the thalamus (white) in the middle. The corpus callosum bridges the two hemispheres. (c) Shows the two hemispheres of the brain viewed from the top. (a) Ramachandran; (b) and (c) redrawn from Zeki, 1993.*

First, prosopognosia, or face blindness. When a structure called the fusiform gyrus in the temporal lobes is damaged on both sides of the brain, the patient can no longer recognize people's faces (Figure 1.3). The patient can still read a book, so is not blind, and is not psychotic or mentally disturbed in any way but is simply no longer able to recognize people by just looking at the face.

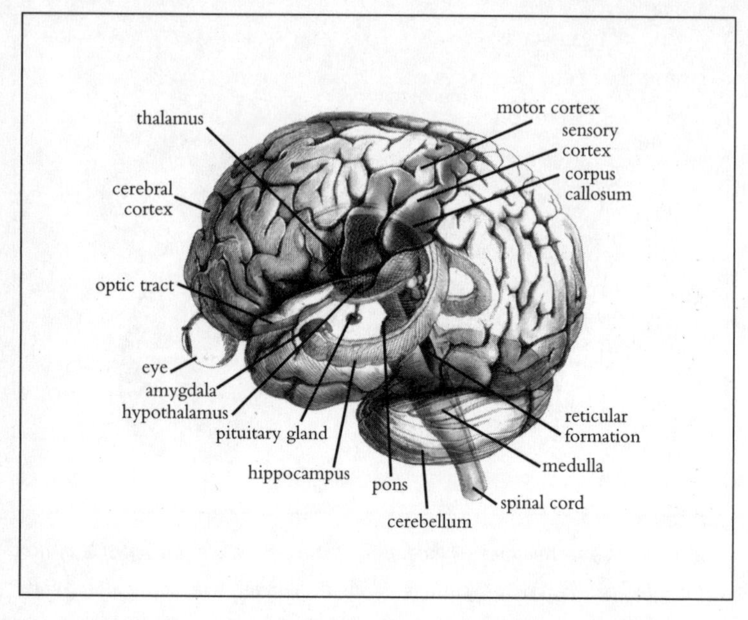

Figure 1.3 *Artist's rendering of a brain with the outer convoluted cortex rendered partially transparent to allow inner structures to be seen. The thalamus (dark) can be seen in the middle, and interposed between it and the cortex are clusters of cells called the basal ganglia (not shown). Embedded in the front part of the temporal lobe you can also see the hippocampus (concerned with memory). In addition to the amygdala, other parts of the limbic system such as the hypothalamus can be seen. The limbic structures mediate emotional arousal. The hemispheres are attached to the spinal cord by the brain stem (consisting of medulla, pons and midbrain), and below the occipital lobes is the cerebellum, concerned mainly with coordination of movements and timing. The fusiform gyrus – concerned with processing faces – is in the inner side of the temporal lobe in the bottom. The amygdala – which receives signals from the fusiform – can be seen clearly in the diagram. From* Brain, Mind and Behaviour *by Bloom and Laserson (1988) by Educational Broadcasting Corporation. Used with permission from W. H. Freeman and Company.*

Prosopognosia is very well known but there is another syndrome that is quite rare – the Capgras syndrome. A patient I saw not long ago had been in a car accident, sustaining a head injury, and was in a coma. He came out of the coma after a couple of weeks and was quite intact neurologically when I examined him. But he had one profound delusion – he would look at his mother and say, "Doctor, this woman looks exactly like my mother but she isn't, she is an impostor." Why would this happen? Bear in mind that this patient, who I will call David, is completely intact in other respects. He is intelligent, alert, fluent in conversation and not emotionally disturbed in any other way.

The standard explanation for the Capgras delusion—found in older psychiatry textbooks—is a Freudian one. According to this view, during infancy all of us men have a strong sexual attraction toward our mothers—the so-called Oedipus complex. But as we grow up the cortex becomes more highly developed and inhibits or "represses" these latent urges originating in the limbic emotional core of the brain. Thank God for that, otherwise we would all be permanently sexually aroused by our mothers! But then, along comes a blow to the head and these repressed sexual urges come flaming to the surface, so that suddenly the patient finds himself sexually aroused by his mother. The only way he can rationalize these forbidden feelings is to say, "If it's mom . . . why am I sexually turned on? It must be an impostor." Hence the delusion. This is an ingenious explanation—as indeed all Freudian explanations are—but it doesn't work. I have seen a patient having the same delusion not only about his mother but also about his pet poodle! "This isn't Fifi, doctor . . . it's some other dog that looks like Fifi." Clearly the Freudian explanation cannot explain this without invoking something absurd like "the latent bestiality of all humans," which would be too far-fetched even by the notoriously lax intellectual standards of Freudian psychology.

So, what really causes Capgras syndrome? To understand this disorder, you have to first realize that vision is not a simple process. When you open your eyes in the morning, it's all out there in front of you and so it's easy to assume that vision is effortless and instantaneous. But in fact within each eyeball, all you have is a tiny distorted upside-down image of the world. This excites the photoreceptors in the retina and the messages then go through the optic nerve to the back of your brain, where they are analyzed in thirty different visual areas. Only after that do you begin to finally identify what you're looking at. Is it your mother? Is it a snake? Is it a pig? And that process of identification takes place partly in a small brain region called the fusiform gyrus – the region which is damaged in patients with face blindness or prosopognosia. Finally, once the image is recognized, the message is relayed to a structure called the amygdala, sometimes called the gateway to the limbic system, the emotional core of your brain, which allows you to gauge the emotional significance of what you are looking at. Is this a predator? Is it prey which I can chase? Is it a potential mate? Or is it my departmental chairman I have to worry about, a stranger who is not important to me, or something utterly trivial like a piece of driftwood? What is it?

In David's case, perhaps the fusiform gyrus and all the visual areas are completely normal, so his brain tells him that the woman he sees looks like his mother. But, to put it crudely, the "wire" that goes from the visual centers to the amygdala, to the emotional centers, is cut by the accident. So he looks at his mother and thinks, "She looks just like my mother, but if it's my mother why don't I feel anything toward her? No, this can't possibly be my mother, it's some stranger pretending to be my mother." This is the only interpretation that makes sense to David's brain, given the peculiar disconnection.

How can an outlandish idea like this be tested? My student Bill Hirstein and I in La Jolla, and Haydn Ellis and Andrew Young in England, did some very simple experiments measuring galvanic skin response (see chapter 5).[1] We found – sure enough – that in David's brain there was a disconnection between vision and emotion as predicted by our theory. Even more amazing is that when David's mother phones him he instantly recognizes her from her voice. There is no delusion. Yet if an hour later his mother were to walk into the room he would tell her that she looked just like his mother but was an impostor. The reason for this anomaly is that a separate pathway leads from the auditory cortex in the superior temporal gyrus to the amygdala, and that pathway perhaps was not cut by the accident. So auditory cognizance remains intact while visual cognizance has disappeared. This is a lovely example of the sort of thing we do: of cognitive neuroscience in action; of how you can take a bizarre, seemingly incomprehensible neurological syndrome – a patient claiming that his mother is an impostor – and then come up with a simple explanation in terms of the known neural pathways in the brain.

Our emotional response to visual images is obviously vital to our survival, but the existence of connections between visual brain centers and the limbic system or emotional core of the brain also raises another interesting question: What is art? How does the brain respond to beauty? Given that these connections are between vision and emotion, and that art involves an aesthetic emotional response to visual images, surely these connections must be involved, and this forms the subject of a later chapter.

Are these intricate connections in the brain laid down by the genome in the fetus, or are they acquired in early infancy as we begin to interact with the world? This is the so-called nature/nurture debate, and is central to my next example: phantom limbs. Most people know what is meant by a phantom limb. A patient has an arm amputated because it has a malignant tumor or has been irreparably damaged in an accident but continues to feel the presence of the amputated arm. A famous example concerns Lord Nelson, who vividly felt a phantom arm long after his real one had been lost in battle. (He actually used it in a somewhat flawed argument for the existence of a non-corporeal soul. For if an arm can survive physical annihilation, he asked, why not the whole body?)

I once had a patient whose arm had been amputated above the left elbow. He sat in my office blindfolded while I gently touched different areas of his body and asked him to say where I was touching him. All went as expected until I touched his left cheek, at which point he exclaimed, "Oh my God, you're touching my left thumb," his missing phantom thumb, in other words. He seemed as surprised as I was. Touching his upper lip produced sensation in his phantom index finger, and touching his lower jaw provoked sensations in his phantom little finger. There was a complete, systematic map of the missing phantom hand draped on his face (Figure 1.4).

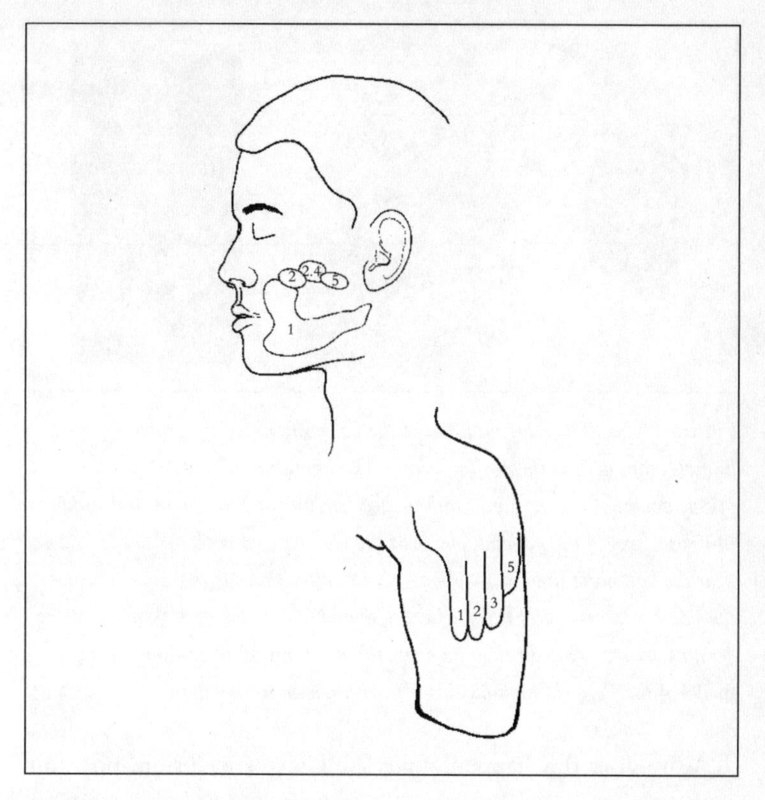

Figure 1.4 *Points on the body surface that yielded referred sensations in the phantom hand (this patient's left arm had been amputated ten years prior to our testing him). Notice that there is a complete map of all the fingers (labelled 1 to 5) on the face and a second map on the upper arm. The sensory input from these two patches of skin is now apparently activating the hand territory of the brain (either in the thalamus or in the cortex). So when these points are touched, the sensations are felt to arise from the missing hand as well.*

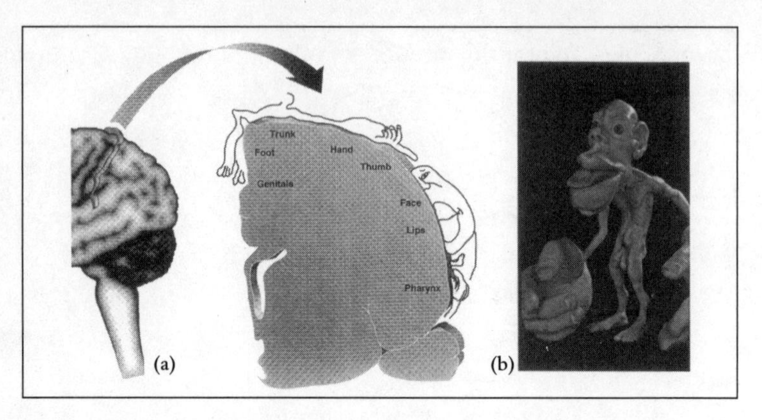

Figure 1.5 *(a) The representation of the body surface on the surface of the human brain behind the central sulcus. The homunculus ("little man") is upside down for the most part and his feet are tucked onto the medial surface (inner surface) of the parietal lobe near the very top, whereas the face is down near the bottom of the outer surface. Notice, also, that the face area is below the hand area instead of being where it should – near the neck – and that the genitals are represented below the foot. (b) A whimsical three-dimensional model of the Penfield homunculus – the little man in the brain.*

Why does this happen? Like the Capgras delusion, phantom limbs are a mystery that would have intrigued Sherlock Holmes. What on earth is going on? The answer lies, again, in the anatomy of the brain. The touch signals from the entire skin surface on the left side of the body are mapped on to the right cerebral hemisphere on a vertical strip of cortical tissue called the post-central gyrus. Actually, there are several maps but for ease of representation we can assume there is only one map, called SI, on the post-central gyrus. This is a faithful representation of the entire body surface – almost as if there were a little person draped on the surface of the brain (Figure 1.5). We call this the Penfield homunculus, and for the most part it is continuous – which, after all, is what one means by a map. But there is one peculiarity: the representation of the face on this map on the surface of the brain is right next to the representation of the hand, not near the neck in its expected position. The head is dislocated. (Why this is so is unclear; perhaps it has something to do with the phylogeny or the way in which the brain develops in early fetal life or in early infancy, but dislocated it is.) This gave me the clue to what was happening. When an arm is amputated, no signals are received by the part of the brain's cortex corresponding to the hand. It becomes hungry for sensory input and the sensory input from the facial skin now invades the adjacent vacated territory corresponding to the missing hand. Signals from the face are then misinterpreted by higher centers in the brain as arising from the missing hand.[2] The specificity of these signals is so compelling that an ice cube or warm water applied to the face will produce a cold or warm phantom digit. One patient, Victor, when the water began to trickle down his face, also felt it trickling down his phantom arm. When he raised his arm he was amazed to feel the trickle going up his phantom, contrary to the laws of physics.

To test our "remapping" or "cross-wiring" hypothesis directly we used the brain imaging technique called MEG or magnetoencephalography. This shows which parts of the brain are stimulated when various parts of the body are touched. Sure enough, we found that in Victor (and other arm amputees like him), touching his face activated not only the face area in the brain but also the hand region of the Penfield map (Figure 1.6). This is very different from what is seen in a normal brain, where touching the face activates only the facial region of the cortex.

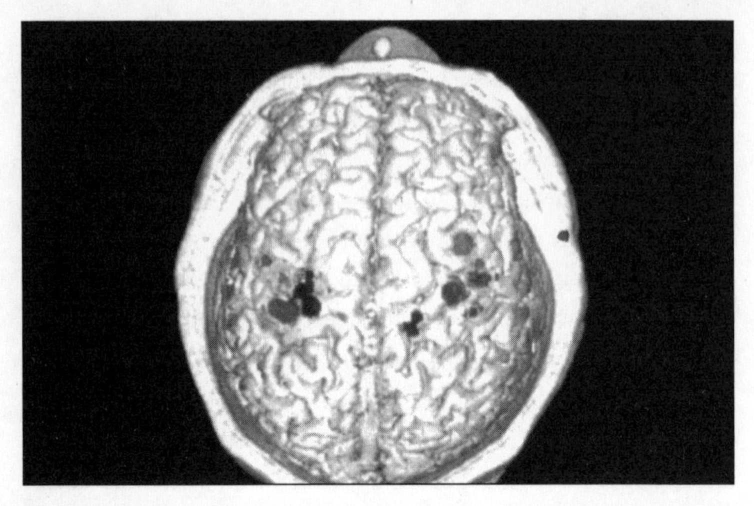

Figure 1.6 *Magnetoencephalography (MEG) image superimposed on a magnetic resonance (MR) image of the brain in a patient whose right arm was amputated below the elbow. The brain is viewed from the top. The right hemisphere shows normal activation of the right hand (hatched), face (black) and upper arm (white) areas of the cortex corresponding to the Penfield map. In the left hemisphere there is no activation corresponding to the missing right hand, but the activity from the face and upper arm has now spread to this area.*

There has obviously been some cross-wiring in Victor's brain, and this is important because it permits us to correlate the changes in brain anatomy, changes in brains' sensory maps, with the phenomenology. This link between physiology and psychology is one of the major goals of cognitive neuroscience.[3]

The discovery also has broader implications. One of the things all medical students learn is that connections in the brain are laid down in the fetus or in early infancy, and that once they are laid down, there is nothing much that can be done to change these connections in an adult. That's why when there's damage to the nervous system, such as is caused by a stroke, there is so little recovery of function. It is also why neurological ailments are notoriously difficult to treat ... or at least that's what we were taught. What I have seen flatly contradicts this view and suggests that there is a tremendous amount of plasticity or malleability even in the adult brain, and this can be demonstrated in a five-minute experiment on a patient with a phantom limb.

It isn't yet clear how this "plasticity" of body maps can be harnessed in the clinic, but I'll mention another example to show how some of these ideas can be clinically useful. Some patients can "move" their phantom limbs, and will say, "It's waving goodbye," or, "It's shaking hands with you."[4] But in many other patients the phantom arm feels "paralyzed," "frozen stiff," "in cement," or "won't budge an inch." Often the phantom hand goes into painful involuntary clenching spasms or is fixed in an awkward painful position which the patient is unable to change. We have discovered that some of these patients had pre-existing nerve damage before the amputation, for example the arm had been paralyzed and lying in a sling. After amputation the patient is stuck with a paralyzed phantom ... as if the paralysis is "carried over" into the phantom. Perhaps when the arm was intact but paralyzed, every time the front of the brain sent a command to the arm saying "move," it was getting visual feedback saying "no, it won't move." Somehow this feedback becomes imprinted on the circuitry in the parietal lobe or somewhere else in the brain. (We call this "learned paralysis.") How could this highly speculative idea be tested? Perhaps if a patient were given visual feedback that the phantom was obeying the brain's com-

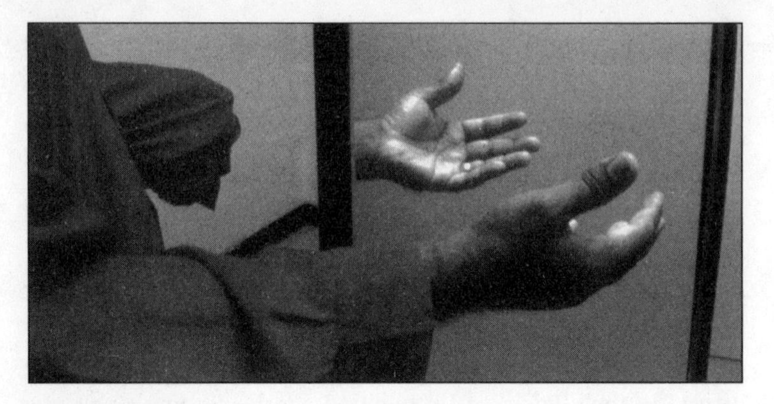

Figure 1.7 *Illustrates the "mirror box" arrangement used for resurrection of the phantom.*

mands the learned paralysis could be "unlearned." We propped up a mirror vertically on a table in front of a prone patient, so that it was at right angles to his chest, and asked him to position his paralyzed phantom left arm on the left of the mirror and mimic its posture with his right hand, which was on the right side of the mirror. We then asked him to look into the right-hand side of the mirror so that he saw the mirror reflection of his intact hand optically superimposed on the felt location of the phantom (Figure 1.7). We then asked him to try to make symmetrical movements of both hands, such as clapping or conducting an orchestra, while looking in the mirror. Imagine his amazement and ours when suddenly he not only saw the phantom move but felt it move as well. I have repeated this experiment with several patients, and it seems that the visual feedback animates the phantom so that it begins to move as never before, often for the first time in years. Many patients have found that this sudden sense of voluntary control and movement in the phantom produces relief from the spasm or awkward posture that was causing much of the agonizing pain in the phantom.[5]

Relief from a phantom pain using a mirror is surprising enough, but can the same trick be applied to real pain in an intact arm or leg? Even though we usually think of pain as one thing, there are at least two different types which may have evolved for different functions. Acute pain evolved to allow reflexive withdrawal from, for example, fire, and probably also to teach avoidance of harmful, pain-producing objects such as thorns. Chronic pain – as in a fracture or a gangrene – is a different thing altogether: it evolved to reflexively immobilize the arm, so letting it rest and remain out of harm's way until fully healed. Ordinarily, pain is a very useful adaptive mechanism – a gift, not a curse. But sometimes the mechanism backfires. We often see patients with a condition called "complex regional pain type 1," which includes the bizarre clinical syndrome of "reflex sympathetic dystrophy" or RSD. In RSD patients, what begins as a minor injury – a bruise or insect sting or fracture of a fingertip – leads to the entire arm becoming excruciatingly painful, completely immobilized, inflamed and swollen – grossly out of proportion to the inciting event. And it lasts forever.

The evolutionary framework helps us to understand how this might come about. Remember, the original purpose of chronic pain is a temporary immobilization to allow recovery, so when the brain sends a motor command to the arm there is intense pain preventing further movement. This is ordinarily adaptive, but I suggest that it sometimes malfunctions and leads to what I call "learned pain": the very act of attempting to move the arm – the motor command signal itself – becomes pathologically associated with excruciating pain. As a result, even after the inciting event has long disappeared, the patient still has a pseudoparalysis caused by learned pain. In 1995 I suggested that this type of pathological chronic pain may also benefit from mirror visual feedback. Imagine the patient sees the reflection of a normal hand superimposed optically on the painfully immobilized abnormal one. If the normal hand is now moved (while partially attempting to move the painful one) the patient will see the bad arm suddenly springing to life and moving quite freely! Perhaps this would help RSD patients to "unlearn" the spurious connection in their brains between arm movement and pain – thereby eliminating the pain and returning mobility. In 1995 this was no more than a far-fetched idea but recently McCabe et al. (2003) tried the mirror procedure on nine patients in placebo-controlled clinical trials. The pain went away completely and mobility returned in many of the patients who used mirrors, whereas the control group, who used Plexiglas, experienced no benefit at all. This result is so surprising that I would have been skeptical had not Patrick Wall – arguably the world's leading expert on both pain and placebos – been one of the authors. If confirmed, this result promises a new and effective treatment for at least some patients with chronic pain.[6]

The cross-wiring in the brain that sometimes results from amputation can also occur owing to a gene mutation. Instead of the brain modules remaining segregated, they become accidentally cross-wired, resulting in a curious condition called synesthesia, first clearly documented by Francis Galton in the nineteenth century. Synesthesia, which appears to be genetically transmitted, results in a mingling of the senses. For example hearing a particular musical note might invoke a particular color: C sharp is red, F sharp is blue, etc. Visually perceived numbers can produce a similar effect: 5 might always be seen as red, 6 always green, 7 always indigo, 8 always yellow ... Synesthesia is surprisingly common, affecting about one in two hundred people. What causes this mixing of signals? A student of mine, Ed Hubbard, and I were looking at brain atlases – specifically at the fusiform gyrus, where color information is analyzed. We saw that the number area of the brain, which represents visual graphemes of numbers, is also in the fusiform gyrus, almost touching the color area. It seems likely that, just as amputation can produce cross-wiring between the face and the hand, synesthesia is caused by cross-wiring between the number and color areas in the fusiform gyrus due to an inherited genetic abnormality.

Even though synesthesia was described by Galton a hundred years ago, the phenomenon never made it into mainstream neuroscience. It was often assumed that these people were either just crazy or simply trying to draw attention to themselves. Or maybe it had something to do with childhood memory: refrigerator magnets or a learning book in which 5 was red, 6 was blue, 7 was green ... but if that were the case, how could it run in families? My colleagues and I wanted to show that synesthesia is a real sensory phenomenon, not mere imagination or memory. We devised a simple computer display, a number of black 5s scattered on a white background. Embedded among those 5s were a number of 2s forming a hidden shape (Figure 1.8). Since these are computer-generated, the 2s are just mirror images of the 5s. Most people looking at this pattern see only a random jumble of numbers, but a synesthete sees the 5s as green and the 2s as forming a red shape

Figure 1.8 *A "clinical test" for synesthesia. The display consists of 2s embedded in a matrix of randomly placed 5s. Non-synesthetes find it very hard to discern the embedded shape (in this case a triangle). Synesthetes who see the numbers as colored can detect the triangle much more easily. (Depicted schematically in Figure 4.1.)*

conspicuously visible against a forest of green (shown schematically in Figure 4.1). The fact that synesthetes can more easily identify these shapes than normal people suggests that they are not crazy but experiencing a genuine sensory phenomenon. It also rules out memory association or some high-level cognitive phenomenon. Our group in La Jolla as well as Jeffrey Gray and Mike Morgan and others in London have conducted experiments to test the idea that there is cross-wiring in the brain. We have shown that there is activation of the fusiform gyrus in the color area when these people are shown numbers in black and white. (In normal people the color area is activated only if they are shown colored numbers.)

Phantom limb, synesthesia and Capgras' syndrome can at least be partly explained in terms of neural circuitry. But I once encountered someone with an even more bizarre syndrome called pain asymbolia. To my amazement, this patient responded to a pain stimulus not with an "ouch" but with laughter. Here is the ultimate irony – a human being laughing in the face of pain. Why would anyone do this? First, we need to answer an even more basic question: why does anybody laugh? Clearly, laughter is hard-wired, it's a trait in all humans. Every society, every civilization, every culture, has some form of laughter and humor (except for Germans). But why did laughter evolve through natural selection? What biological purpose does it serve?

The common denominator of all jokes is a path of expectation that is diverted by an unexpected twist necessitating a complete reinterpretation of all the previous facts – the punch-line. Obviously a sudden twist per se is not sufficient for laughter, otherwise every great scientific discovery that generates a "paradigm shift" would be greeted with hilarity, even by those whose theory had just been disproved. (No scientist would be amused if you disproved his theory; believe me, I've tried!) Reinterpretation alone is insufficient. The new model must be inconsequential. For example, a portly gentleman walking toward his car slips on a banana peel and falls. If he breaks his head and blood spills out, obviously you are not going to laugh. You are going to rush to the telephone and call an ambulance. But if he simply wipes off the goo from his face, looks around him, and then gets up, you start laughing. The reason is, I suggest, because now you know it's inconsequential, no real harm has been done. I would argue that laughter is nature's way of signaling that "it's a false alarm." Why is this useful from an evolutionary standpoint? I suggest that the rhythmic staccato sound of laughter evolved to inform our kin who share our genes: don't waste your precious resources on this situation; it's a false alarm. Laughter is nature's OK signal.

But what does this have to do with my pain asymbolia patient? Let me explain. When we examined his brain using a CT scan we found there was damage close to the region called the insular cortex on the sides of the brain. The insular cortex receives pain signals from the viscera and from the skin. That's where the raw sensation of pain is experienced, but there are many layers to pain – it is not just a unitary thing. From the insular cortex the message goes to the amygdala, encountered earlier in the context of the Capgras syndrome, and then to the rest of the limbic system, and especially the anterior cingulate, where we respond emotionally to the pain. We experience the agony of pain and take the appropriate action. So perhaps in this patient the insular cortex was normal, so he could feel the pain, but the wire that goes from the insula to the rest of the limbic system and the anterior cingulate was cut: a disconnection similar to that seen in the Capgras patient. Such a situation would produce the two key ingredients required for laughter and humor: one part of the brain signals a potential danger but the very next instant another part – the anterior cingulate – does not receive a confirmatory signal, thereby leading to the conclusion "it's a false alarm." Hence the patient starts laughing and giggling uncontrollably. The same sort of thing happens in tickling, which may be a sort of crude "playtime" rehearsal for adult humor. An adult approaches a child with his hands extending menacingly toward vulnerable parts of the child's body but then anti-climactically deflates the potential threat with gentle stimulation and a "Koochy, koochy, koo!" This takes the same form as mature adult humor: a potential threat followed by a deflation.

The examples we have considered so far suggest that syndromes such as phantom limb, the Capgras delusion, or pain asymbolia can help us understand the neural basis, functional logic and evolution of many otherwise mysterious aspects of our minds. This is true whether you are considering little quirks of human behavior, such as our response to tickling, or the loftiest question of all – the neural basis of the self. There is something distinctly odd about a hairless, neotenous primate that has evolved into a species that can look back over its own shoulder to ponder its origins. Odder still, the brain can not only discover how other brains work but also ask questions about itself: Who am I? What is the meaning of my existence? Why do I laugh? Why do I dream? Why do I enjoy art, music and poetry? Does my mind consist entirely of the activity of neurons in my brain? If so, what scope is there for free will? It is the peculiar recursive quality of these questions as the brain struggles to understand itself that makes neurology so fascinating.

The prospect of answering these questions in the millennium is both exhilarating and disquieting, but it's surely the greatest adventure that our species has ever embarked upon.

2

Believing Is Seeing

Our ability to perceive the world around us seems so effortless that we tend to take it for granted. But just think of what's involved. You have two tiny, upside-down distorted images inside your eyeballs but what you see is a vivid three-dimensional world out there in front of you. This transformation – as Richard Gregory has said – is nothing short of a miracle. How does it come about? What is perception?

One common fallacy is to assume there is an image inside your eyeball, the optical image, exciting photoreceptors on your retina and then that image is transmitted faithfully along a cable called the optic nerve and displayed on a screen called the visual cortex. This is obviously a logical fallacy because if you have an image displayed on a screen in the brain, then you have to have someone else in there watching that image, and that someone needs someone else in *his* head, and so on ad infinitum.

The first step we must take toward understanding perception

is to forget the idea of images in the brain and think instead of transforms or symbolic representations of objects and events in the external world. Just as little squiggles of ink called writing can symbolize or represent something they don't physically resemble, so the action of nerve cells in the brain, the patterns of firing, represent objects and events in the external world. Neuroscientists are like cryptographers trying to crack an alien code, in this case the code used by the nervous system to represent the external world.

This chapter concerns the process we call seeing – and how we become consciously aware of things around us. As in chapter 1, I'll begin with some examples of patients with strange visual defects and then explore the wider implications of these syndromes for understanding the nature of conscious experience. How does the activity of neurons – mere wisps of protoplasm – in the visual areas of the brain give rise to all the richness of conscious experience, the redness of red or blueness of blue? Or the ability to tell a burglar from a lover?

We primates are highly visual creatures. We have not just one visual area, the visual cortex, but thirty areas in the back of our brains which enable us to see the world. It's not clear why we need thirty areas and not just one. Perhaps each of these areas is specialized for a different aspect of vision. For example, one area called V4 seems to be concerned mainly with processing color information, seeing colors, whereas another area in the parietal lobe called MT, or the middle temporal area, is concerned mainly with seeing motion.

The most striking evidence for this comes from patients with tiny lesions that damage just V4, the color area, or just MT, the motion area. For example, if V4 is damaged on both sides of the brain a syndrome called cortical color blindness or achromatopsia results. Patients with cortical achromatopsia see the world in shades of grey, like a black and white film, but they have no problem reading a newspaper or recognizing people's faces or seeing direction of movement. Conversely if MT, the middle temporal area, is damaged, the patient can still read books and see colors but can't tell you which direction something is moving, or how fast.

A woman in Zurich who had this problem was terrified to cross the street because she saw the cars there not as moving but as a series of static images as though lit by a strobe light in a discotheque. She couldn't tell how fast a car was approaching even though she could read its number plate and tell you what color it was. Even pouring wine into a glass was an ordeal: she could not gauge the rate at which the wine level was rising and so the wine would always overflow. Most of us cross the road or pour a drink without even thinking about it – only when something goes wrong do we realize how extraordinarily subtle the mechanisms of vision really are and how complex a process it is.

Although the anatomy of these thirty "visual" areas in the brain seems bewildering at first, there is an overall plan of organization. The message from the eyeball on the retina goes though the optic nerve to two major visual centers in the brain. One of these, which I'll call the old system, is the evolutionary ancient pathway that includes a structure in the brain stem called the superior colliculus. The second pathway, the new pathway, goes to the visual cortex in the back of the brain (Figure 2.1). The new pathway in the cortex is doing most of what we usually think of

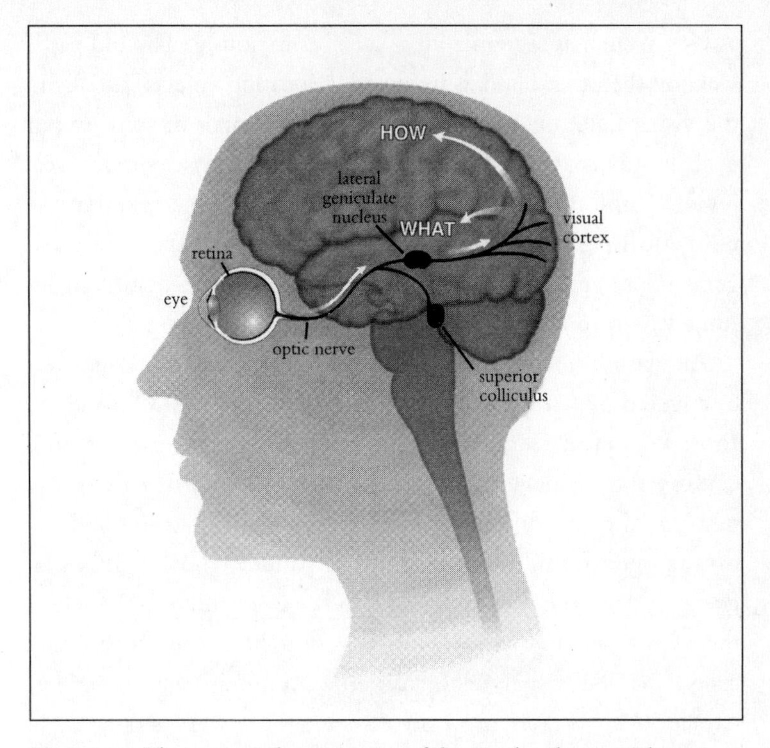

Figure 2.1 *The anatomical organization of the visual pathways. Schematic diagram of the left hemisphere viewed from the left side. The fibers from the eyeball diverge in two parallel "streams": a new pathway that goes to the lateral geniculate nucleus (shown here on the surface for clarity, though it is actually inside the thalamus, not the temporal lobe) and an old pathway that goes to the superior colliculus in the brain stem. The "new" pathway goes to the visual cortex and diverges again (after a couple of relays) into two pathways (white arrows) – a "how" pathway in the parietal lobes that is concerned with grasping, navigation and other spatial functions, and the second, "what" pathway in the temporal lobes concerned with recognizing objects. These two pathways were discovered by Leslie Ungerleider and Mortimer Mischkin of the National Institutes of Health. The two pathways are shown by white arrows.*

as vision, such as recognizing objects, consciously. The old pathway, on the other hand, is involved in locating objects spatially in the visual field, enabling you to reach out for it or swivel your eyeballs toward it. This allows the high-acuity central foveal region of the retina to be directed toward the object so that the new visual pathway can then proceed to identify the object and generate an appropriate behavior toward it: eat it, mate with it, run away from it, name it, etc.

An extraordinary neurological syndrome called blindsight was discovered by Larry Weiskrantz and Alan Cowey at Oxford and Ernst Poppel in Germany. It has been known for more than a century that damage to the visual cortex (which is part of the new visual pathway) on one side of the brain results in blindness on the opposite side. For example, a patient whose right visual cortex is damaged is completely blind to everything to the left of their nose when they are looking straight ahead (technically called the left visual field). When examining such a patient, named GY, Weiskrantz noticed something very strange. He showed the patient a little spot of light in the blind region and asked what he saw. The patient said "nothing," as would be expected. But then he asked the patient to reach out and touch the light, even though he couldn't see it.

"But I can't see it," said the patient. "How do you expect me to point to it?" Weiskrantz said to try anyway; take a guess. To the researcher's surprise, the patient reached out and pointed accurately to the dot that he could not consciously perceive. After hundreds of trials it became obvious that he could point with 99 percent accuracy, even though he claimed on each trial that he was just guessing and didn't know if he was getting it right or not. The implications of this are staggering. How can someone reach out and touch something he cannot see?

The answer is, in fact, obvious. GY has damage to his visual cortex – the new pathway – which is why he is blind. But remember, he still has the other old pathway, the other pathway going through his brain stem and superior colliculus as a backup. So even though the message from the eyes and optic nerves doesn't reach the visual cortex, given that the visual cortex is damaged, it takes the parallel route to the superior colliculus which allows him to locate the object in space. The message is then relayed to higher brain centers in the parietal lobes that guide the hand movement accurately to point to the invisible object! It's as if even though GY the person, the human being, is oblivious to what's going on, there's another unconscious being – a "zombie" – inside him who can guide his hand with uncanny accuracy.

This explanation suggests that only the new pathway is conscious – events in the old pathway, going through the colliculus and guiding the hand movement, can occur without a person being conscious of it! Why? Why should one pathway alone, or its computational style, perhaps, lead to conscious awareness, whereas neurons in a parallel part of the brain, the old pathway, can carry out even complex computations without being conscious? Why should *any* brain event be associated with conscious awareness given the "existence proof" that the old pathway through the colliculus can do its job perfectly well without being conscious? Why can't the rest of the brain do without consciousness? Why can't it all be blindsight, in other words?

We can't yet answer this question directly but as scientists the best we can do is to establish correlations and try to home in on the answer. We can make a list of all brain events that reach consciousness and a list of those brain events that don't. We can then compare the two lists and ask whether there is a common denominator in each list that distinguishes it from the other. Is it only certain styles of computation that lead to consciousness? Or perhaps certain anatomical locations that are linked to being conscious? That is a tractable empirical question which, once tackled, might get us closer to answering what the function of consciousness, if any, might be and why it evolved (just as knowing that heredity was embodied in DNA allowed us to crack the genetic code).

To compile these two lists we do need to know a great deal more about what the limits of blindsight are. How sophisticated is it? This has yet to be studied in detail. We already know from Alan Cowey and Petra Stoerig that it is capable of some degree of wavelength (color) discrimination. We know it cannot recognize faces, but can it correctly "guess" someone's expression?

One claim made for visual awareness and consciousness is that it is required for *binding* different features of an object together. If you are shown a red object moving right while a green one is simultaneously moving left, and if your color area and motion area in your brain are simultaneously signaling these attributes, how do you know which direction goes with which color? It has been suggested that consciousness is not required for the initial stage of extracting the attributes (red vs. green, left vs. right) but it *is* required to solve "the binding problem" – i.e., to know which color goes with which direction. Patients like GY can help us test this theory. The prediction would be that if he were simultaneously shown two balls on his blind side – a red one moving right and green one moving left – he should be able to tell you there is a red object and a green object and that one of them is moving right and the other left, but he won't know which is which. Or we could dispense with color and simply have two objects one below the other (both in the blind left visual field) simultaneously moving in opposite directions. Could GY tell which is which?

I should point out that the blindsight syndrome in GY seemed so bizarre that it was (and still is) greeted with skepticism by some of my colleagues. This is in part because the syndrome is very rare but also partly because it seems to violate common sense. How can you point to something you don't see? However, that is not a good reason for rejecting it because in a sense we all suffer from blindsight. Let me explain.

Imagine you are driving your car and having an animated conversation with your friend sitting next to you. Your attention is entirely on the conversation, it's what you're conscious of. But in parallel you are negotiating traffic, avoiding the pavement, avoiding pedestrians, obeying red lights and performing all these very complex elaborate computations without being really conscious of any of it unless something strange happens, such as a leopard crossing the road. So in a sense you are no different from GY: you have "blindsight" for driving and negotiating traffic. What we see in GY is simply an especially florid version of blindsight unmasked by disease, but his predicament is not fundamentally different from that of us all.[1]

Intriguingly, it is impossible to imagine the converse scenario: paying conscious attention to driving and negotiating traffic while unconsciously having a creative conversation with your friend. This may sound trivial but it is a thought experiment and it is already telling you something valuable: that computations involved in the meaningful use of language require consciousness but those involved in driving, however complicated, do not involve consciousness. It is true that sleepwalkers sometimes "talk" without (presumably) being conscious – but their mumblings are hardly like the two-way exchange of normal open-ended conversation. The link between language and consciousness is a topic we will explore further in chapter 5.

I believe this approach to consciousness will take us a long way toward answering the riddle of the benefits of consciousness and why it evolved. My own philosophical position about consciousness accords with the view proposed by the first Reith lecturer, Bertrand Russell, that there is no separate "mind stuff" and "physical stuff" in the universe: the two are one and the same. (The formal term for this is neutral monism.) Perhaps mind and matter are like the two sides of a Möbius strip that appear different but are in fact the same.

So much for the messages in the new visual pathway. Now let us turn to the other pathway, the old pathway which goes to the colliculus and which mediates blindsight and projects to the parietal lobe in the sides of the brain. The parietal lobes are concerned with creating a symbolic representation of the spatial layout of the external world. The ability that we call spatial navigation—avoiding obstructions, dodging a snowball, catching a football—all of these abilities depend crucially on the parietal lobes.

Damage to the right parietal lobe produces a fascinating syndrome called neglect, in a sense the converse of blindsight. The patient no longer moves his eyes toward the object, which is looming toward him from the left, and he can no longer reach out and point to it or grab it. But he is not blind to events on the left side of the world because if you draw his attention to an object there he can see it perfectly clearly and can identify it. The best description of the neglect syndrome is an indifference to the left side of the world. A patient suffering from this syndrome will eat only from the right side of a plate and leave the food on the left side uneaten. Only if the patient's attention is drawn to the uneaten food will he eat what's there. A man will shave only the right side of his face; a woman will apply makeup similarly. And if you give the patient a sketchpad and ask her to draw a flower the result will be only half a flower – the right half (Figure 2.2).

Neglect is caused by damage to the right hemisphere, and the patient is also usually paralyzed on the left side because the right hemisphere of the brain controls the left side of the body. I wondered if it would be possible to "cure" neglect. Could patients be treated by making them pay attention to the left side of the world which they are ignoring?

I again hit on the idea of using a mirror, as in the case of phantom limbs in chapter 1. I had the patient sit on a chair and then I stood to her right side and held a mirror so that when she rotated her head to the right she would be looking directly into it. What she would see was a reflection of the left side of the world which she had previously ignored. Would this make her suddenly realize that there's a whole left side to the world she had been ignoring so that she would turn to the left and look at it? If so we would have cured her neglect by merely using a

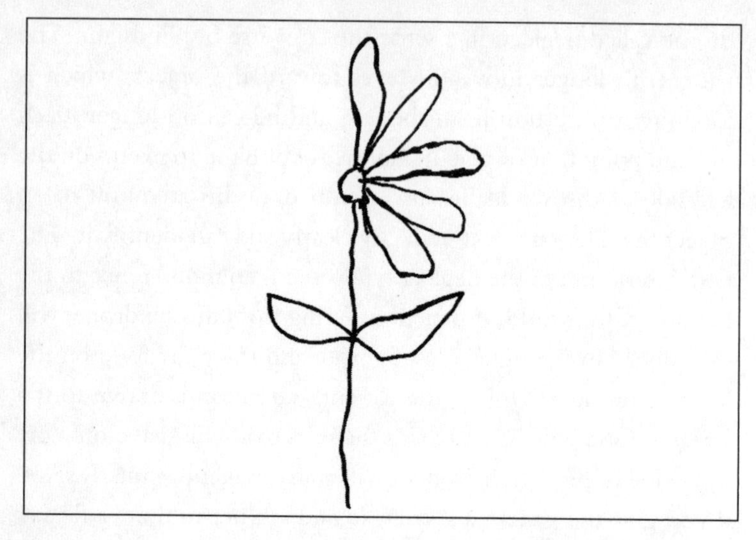

Figure 2.2 *Drawing made by a neglect patient. Notice that the left half of the flower is missing. Many neglect patients will also only draw half of the flower when drawing from memory – even with their eyes closed. This implies that the patient has also lost the ability to "scan" the left side of the internal mental picture of the flower.*

mirror! Or would her brain say (in effect) "Well, that's on my left, which doesn't really exist for me, so I'll continue to ignore it"?

The answer was, as often happens in science – neither! Before holding up the mirror I positioned John, my student, on her left side, holding a pen. I raised the mirror and asked the patient what she saw, what I was holding. The patient identified the mirror, saying that she could see her reflection in it and that it was cracked on the top – which it was.

She said she could also see John, and that he was holding a pen. I asked her to use her right hand – the hand that was not paralyzed – to reach out, take the pen and write her name with it. Of course, any normal person would turn to their left for the pen, but my patient began clawing the surface of the mirror, even reaching behind it, pulling my tie, grabbing my belt buckle … I explained that I didn't want her to reach for the reflection but for the real pen. The patient replied: "The real pen is inside the mirror, Doctor," or, on another occasion: "The pen is behind the darn mirror, Doctor."

This is a problem that can be solved without difficulty by a three-year-old child. Even a chimpanzee doesn't confuse a mirror image for a real object. But the wise Mrs. D – in spite of seventy years of experience with reflections – reaches straight into the mirror. We call this "mirror agnosia" or "looking-glass syndrome" in honor of Alice, who actually walked into the mirror thinking it was a real world.

What causes mirror agnosia? I think what happens is the patient knows she's looking at a reflection, therefore the object is on her left. But because left doesn't exist in her universe the only possible explanation, however improbable, is that the object is inside the mirror. Remarkably, all her abstract knowledge about the laws of optics and mirrors is distorted to accommodate this strange new sensory world in which the patient finds herself trapped.[2] This isn't just some absent-minded or impulsive response to the mirror image: she actually recognizes the presence of the mirror and starts groping behind it, regarding it as an obstacle.

Another even more extraordinary disorder which is also caused by damage to the right parietal is denial or anosognosia. Remember, most patients with right parietal damage also have some damage to the internal capsule and so are completely paralyzed on the left side of the body. This is what is meant by a stroke, a complete paralysis of one side of the body. Most of them complain about this, as indeed they should, but a small percentage vehemently deny that their left arm is paralyzed, and some of these patients don't have any neglect. They keep saying their arm is moving correctly.

The fact that this behavior is usually seen only when the right parietal is damaged, but rarely when the left parietal is damaged, gives us a clue to what is happening. It seems that the denial syndrome has something to do with hemispheric specialization: the manner in which the two cerebral hemispheres deal with the external world, especially the manner in which they deal with discrepancies in sensory input and discrepancies in beliefs. Specifically, when confronted with a discrepancy, the left hemisphere's coping style is to smooth over it, pretend it doesn't exist and forge ahead. (Freudian defense mechanisms are an example of this.) The right hemisphere's coping style is the exact opposite. It is highly sensitive to discrepancies, so I call it the anomaly detector.

A patient with a right hemisphere stroke (left side paralyzed) sending a command to move his arm receives a visual feedback signal saying it is not moving, so there is a discrepancy. His right hemisphere is damaged, but his intact left hemisphere goes about its job of denial and confabulation, smoothing over the discrepancy and saying, all is fine, don't worry. On the other hand, if the left hemisphere is damaged and the right side is paralyzed, the right hemisphere is functioning as it should, so it notices the discrepancy between the motor command and the lack of visual feedback and recognizes the paralysis.[3] This was an outlandish idea but it's now been tested with brain imaging experiments and shown to be essentially correct.[4]

For a person to deny that he or she is paralyzed is quite bizarre, but seven or eight years ago we found something even more amazing. Some patients will also deny that *another* patient is paralyzed. I tell patient B, who is paralyzed, to move his arm. Patient B, of course, doesn't move, but if I ask my patient A who has anosognosia whether B moved his arm, A says yes, he did. Patient A is engaging in denial of another person's disabilities.[5]

At first this made no sense to me, but then I came across some studies by Giaccomo Rizzollati of experiments done on monkeys. It is well known that parts of the frontal lobes which are concerned with motor commands contain cells which fire when a monkey performs certain specific movements. One cell will fire when the monkey reaches out and grabs a peanut, another cell will fire when the monkey pulls something, yet another cell when the monkey pushes something. These are motor command neurons. Rizzollati found that some of these neurons will also fire when the monkey watches another monkey performing the same action. For example, a peanut-grabbing neuron which fires when the monkey grabs a peanut also fires when the monkey watches another monkey grab a peanut. The same thing happens in humans. This is quite extraordinary, because the visual image of somebody else grabbing a peanut is utterly different from the image of yourself grabbing a peanut – your brain must perform an internal mental transformation. Only then can that neuron fire in response both to its own movements *and* to another person making the same movements. Rizzollati calls these mirror neurons. Another name for them is monkey-see, monkey-do neurons, and these neurons are, I think, the ones that were damaged in our patients.

Consider what's involved in judging somebody else's movements. Maybe you need to do a virtual reality internal simulation of what that person is doing, and that may involve the activity of these very same neurons, these mirror neurons. So mirror neurons, instead of being some kind of curiosity, have important implications for understanding many aspects of human nature, such as interpreting somebody else's actions and intentions. We think it is this system of neurons that is damaged in some patients who have anosognosia. The patient can therefore no longer construct an internal model of somebody else's actions in order to judge whether that person is accurately carrying out a command or not.

I believe that these neurons may have played an important role in human evolution.[6] One of the hallmarks of our species is what we call culture. Culture depends crucially on imitation of parents and teachers, and the imitation of complex skills may require the participation of mirror neurons. I think that, somewhere around 50,000 years ago, maybe the mirror neurons system became sufficiently sophisticated that there was an explosive evolution of this ability to mime complex actions, in turn leading to cultural transmission of information, which is what characterizes us humans.

Mirror neurons also permit a sort of "virtual reality" simulation of other people's actions and intentions, which would explain why we humans are the "Machiavellian" primate – so good at constructing a "theory of other minds" in order to predict their behavior. This is indispensable for sophisticated social interactions, and some of our recent studies have shown that this system may be flawed in autistic children, which would explain their extreme social awkwardness.

But although the studies on patients are intriguing in themselves, our real agenda here is to understand how the *normal* brain works.[7] How does the activity of neurons generate the whole spectrum of abilities that we call human nature, whether it is body image or culture or language or abstract thinking? It is my belief that such a deeper understanding of the brain will have a profound impact not just on the sciences but on the humanities as well. Lofty questions about the mind are fascinating to ask – philosophers have been asking them for three millennia both in my native India and in the West – but it is only in the brain that we can eventually hope to find the answers.

3

The Artful Brain

In this chapter – the most speculative in the book – I consider one of the most ancient questions in philosophy, psychology and anthropology, namely, what is art? When Picasso said, "Art is the lie that reveals the truth," what exactly did he mean?

As we have seen, neuroscientists have made some headway in understanding the neural basis of psychological phenomena such as body image or visual perception. But can the same be said of art – given that art obviously originates in the brain?

In particular we can ask whether there are such things as artistic universals. There is obviously an enormous number of artistic styles across the world: Tibetan art, Classical Greek art, Renaissance art, Cubism, Expressionism, Impressionism, Indian art, pre-Columbian art, Dada … the list is endless. But despite this staggering diversity can we come up with some universal laws or principles that transcend these cultural boundaries and styles?

The question may seem meaningless to many social scientists; after all, science deals with *universal* principles whereas art is the ultimate celebration of human individuality and originality – the ultimate antidote to the homogenizing effects of science. There is some truth to this, of course, but even so I'd like to argue in this chapter that such universals do exist.

First, a note of caution. When I speak of "artistic universals" I am not denying the enormous role played by culture. Obviously, without culture there would be no different artistic styles – but neither does it follow that art is completely idiosyncratic and arbitrary, or that there are no universal laws.

To put it somewhat differently, let us assume that 90 percent of the variance seen in art is driven by cultural diversity or – more cynically – by just the auctioneer's hammer, and only 10 per cent by universal laws that are common to all brains. The culturally driven 90 percent is what most people already study – it's called art history. As a scientist, what I am interested in is the 10 per cent that is universal – not in the endless variations imposed by cultures. The advantage that scientists have today is that unlike philosophers we can now test our conjectures by directly studying the brain empirically. There's even a new name for this discipline. My colleague Semir Zeki calls it neuroaesthetics – just to annoy the philosophers.

I recently started reading about the history of ideas on art – especially Victorian reactions to Indian art – and it's a fascinating story. For example let's go to southern India and look at the famous Chola bronze of the goddess Parvathi dating back to the twelfth century (Figure 3.1). To Indian eyes, she is supposed to represent the very epitome of feminine sensuality,

Figure 3.1 *Parvathi, consort of Lord Shiva; twelfth-century Chola dynasty (replica).*

grace, poise, dignity, elegance: everything that's good about being a woman. And she's of course also very voluptuous.

But the Victorian Englishmen who first encountered these sculptures were appalled. Partly because they were prudish, but partly also because of just plain ignorance.

They complained that the breasts were far too big, the hips were too wide and the waist was too narrow. It didn't look anything like a real woman – it wasn't realistic – it was primitive art. And they said the same thing about the voluptuous nymphs of Kajuraho – even about Rajastani and Mogul miniature paintings. They said the paintings lacked perspective, that they were distorted.

The Victorians were unconsciously judging Indian art using the standards of Western art – especially classical Greek and Renaissance art, where realism is strongly emphasized.

But obviously this is a fallacy. Anyone today will tell you that art has nothing to do with realism. It is not about creating a replica of what's out there in the world. I can take a realistic photograph of my pet cat and no one would give me a penny for it. In fact, art is not about realism at all – it's the exact opposite. It involves deliberate hyperbole, exaggeration, even distortion, in order to create pleasing effects in the brain.

But obviously that can't be the whole story. You can't just take an image and randomly distort it and call it art. (Although in California, where I come from, many do!) The distortion has to be "lawful." The question then becomes, what kinds of distortion are effective? What are the laws?

I was sitting in a temple in India and in a whimsical frame of mind when I just jotted down what I think of as the universal laws of art, the ten laws of art which cut across cultural boundaries (see box).[1] The choice of 10 is arbitrary ... but it's a place to start.

The first law I call peak shift and to illustrate this I'll use a hypothetical example from animal behavior, from rat psychology.

Imagine you're training a rat to discriminate a square from a rectangle by giving it a piece of cheese every time it sees a particular rectangle. When it sees a square it receives nothing. Very soon it learns that the rectangle means food; it starts liking the rectangle – although a behaviorist wouldn't put it that way. Let's just say it starts going toward the rectangle because it prefers the rectangle to the square.

But if you take a longer, thinner rectangle and show it to the rat, it actually prefers the second rectangle to the first. This is because the rat is learning a rule – rectangularity. Longer and thinner equals more rectangular and, so far as the rat is concerned, the more rectangular, the better.

Professor Ramachandran's suggested
10 universal laws of art

1. Peak shift
2. Grouping
3. Contrast
4. Isolation
5. Perceptual problem solving
6. Symmetry
7. Abhorrence of coincidence/generic viewpoint
8. Repetition, rhythm and orderliness
9. Balance
10. Metaphor

And what has that to do with art?

Think about caricature. To produce a caricature of, say, Richard Nixon an artist must first ask: What's special about his face? What makes him different from other people? The artist will take the mathematical average, so to speak, of all male faces and subtract it from Nixon's face, leaving a big bulbous nose and shaggy eyebrows. These are then amplified to produce an image that looks even more like Nixon than Nixon himself. Skilled artists work this way to produce great portraiture;[2] take it a step further and you get caricature. It looks comical, but it still looks even more like Nixon than the original Nixon. So you're behaving exactly like that rat.

What has all this to do with the rest of art? Let's go back to the Chola bronze of Parvathi, where the same principle applies.

How does the artist convey the very epitome of feminine sensuality? He simply takes the average female form and subtracts the average male form – leaving big breasts, big hips and a narrow waist. And then amplifies the difference. The result is one anatomically incorrect but very sexy goddess.

But that's not all there is to it – what about dignity, poise, grace?

Here the Chola bronze artist has done something quite clever. There are some postures that are impossible for a male owing to the constraints imposed by pelvic anatomy, curvature of the lumbar spine and angle between the neck and shaft of the femur. I can't stand like that even if I want to. But a woman can do it effortlessly. So the artist visits an abstract space I call "posture space," subtracts the average male posture from the average female and then exaggerates it. Doing this produces the elegant triple flexion – or tribhanga – pose, where the head is tilted one way, the body is tilted exactly the opposite way, and the hips again the other way. And again the viewer's reaction is not that the figure is anatomically inappropriate because nobody can stand like that. What the viewer sees is a gorgeous, beautiful, celestial goddess. This extremely evocative image is an example of the peak shift principle in Indian art.

So much for faces and caricatures and bodies and Chola bronzes. But what about the rest of art? What about abstract art, semi-abstract art, Impressionism, Cubism? What about Picasso, Van Gogh, Monet, Henry Moore? How can my ideas even begin to explain the appeal of some of those artistic styles?

To answer this question, we need to look at evidence from

ethology, especially the work of Niko Tinbergen at Oxford more than fifty years ago, who was doing some very elegant experiments on herring-gull chicks.

As soon as the herring-gull chick hatches, it sees its mother's long yellow beak with a red spot on it. It starts pecking at the red spot, begging for food. The mother then regurgitates half-digested food into the chick's gaping mouth, the chick swallows the food and is happy. Tinbergen asked himself: "How does the chick recognize its mother? Why doesn't it beg for food from a person who is passing by or from a pig?"

And he found that you don't need a mother. A hatchling would react in exactly the same way to a disembodied beak with no mother attached.

Why does a chick think a scientist waving a beak is a mother seagull? Well, the goal of vision is to do as little processing or computation as is necessary for the job on hand, in this case for recognizing mother. And through millions of years of evolution, the chick has acquired the wisdom that this long thing with a red spot always has a mother attached to it, rather than a mutant pig or a malicious ethologist. So it can take advantage of the statistical redundancy in nature and assume: "Long yellow thing with a red spot equals mother," thereby simplifying the processing and saving a lot of computational labor.

That seems fair enough. But what Tinbergen found next is that he didn't need even a beak. He took a long yellow stick with three red stripes, which looked nothing like a beak – and that's important – and the chicks pecked at the stick even more than they would have pecked at a real beak. They preferred it to a real beak, even though it didn't resemble a beak. Tinbergen had stumbled on a superbeak – an ultrabeak. So the chick's brain goes: "Wow – what a sexy beak!"

Why does this happen? We don't know exactly, but obviously there are neural circuits in the visual pathways of the chick's brain that are specialized to detect a beak as soon as the chick hatches. They fire upon seeing the beak. Perhaps because of the way they are wired up, they may actually respond more powerfully to the stick with three stripes than to a real beak. Maybe the neurons' receptive field embodies a rule such as "the more red contour the better." And so even though the stick doesn't look like a beak – maybe not even to the chick – this strange object is actually more effective in driving beak detectors than a real beak. And a message from this beak-detecting neuron goes to the emotional limbic centers in the chick's brain, giving it a big jolt and the message: "Here is a superbeak." The chick is absolutely mesmerized.

All of which brings me to my punch-line about art. If herring-gulls had an art gallery, they would hang a long stick with three red stripes on the wall; they would worship it, pay millions of dollars for it, call it a Picasso, but not understand why – why they are mesmerized by this thing even though it doesn't resemble anything. That's all any art lover is doing when buying contemporary art: behaving exactly like those gull chicks.

In other words human artists through trial and error, through intuition, through genius, have discovered the figural primitives of our perceptual grammar. They are tapping into these and creating for the human brain the equivalent of the long stick with three stripes. And what emerges is a Henry Moore or a Picasso.

The advantage with these ideas is that they can be tested experimentally. It is possible to record from cells in the fusiform gyrus of the brain that respond powerfully to individual faces. Some of them will fire only to a particular view of a face, but higher up are found neurons each of which will respond to *any* view (profile vs. full frontal) of a given face. And I predict that if you present a monkey with a Cubist portrait of a monkey's face – two different views of a monkey's face superimposed in the same location in the visual field – then that cell in the monkey's brain will be hyperactivated just as a long stick with three stripes hyperactivates the beak-detecting neurons in the chick's brain. So what we have here is a neural explanation for Picasso – for Cubism.[3]

I've discussed one of my universal laws of art so far – peak shift and the idea of ultra-normal stimuli – and have borrowed insights from ethology, neurophysiology and rat psychology to account for why people like non-realistic art.[4, 5]

The second law is more familiar. It's called grouping.

Most of us are familiar with puzzle pictures, such as Richard Gregory's Dalmatian dog. At first sight you see nothing but a bunch of splotches, but you can sense your visual brain trying to solve a perceptual problem, trying to make sense of this chaos. And then after 30 or 40 seconds suddenly everything clicks in place and you group all the correct fragments together to see a Dalmatian dog (Figure 3.2).

You can almost sense your brain groping for a solution to the perceptual riddle and as soon as you successfully group the correct fragments together to see the object, what I suggest is that a message is sent from the visual centers of the brain to the limbic-emotional centers of the brain, giving it a jolt and saying: "A-ha, there is an object – a dog," or "A-ha, there is a face."

The Dalmatian example is very important because it reminds us

Figure 3.2 *Gregory's Dalmatian dog (photo by Ron James).*

that vision is an extraordinarily complex and sophisticated process. Even looking at a simple scene involves a complex hierarchy, a stage-by-stage processing. At each stage in the hierarchy of processing, when a partial solution is achieved – when a part of the dog is identified – there is a reward signal "a-ha," a partial "a-ha," and a small bias is sent back to earlier stages to facilitate the further binding of the features of the dog. And through such progressive bootstrapping the final dog clicks in place to create the final big "A-HA!" Vision has much more in common with problem solving – like a twenty-questions game – than we usually realize.

The grouping principle is widely used in both Indian and in Western art – and even in fashion design. For example, you go shopping and pick out a scarf with red splotches on it. Then you look for a skirt which has also got some red splotches on it.

Why? Is it just hype, just marketing, or is it telling you something very deep about how the brain is organized? I believe it is telling you something very deep, something to do with the way the brain evolved.

Vision evolved mainly to discover objects and to defeat camouflage. You don't realize this when you look around you and you see clearly defined objects, but imagine your primate ancestors scurrying up in the treetops trying to detect a lion seen behind fluttering green foliage. What you get inside the eyeball on the retina is just a mass of yellow lion fragments obscured by all the leaves. But the visual system of the brain "knows" that the likelihood that all these different yellow fragments being exactly the same yellow simply by chance is zero. They must all belong to one object. It links them together, decides it's a lion (based on the overall shape) and sends a big "a-ha" signal to the limbic system telling you to run.

Arousal and attention culminate in titillating the limbic system. Such "a-has" are created, I maintain, at every stage in the visual hierarchy as partial object-like entities are discovered that draw our interest and attention. What an artist tries to do is to generate as many of these "a-ha" signals in as many visual areas as possible by more optimally exciting these areas with painting or sculpture than could be achieved with natural visual scenes or realistic images. Not a bad definition of art, if you think about it.

That brings me to my third law – the law of perceptual problem solving or visual peek-a-boo.

As anyone knows, a nude seen behind a diaphanous veil is much more alluring and tantalizing than a full-color *Playboy* photo or a Chippendale pin-up. Why? (This question was first raised by the Indian philosopher Abhinavagupta in the tenth century.) After all, the pin-up is much richer in information and should excite many more neurons.

As I have said, our brains evolved in highly camouflaged environments. Imagine you are chasing your mate through dense fog. Then you want every stage in the process – every partial glimpse of her – to be pleasing enough to prompt further visual search – so you don't give up the search prematurely in frustration. In other words, the wiring of your visual centers to your emotional centers ensures that the very act of searching for the solution is pleasing, just as struggling with a jigsaw puzzle is pleasing long before the final "a-ha." Once again it's about generating as many "a-has" in your brain as possible.[6] Art may be thought of as a form of visual foreplay before the climax.

We have discussed three laws so far: peak shift, grouping and perceptual problem solving. Before I go any further I'd like to emphasise that looking for universal laws of aesthetics does not negate the enormous role of culture, nor the genius and originality of an individual artist. Even if the laws are universal, which particular law (or combination of them) an artist chooses to deploy depends entirely on his or her genius and intuition. Thus, while Rodin and Henry Moore were mainly tapping into "form," Van Gogh and Monet were mainly introducing peak shifts in an abstract "color space" – brain maps in which adjacent points in color space rather than Cartesian space are mapped adjacently. Hence the effectiveness of artificially heightened "non-realistic" colors of their sunflowers or water lilies. These two artists also deliberately blurred the outlines to avoid distracting attention from the colors where it was needed most. Other artists may choose to emphasise even more abstract attributes such as shading or illumination (Vermeer).

And that brings us to my fourth law – the law of isolation or understatement.

A simple outline doodle of a nude by Picasso, Rodin or Klimt can be much more evocative than a full-color pin-up photo. Similarly the cartoon-like outline drawings of bulls in the Lascaux Caves are much more powerful and evocative of the animal than a *National Geographic* photograph of a bull. Hence the famous aphorism: "Less is more."

But why should this be so? Isn't it the exact opposite of the first law, the idea of hyperbole, of trying to excite as many "a-has" as possible? A pin-up or a Page Three photo has, after all, much more information. It's going to excite many more areas in the brain, many more neurons, so why isn't it more beautiful?[7]

The answer to this paradox lies in another visual phenomenon: "attention." It is well known that there cannot be two overlapping patterns of neural activity simultaneously. Even though the human brain contains a hundred billion nerve cells, no two patterns may overlap. In other words, there is a bottleneck of attention. Attentional resources may be allocated to only one entity at a time.

The main information about the sinuous, soft contours of a Page Three girl is conveyed by her outline. Her skin tone, hair color, etc. are irrelevant to her beauty as a nude. All this irrelevant information clutters the picture and distracts attention from where it needs critically to be directed – to her contours and outlines. By omitting such irrelevant information from a doodle or sketch the artist is saving your brain a lot of trouble. And this is especially true if the artist has also added some peak shifts to the outline to create an "ultra nude" or a "super nude."

This theory can be tested by doing brain imaging experiments comparing neural responses to outline sketches and caricatures versus full-color photos. But there is also some striking neurological evidence from children with autism. Some of these children have what is known as the savant syndrome. Even though they are retarded in many respects, they have one preserved island of extraordinary talent.

For example, a seven-year-old autistic child, Nadia, had exceptional artistic skills. She was quite retarded mentally, could barely talk, yet she could produce the most amazing drawings of horses and roosters and other animals. A horse drawn by Nadia would almost leap out at you from the canvas (Figure 3.3 left). Contrast this with the lifeless, two-dimensional, tadpole-like sketches drawn by most normal eight- or nine-year-olds (right) – or even a very good one by Leonardo da Vinci (center).

So we have another paradox. How can this retarded child produce a drawing that is so incredibly beautiful? The answer, I maintain, is the principle of isolation.

In Nadia, perhaps many or even most of her brain modules are damaged because of her autism, but there is a spared island of cortical tissue in the right parietal. So her brain spontaneously allocates all her attentional resources to the one module that's still functioning, her right parietal. The right parietal is the part of the brain concerned with our sense of artistic proportion. We know this because when it's damaged in an adult, artistic sense is lost. Stroke patients with right parietal damage produce drawings that are often excessively detailed but lack the vital essence of the

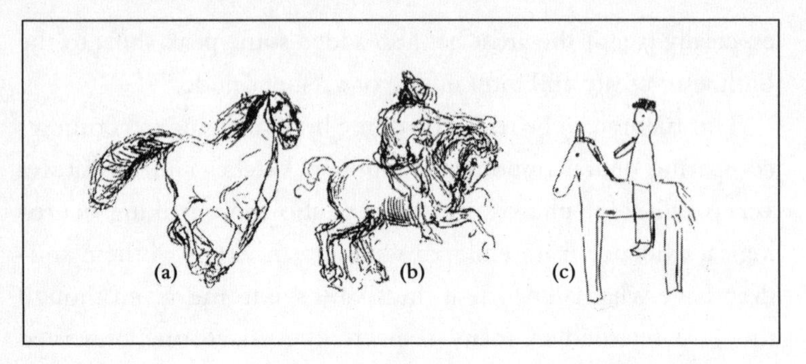

Figure 3.3 *(a) A drawing of a horse made by Nadia, the autistic savant, when she was five years old. (b) A horse drawn by Leonardo da Vinci. (c) A drawing of a horse by a normal eight-year-old. Notice Nadia's drawing is vastly superior to that of the normal eight-year-old and almost as good as (or perhaps better than!) Leonardo's horse. (a) and (c) reprinted from* Nadia, *by Lorna Selfe, with permission from Academic Press (New York).*

picture they are trying to depict. They have lost their sense of artistic proportion. Nadia, since everything else is damaged in her brain, spontaneously allocates all her attention to the right parietal – so she has a hyperfunctioning art module in her brain which is responsible for her beautiful renderings of horses and roosters. What most of us "normals" have to learn to do through years of training – ignoring irrelevant variables – she does effortlessly. Consistent with this idea, Nadia lost her artistic sense once she grew up and improved her language skills.

Another example is equally striking. Steve Miller, of the University of California, has studied patients who start developing rapidly progressing dementia in middle age, a form of dementia called the fronto-temporal dementia. This affects the frontal and temporal lobes, but spares the parietal lobe. Some of these patients suddenly start producing the most amazingly beautiful paintings and drawings, even though they had no artistic talent before the onset of their dementia. Again, the isolation principle is at work. With all other modules in the brain not working, the patient develops a hyper-functioning right parietal. There are even reports from Alan Snyder in Australia that it is possible to unleash such hidden talents by temporarily paralyzing parts of the brain in normal volunteers. If his findings are confirmed, it will truly be a brave new world.

That brings me to another question: why do humans even bother creating and viewing art?[8] I've already hinted at some possible answers but let me spell them out more explicitly. There are at least four possibilities – none mutually exclusive.

First, it is possible that once laws of aesthetics have evolved (for reasons such as discovering, attending to and identifying objects) then they may be artificially hyperstimulated even though such titillation serves no direct adaptive purpose, just as saccharin tastes "hypersweet" even though it provides zero energy and zero nutrition.

Second, as suggested by Miller, artistic skill may be an index of skillful eye–hand coordination and, therefore, an advertisement of good genes for attracting potential mates (the "come and see my etchings" theory). This is a clever idea that I don't find convincing. It doesn't explain why the so-called "index" takes the particular form that it does: art. After all, few women – not even feminists! – find the ability to knit or embroider attractive in a man, even though these demand excellent eye–hand coordination. My point is, why not use a much more straightforward "index" such as proficiency in archery or javelin throwing (which, to be sure, *are* attractive in a man)?

Third, there is Steve Pinker's idea that people acquire art as a status symbol to advertise their wealth: the "I own a Picasso, so help me spread our genes together" theory. Anyone who has been to a cocktail reception at an art gallery knows there's some truth to this.

Fourth – the idea I favor – art may have evolved as a form of virtual reality simulation. When you imagine something – as when rehearsing a forthcoming bison hunt or amorous encounter – many of the same brain circuits are activated as when you really do something. This allows you to practice scenarios in an internal simulation without incurring the energy cost or risks of a real rehearsal.

But there are obvious limits. Evolution has seen to it that our imagery – internal simulation – isn't perfect. A hominid with mutations that enabled it to perfectly imagine a feast instead of having one, or imagine orgasms instead of pursuing mates, is unlikely to spread its genes. This limitation in our ability to create internal simulations may have been even more apparent in our ancestors. For this reason they may have created real images ("art") as "props" to rehearse real bison hunts or to instruct their children. If so, we could regard art as Nature's own "virtual reality" (just as my mirror box allows patients to actually see their phantom arm and move it – whereas they couldn't do so just using imagination).

Limitations of space prevent the discussion of all my other laws in detail, but I will mention the last on my list. In many ways it is the most important, yet the most elusive: visual metaphor. A metaphor in literature juxtaposes two seemingly unrelated things to highlight certain important aspects of one of them (as when the Indian poet Rabindranath Tagore referred to the Taj Mahal as "A teardrop on the cheek of time"). The same thing is possible in visual art. For example, the multiple arms on the Chola bronze of the dancing Shiva or Nataraja (Figure 3.4) are not meant to be taken literally, as they were by the Victorian art critic Sir George Birdwood, who called it a multi-armed monstrosity. (Funnily enough, he didn't think that angels sprouting wings were monstrosities – although I can tell you as a medical man that to possess multiple arms is anatomically possible, but wings on scapulae are not!)

The multiple arms are meant to symbolize multiple divine attributes of God and the ring of fire that Nataraja dances in – indeed his dance itself – is a metaphor of the dance of the cosmos and of the cyclical nature of creation and destruction, an idea championed by the late Fred Hoyle. Most great works of art – be they Western or Indian – are pregnant with metaphor and have many layers of meaning.[9]

Everyone knows that metaphors are important, yet we have no idea why. Why not just say "Juliet is radiant and warm" instead of saying "Juliet is the sun"? What is the neural basis for metaphor? We don't know, but I will attempt some answers in chapter 4.

Many social scientists feel rather deflated when informed that beauty, charity, piety and love are the result of the activity of neurons in the brain, but their disappointment is based on the false assumption that to explain a complex phenomenon in terms of its component parts ("reductionism") is to explain it away. To understand why this is a fallacy imagine it's the twenty-second century and I am a neuroscientist watching you and your partner

Figure 3.4 *Nataraja or dancing Shiva. Chola dynasty copper alloy, thirteenth century.*

(Esmeralda) making love. I scan Esmeralda's brain and tell you everything that's going on in it when she is in love with you and is making love to you. I tell you about activity in her septum, in her hypothalamic nuclei, and how certain peptides are released along with the affiliation hormone prolactin, etc. You might then turn to her and say, "You mean that's all there is to it? Your love isn't real? It's all just chemicals?" To which Esmeralda should respond, "On the contrary, all this brain activity provides hard evidence that I *do* love you, that I'm not just faking it. It should increase your confidence in the reality of my love." And the same argument holds for art or piety or wit.

Do these laws of neuroaesthetics encompass everything there is to know about art? Of course not; I have barely scratched the surface. But I hope that these laws have provided some hints about the general form of a future theory of art, and about how a neuroscientist might try to approach the problem.

The solution to the problem of aesthetics, I believe, lies in a more thorough understanding of the connections between the thirty visual centers in the brain and the emotional limbic structures (and of the internal logic and evolutionary rationale that drives them). Once we have achieved a clear understanding of these connections, we will be closer to bridging the huge gulf that separates C. P. Snow's two cultures – science on the one hand and arts, philosophy and humanities on the other.

We could be at the dawning of a new age where specialization becomes old-fashioned and a twenty-first-century version of the Renaissance man is born.

4

Purple Numbers and Sharp Cheese

> You know my method, Watson. It is founded upon the observation of trifles.
>
> SHERLOCK HOLMES

In the nineteenth century the Victorian scientist Francis Galton, who was a cousin of Charles Darwin, noticed something very odd. He found that certain people in the population who were otherwise perfectly normal would experience a specific color every time they heard a specific tone. For example, C sharp might be red, F sharp might be blue, another tone might be indigo. He called this curious mingling of the senses synesthesia. Some of these people also see colors when they see numbers. Every time they see a black number, say the number five, printed on a white page (or a white five on a black page, for that matter), they would see it tinged, say, red. Six might be green, seven indigo, eight yellow and so on. Galton also asserted that this condition runs in families, something which Simon Baron-Cohen in Cambridge has recently confirmed.

It is fair to say that, even though synesthesia has been known about for over a hundred years, it has been by and large regarded as a curiosity and has never made it into mainstream neuroscience and psychology. But such "anomalies" (to use Thomas Kuhn's phrase) can be extremely important in science. Of course, most anomalies turn out to be false alarms, such as telepathy, spoon bending or cold fusion, but if you pick the right one you can completely change the direction of research in a field and generate a scientific revolution.

But first let's look at the most common explanations that have been proposed to account for synesthesia.

There are four. The first explanation is the most obvious: that these people are just crazy. It's the common reaction of scientists: when something does not fit the accepted "big picture" it just gets brushed under the carpet. The second explanation is that they've been on drugs. This is not an entirely inappropriate criticism because synesthesia is more common among people who use LSD, but in my view that makes it more interesting, not less so. Why should some chemicals cause synesthesia (if indeed they do)?

The third idea is that synesthetes are just recalling childhood memories. For example, maybe they saw refrigerator magnets where five was red and six was blue and seven was green, and for some reason they're stuck with these memories. This has never made much sense to me because if it were true why should the condition run in families? (Unless the same magnets are passed down, or the propensity to play with magnets runs in families ...) The fourth explanation is more subtle and it invokes sensory metaphors. Our everyday language is replete with synesthetic metaphors, cross-sensory metaphors: for example "Cheddar cheese is sharp." Well, cheese isn't sharp, it's soft. What we mean is it tastes sharp; "sharp" is a metaphor. But this argument is circular – why use a tactile adjective, sharp, for a taste sensation?

In science, one mystery cannot be explained with another mystery. Saying that synesthesia is just a metaphor explains nothing because we don't know what a metaphor is or how it's represented in the brain. Indeed, I'd suggest the very opposite, that synesthesia is a sensory phenomenon whose neural basis can be discovered in the brain and that in turn can provide an experimental foothold for understanding more elusive aspects of the mind, such as metaphor.

Why has synesthesia been ignored for so long? There's an important lesson here in the history of science. I think in general it is fair to say that for a curious phenomenon, an anomaly, to make it into mainstream science and have an impact, it has to fullfil three criteria. First, it must be a demonstrably real phenomenon ... it has to be reliably repeatable under controlled conditions. Second, there must be a candidate mechanism that explains the phenomenon in terms of previously known principles. And third, it has to have significant implications beyond the phenomenon itself. Take, for example, telepathy. Telepathy has vastly significant implications if real, so the third criterion is fulfilled, but the first criterion is not fulfilled; it's not reliably repeatable. We don't even know if it's a real phenomenon.

Indeed, the more you measure it the smaller it becomes and that's always a bad sign. Another example is bacterial transformation. Some years ago it was discovered that incubating one species of the bacteria pneumococcus with another species resulted in the second species becoming transformed into the first. In fact, the transformation could be induced simply by extracting the chemical we would today call the bacteria's DNA. This was published in a prestigious journal and was reliably repeatable. Yet people ignored it because nobody could think of a candidate mechanism. How could you possibly encode heredity in a chemical? Then Watson and Crick described the double-helical structure of DNA and cracked the genetic code. Once that happened, the scientific community perked up and recognized the importance of bacterial transformation.

I have tried to do something similar with synesthesia. First of all, I will try to show it's real, not bogus. Second, I will suggest candidate mechanisms, what is going on in the brain. And third, I will argue that synesthesia has very broad implications. It might tell us about things like metaphor and how language evolved in the brain, maybe even the emergence of abstract thought that we human beings are very good at.

In an attempt to show that synesthesia is a real phenomenon, my colleagues and I essentially developed a clinical test to identify closet synesthetes. First, we found two synesthetes who saw numbers as colored; in their case 5 was green and 2 was red. We produced a computerized display which had a random jumble of 5s on the screen and embedded among these 5s a number of 2s, arranged to form a geometric shape. A non-synesthete would take as long as twenty seconds to see the arrangement of all the

Figure 4.1 *A "clinical test" for synesthesia. The display consists of 2s embedded in a matrix of randomly placed 5s. Non-synesthetes find it very hard to discern the embedded shape (in this case a triangle). Synesthetes who see the numbers as colored can detect the triangle much more easily. (Depicted schematically in this figure; compare with Figure 1.8.)*

2s (Figure 1.8), but the two synesthetes immediately or very quickly saw the shape formed by the red 2s against the background of contrasting green 5s (shown schematically in Figure 4.1). They are obviously not crazy: how can someone crazy outperform normals? And the effect must be sensory rather than based on memory, or they wouldn't be able to literally see the geometric shape. Using this and other similar tests, we have discovered that synesthesia is much more common than has been assumed in the past. In fact, we found that one in two hundred people is a synesthete.

What causes synesthesia? In 1999, my student Ed Hubbard and I were looking at a structure called the fusiform gyrus in the temporal lobes. The fusiform gyrus contains the color area V4 which was described by Semir Zeki (Zeki and Marini, 1998). This is the area which processes color information, but we were struck by the fact that the number area of the brain, which represents visual numbers as shown by brain imaging studies, is right next to it, almost touching the color area of the brain (Figure 4.2). This is an unlikely coincidence; the most common type of synesthesia is number/color synesthesia *and* the number area and color areas are right next to each other in the same part of the brain. It seemed likely that these people had some accidental cross-talk, or cross-wiring, just as in my phantom limb patients (see chapter 1). The difference here is that it happens not because of amputation but because of some genetic change in the brain. Imaging experiments on people with synesthesia (Figure 4.3) suggest that showing black and white numbers to a synesthete produces activation in the color area of the fusiform gyrus.[1]

Further evidence for this "cross-activation" theory came from a most unexpected observation. We recently came across a man who was partially color-blind but had full-blown synesthesia. Because of a deficiency in his cone pigments (in the retina) he couldn't see the full range of colors in the world. Yet when looking at numbers he could see colors that he could never experience otherwise. He referred to them charmingly as "Martian colours." We suggest that this occurs because, even though

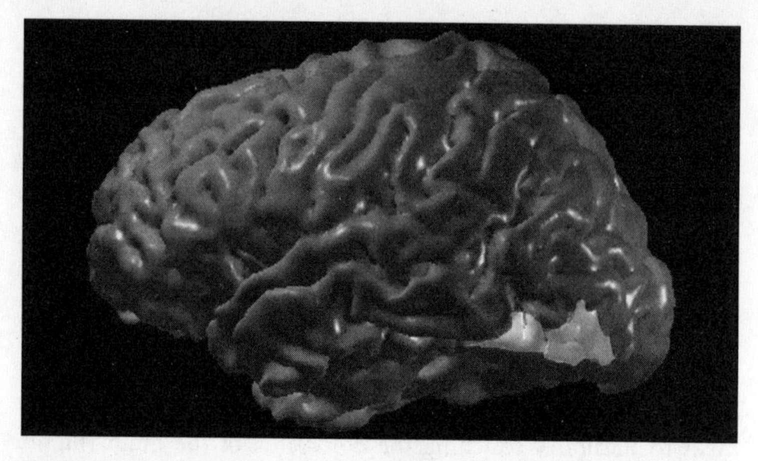

Figure 4.2 *Shows that the color area V4 and the so-called "number-grapheme area" are close to each other in the fusiform gyrus of the temporal lobes. A "cross-activation" between these adjacent areas may provide a neural substrate for synesthesia. Area V4 is indicated by diagonal lines while the number-grapheme area is indicated by cross-hatching.*

his receptors are deficient in the eye, the color areas in his brain are normal and can be accessed indirectly through cross-activation by numbers. The observation provides strong evidence against the memory association hypothesis: how could he remember something he had never actually seen?

Intriguingly, in some synesthetes even an invisible number can evoke color. If a number (say, 5) is presented off to one side and flanked by two other numbers, which we call distractors, then normal people find it hard to discern the middle number – an effect called crowding. This isn't caused by a decline of visual acuity in peripheral vision because the same number is readily seen if the two flanking distractors are removed (Figure 4.4).

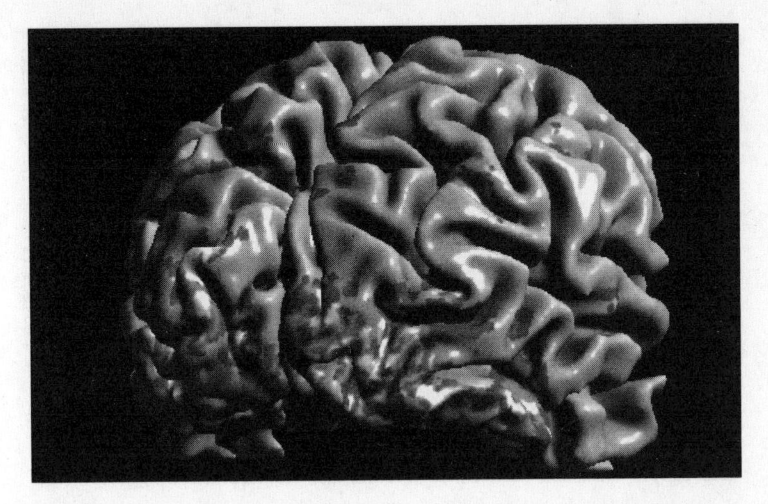

Figure 4.3 *Rear view of a synesthete's brain; fMR (functional magnetic resonance) image showing high activity in V4 – a color processing area – as the subject looks at white numbers on a grey background. This area is not active in normal people viewing the same figures.*

Crowding occurs because the flanking numbers distract attention from the middle number.

But a synesthete, unable to discern the number itself, will still identify a 5 "because it looks red" – implying that even a number that is not consciously visible can nevertheless evoke color! I suggest that the cross-activation occurs before the stage where the number reaches conscious awareness and the color evoked is then relayed to higher centers in the brain where it is consciously perceived and used to deduce intellectually what the number must have been.[2] This phenomenon bears an uncanny resemblance to what we called blindsight in chapter 2. It would also explain why many synesthetes actually use their colors as a

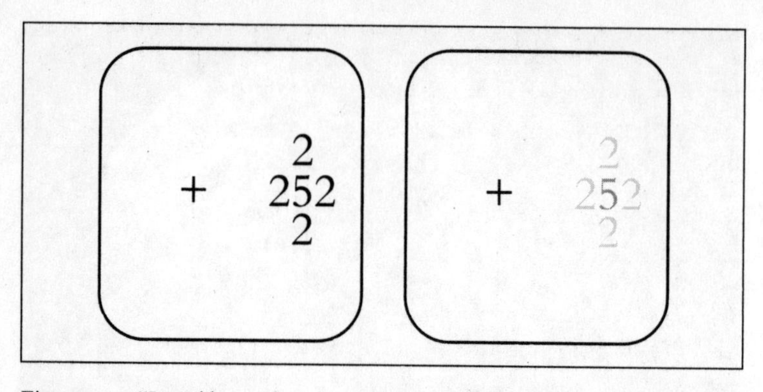

Figure 4.4 *"Invisible numbers." When a normal person stares at the central fixation sign (here a plus sign, +), a single digit off to one side is easy to see with peripheral vision (left). But, if the number is surrounded by other flanking numbers it appears indiscernible to the average person. In contrast a synesthete could deduce the central number by the color it evokes.*

mnemonic aid, e.g., for learning phone numbers and musical scales.

Why does this cross-wiring or cross-activation occur? The fact that it runs in families suggests that there is a gene, or set of genes, involved. What might this bad gene be doing? One possibility is that we are all born with excess connections in the brain. In the fetus there are many redundant connections which get pruned away to produce the modular architecture characteristic of the adult brain. What I think has happened in these people is that the "pruning" gene is defective, which has resulted in cross-activation between areas of the brain.[3] Or perhaps some kind of chemical imbalance has caused cross-activation between adjacent parts of the brain that are normally only loosely connected.

What we found next was even more amazing. We showed our two subjects Roman numbers V and VI instead of the Indian/Arabic numbers 5 and 6. They knew it was a five or six, but saw no color. This result is very important because it demonstrates that it is not the numerical concept that drives the color but the visual appearance of the number. This fits my argument because the fusiform gyrus represents the *visual* appearance of numbers and letters, not the abstract concept of sequence or ordinality.[4]

We don't know where in the brain the abstract idea of number is represented, but a good guess is the angular gyrus in the left hemisphere. When that area is damaged, patients can no longer do arithmetic even though they can see and identify numbers correctly. They are fluent in conversation, they remain intelligent, but cannot do even simple calculations such as seventeen minus three. This would indicate that abstract number concepts are represented in the angular gyrus while the fusiform gyrus deals with the visual appearance of a number.

But not all synesthetes are created equal. We soon ran into others in whom not merely numbers but even days of the week and even months of the year evoke colors: Monday is red, Tuesday is indigo, December is yellow. No wonder people thought they were crazy! What days, months and numbers have in common, though, is the abstract idea of sequence or ordinality, which I believe is represented higher up in the temporal parietal occipital (TPO) junction, in the vicinity of the angular gyrus (Figure 4.5). It should by now come as no surprise that the next color area in the color processing hierarchy is higher up in the general vicinity of the TPO junction, not far from the angular

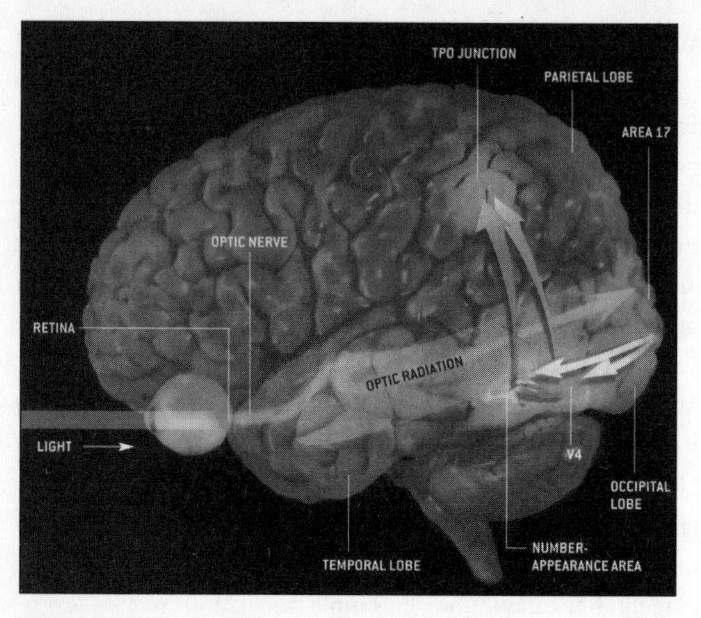

Figure 4.5 *Stages of number and color processing in the human brain. (Illustration by Carol Donner.)*

gyrus. So I think that in these people, those who see colors in days and months, the cross-wiring is higher up in the angular gyrus. For this reason, I call them higher synesthetes. In summary, if the faulty gene is selectively expressed in the fusiform gyrus, at an earlier stage in processing, the result is a lower synesthete driven by visual appearance. If the gene is expressed selectively higher up, in the vicinity of the angular gyrus, the result is a higher synesthete driven by numerical concept rather than visual appearance. Such selective gene expression can occur as a result of transcription factors.[5]

One in two hundred people has this completely useless peculiarity of seeing colored numbers – why did this gene survive? I would suggest it's a bit like sickle-cell anemia – there's a hidden agenda. These genes are doing something else important.[6]

One of the odd facts about synesthesia, which has been known, and ignored, for a long time, is that it is seven times more common among artists, poets, novelists – in other words, flaky types! Is this because artists are crazy? Or just willing to report their experiences unselfconsciously? Or maybe even trying to attract attention to themselves? (It's sexy to be a synesthete, given that many eminent artists are or were.) But I'd propose a very different view. What artists, poets and novelists all have in common is their skill at forming metaphors, linking seemingly unrelated concepts in their brain, as when Macbeth said "Out, out brief candle," talking about life. But why call it a candle? Is it because life is like a long white thing? Obviously not. Metaphors are not to be taken literally (except by schizophrenics, which is another story altogether). But in some ways life *is* like a candle: it's ephemeral, it can be snuffed out, it illumines only very briefly. Our brains make all the right links, and Shakespeare, of course, was a master at doing this. Now, let's make one further assumption – that this "cross-activation" or "hyperconnectivity" gene is expressed more diffusely throughout the brain, and not just in the fusiform or in the angular. As we have seen, if the gene is expressed in the fusiform you get a lower synesthete, and if expressed in the angular gyrus/TPO junction you get a higher synesthete. But if it's expressed everywhere there is greater hyperconnectivity throughout the brain, making that person more prone to metaphor, the ability to link seemingly unrelated things. (After all, even so-called abstract concepts are also represented in brain maps.) This may seem counter-intuitive, but just think of something like a number. There is nothing more abstract than a number. Five pigs, five donkeys, five chairs, even five tones – all very different, but with fiveness in common. That fiveness is represented in a fairly small region, the angular gyrus. So it's possible that other high-level concepts are also represented in brain maps and that artistic people, with their excess connections, can make these associations much more fluidly and effortlessly than less gifted people.[7]

So far we have shown that synesthesia is a genuine sensory effect and proposed a candidate mechanism, so satisfying the first two criteria outlined above for its entry into mainstream science. All that remains is to show that synesthesia is not just some oddity – something weird – but has implications beyond the confines of a narrow speciality. In my view, synesthesia is far more than just a quirk in some people's brains. In fact, most of us are synesthetes, as I will now demonstrate. Imagine that in front of you is a bulbous amoeboid shape on which are many undulating curves. And right next to it imagine a jagged shape – like a piece of shattered glass with sharp jagged edges (Figure 4.6). These shapes are the first two letters of the Martian alphabet. One of these shapes is kiki and the other is booba, and you have to decide which is which. Look at the figure now and decide which is kiki. In experiments, 98 percent of people say the jagged shape is kiki and the bulbous amoeboid shape is booba. If you are among them you're a synesthete. Let me explain. Look at the letter kiki and compare it with the sound "kiki." They both share one property: the kiki visual shape has a sharp inflexion and the sound "kiki" represented in your auditory cortex, in

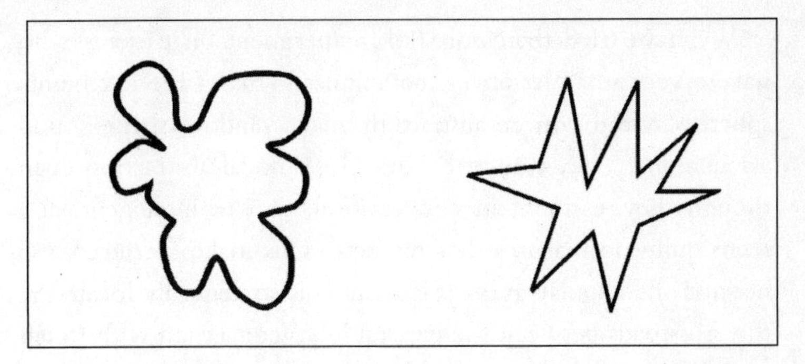

Figure 4.6 *If asked which of these two abstract shapes is "booba" and which "kiki", 95–98 percent of respondents pick the blob as booba and the jagged shape as kiki. This is also true for non-English-speaking Tamillians for whom the shapes bear no resemblance to visual shapes of the Tamil alphabet corresponding to B or K. The effect demonstrates the brain's ability to engage in cross-modal abstraction of properties such as jaggedness or curviness. Our preliminary results suggest that the effect is compromised in patients with left angular gyrus lesions who also have difficulties with metaphor.*

the hearing centers of your brain, also has a sharp sudden inflection. Your brain performs a cross-modal synesthetic abstraction, recognizing that common property of jaggedness, extracting it, and so reaching the conclusion that they are both kiki.

(Interestingly, Tamillians, who don't speak or write English, produce the same results, so this phenomenon is unrelated to the jagged shape resembling the visual appearance of the letter K. Other shapes can also be paired with sounds in this manner: for example, if you show a blurred or smudged line and a sawtooth and ask people which is "rrrrr" and which is "shhhhh" they spontaneously pair the former with "shhhhh" and the latter with "rrrrr.")

We have tried the booba/kiki experiment on patients who have a very small lesion in the angular gyrus of the left hemisphere. Unlike you and me, they make random shape–sound associations. They cannot do this cross-modal abstraction even though they're fluent in conversation, they're intelligent, and seem quite normal in other respects. This makes perfect sense because the angular gyrus (Figure 1.3) is strategically located at the crossroads between the parietal lobe (concerned with touch and proprioception), the temporal lobe (concerned with hearing) and the occipital lobe (concerned with vision). So it is strategically placed to allow a convergence of different sense modalities to create abstract, modality-free representations of things around us. Logically, the jagged shape and the sound "kiki" have nothing in common: the shape comprises photons hitting the retina in parallel, the sound is a sharp air disturbance hitting the hair cells of the inner ear sequentially. But the brain abstracts the common denominator – the property of jaggedness. Here in the angular gyrus are the rudimentary beginnings of the property we call abstraction that we human beings excel in.

Why did this ability evolve in humans in the first place? Why cross-modal abstraction? If we compare the brains of lower mammals, monkeys, great apes and humans, we find a progressive enlargement of the TPO junction and angular gyrus, an almost explosive development. And especially so in humans. I think this ability initially evolved to help us survive in treetops, grabbing handholds, jumping from branch to branch. To do this it is necessary to adjust the angle of the arm and the fingers so that the proprioceptive map (signaled by receptors in muscles and joints) matches the horizontality of a branch's visual appearance – the horizontal array of photons – which is why the angular gyrus became larger and larger. But once this ability to engage in cross-modal abstraction was developed, that structure in turn became an exaptation for the other types of abstraction that modern humans excel in, be it metaphor or any other type of abstraction.[8] Such opportunistic hijacking of a structure for a function other than the one for which it originally evolved is the rule rather than the exception in biology. For example, two of the bones in the lower jaws of reptiles which evolved for chewing became transformed into the tiny middle-ear bones of mammals – used for hearing – simply because these bones were "at the right place at the right time."

I would conjecture that the TPO junction – especially the angular gyri – in the two hemispheres may have also evolved complementary roles in mediating somewhat different types of metaphor: the left one for cross-modal ones (e.g., "loud shirt," "sharp cheese") and the right for spatial metaphors (he "stepped down" from his post). This has not yet been tested systematically. But, as I noted earlier, two patients with left angular gyrus lesions I tested recently were both abysmal at interpreting proverbs and metaphors and also failed the booba/kiki test.

Finally, I would like to turn to the evolution of language. This has always been a very controversial topic. Language is amazing. Its subtleties and nuances (including what's called recursive embedding) combine with an enormous lexicon to produce a highly sophisticated mechanism. It's easy to imagine a single trait ... like a giraffe's long neck arising from the progressive accumulation of chance variations. But how could an extraordinarily complex mechanism like language with so many interlocking components have evolved through the blind workings of chance – through natural selection? How did we make the transition from the grunts and howls and groans of our ape-like ancestors to all the sophistication of a Shakespeare, or a George W. Bush? There have been several theories. Alfred Russell Wallace said the mechanism is so complicated it couldn't have evolved through natural selection at all and must have resulted from divine intervention. The second theory was by Noam Chomsky, the founding father of linguistics. Chomsky said something quite similar, although he didn't invoke God. He said this mechanism is so sophisticated and elaborate that it couldn't have emerged through natural selection, through the blind workings of chance, but that packing one hundred billion nerve cells in such a tiny space may result in some new laws of physics emerging. He almost says it's a miracle, although he doesn't use the word. Unfortunately, neither Wallace's nor Chomsky's theory can be tested. A third idea was proposed by the brilliant MIT psychologist Steve Pinker. Pinker suggests that the evolution of language is no big mystery. What we see now is the final result of evolution, and it seems mysterious only because we don't know what the intermediate steps were. I suspect he is on the right track – natural selection is the only plausible explanation – but he doesn't go far enough. As biologists, we want the details – the devil's in the details. We want to know what those intermediate steps were, not merely that language *could* have evolved through natural selection. (The technical term for this is "trajectory through the fitness landscape.") And the vital clue to discovering those steps comes from the booba/kiki example, from synesthesia, leading me to propose what I call the synesthetic bootstrapping theory of language origins.

Let's begin with the lexicon. How did we evolve a shared vocabulary – such a huge repertoire of thousands of words?[9] Did our ancestral hominids sit around the fire and say, "Everybody call that object an axe?" Of course not. But if they didn't do that, what did they do? The booba/kiki example provides the clue. It shows there is a pre-existing, *non-arbitrary* translation between the visual appearance of an object represented in the fusiform gyrus and the auditory representation in the auditory cortex. In other words, a synesthetic cross-modal abstraction is already going on, a pre-existing translation, if you like, between visual appearance and auditory representation. Admittedly, this is a very small bias, but that is all that is required in evolution to get something started.[10]

But this is only part of the story. Just as there is a pre-existing, built-in cross-activation between sound and vision – the booba/kiki effect – there's also a non-arbitrary cross-activation between the visual area in the fusiform and the Broca's area in the front of the brain that generates programs which control the muscles of vocalization, phonation and articulation – how we move our lips, tongue and mouth. How do we know that? Say the words "teeny weeny," "un peu," "diminutive." Look at what your lips are doing: they are physically mimicking the visual appearance of what you are saying. Now say "enormous," "large" ... Such mimicry indicates a pre-existing bias to systematically map certain visual shapes on to certain "sounds" represented in the motor maps in the Broca's area (Figure 4.7).

The third part of my theory is that there is also a pre-existing cross-activation between the hand area and the mouth area, which are right next to each other in the Penfield motor map in the brain (see Figure 1.6). I can illustrate with an example from

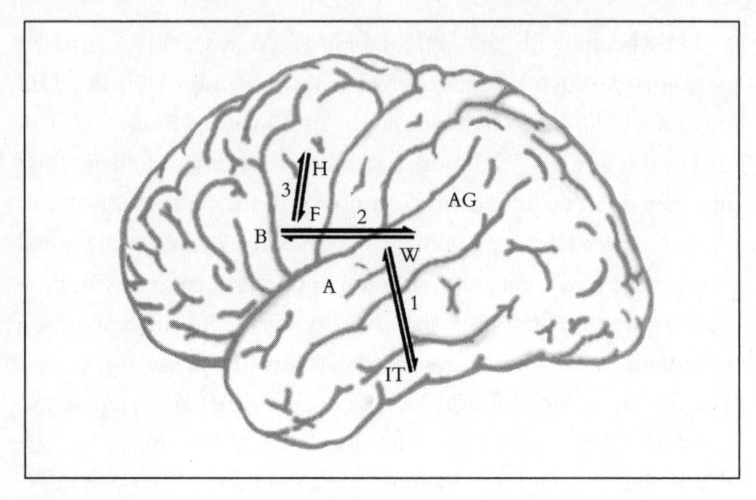

Figure 4.7 *A new synesthetic bootstrapping theory of language origins. Arrows depict cross-domain remapping of the kind we postulate for synesthesia in the fusiform gyrus. (1) A non-arbitrary synesthetic correspondence between visual object shape (as represented in IT and other visual centers) and sound contours represented in the auditory cortex (as in our booba/kiki example). Such synesthetic correspondence could be based on either direct cross-activation or mediated by the angular gyrus – long known to be involved in inter-sensory transformations. (2) Cross-domain mapping (perhaps involving the arcuate fasciculus) between sound contours and motor maps in or close to Broca's area (mediated, perhaps, by mirror neurons). (3) Motor to motor mappings (synkinesia) caused by links between hand gestures and tongue, lip and mouth movements in the Penfield motor homunculus synkinetically mimic the small pincer gesture made by opposing thumb and index finger (as opposed to "large" or "enormous").*

Charles Darwin. He noticed that when people cut something with a pair of scissors they clench and unclench their jaws unconsciously, as if to echo or mimic the movements of the fingers. I call this phenomenon synkinesia, because the hand and mouth areas are right next to each other and perhaps there is some spill-over of signals from gestures to vocalizations (e.g., the oral gestures for "little" or "diminutive" or "teeny weeny") synkinetically mimic the small pincer gesture of opposing the thumb and index finger.

A system of non-verbal communication would have been important to our ancestral hominids unable to engage in loud vocalization when hunting. We also know that the right hemisphere produces guttural emotional utterances along with the anterior cingulate. Combine these with the pre-existing translation of manual gesturing into mouth, lip and tongue movements and you get humankind's first words: proto-words.

Thus we have three things in place – first, hand to mouth; second, mouth in Broca's area to visual appearance in the fusiform and sound contours in the auditory cortex; and third, auditory to visual, the booba/kiki effect. Acting together, these three have a synergistic bootstrapping effect – an avalanche culminating in the emergence of a primitive language (Figure 4.7).

All this is fine, but how do we explain the hierarchical structure of syntax? For example, "He knows that I know that he knows that I had an affair with his wife," or, "She hit the boy who kissed the girl that she disliked." How does this hierarchic embedding in language come about? Partly, I think, from semantics, from the region of the TPO junction which is involved in abstraction (I have already explained how abstraction might have evolved). It is possible that abstraction (and semantics) feeds into syntactic structure and may have played a role in "guiding" its evolution. But partly, the hierarchical "tree" structure of syntax may have evolved from tool use. Early hominids were very good at tool use, especially what is known as the sub-assembly technique: a piece of flint is made into a head – step one – then hafted on to a handle – step two – then the whole thing is used as a tool or weapon – step three. There is a close operational analogy between this function and the embedding of noun clauses within longer sentences. So perhaps what originally evolved for tool use in the hand area is now exapted and assimilated in the Broca's area to be used in aspects of syntax such as hierarchic embedding.

Each of these effects is a small bias, but acting in conjunction they may have paved the way for the emergence of sophisticated language. This is very different from Steve Pinker's idea that language is a specific adaptation which evolved step by step for the sole purpose of communication. I suggest, instead, that it is the fortuitous synergistic combination of a number of mechanisms which evolved for other purposes initially that later became assimilated into the mechanism that we call language. This often happens in evolution, but this style of thinking has yet to permeate neurology and psychology. It seems odd that neurologists often overlook evolution as explanation, given that, as Dobzhansky once said, nothing in biology makes sense except in the light of evolution.[11]

My final point concerns the mirror neurons mentioned in chapter 2 – cells in the parietal and frontal lobes that fire not only when you move your hand but also when you watch someone else move their hand. Similar neurons also exist for orofacial movements: not only do they fire when you stick your tongue out or purse your lips, but also when you see someone else do it, even though you have never directly seen your own lips or tongue. The neurons must be establishing a congruence between (1) the highly specific volitional motor command sequence sent to muscles of phonation and articulation; (2) *felt* lip and tongue position (proprioception) from sensors in your oral muscles; (3) the *seen* image of someone else's lips and tongue; and (4) the *heard* phoneme. This ability is partly innate – if you stick your tongue out at a newborn baby, it will imitate you – but the more complex three-way congruence between the phoneme sound, the appearance of lips and tongue and their felt position that is required for lipreading is probably acquired in childhood. It seems likely that these neurons may have played an important role in developing a shared vocabulary by miming seen vocalizations and establishing a congruence with heard sounds.

As an illustration, consider the fact that if a normal English-speaking person watches me silently pronounce the syllable Rrrrr or Llll he can "lipread" and correctly infer which one I am producing, presumably using his mirror neurons. But when I recently tried this on a native Chinese speaker (who learned English only as an adult) he had considerable difficulty making this judgment; perhaps the mirror neurons required for making this particular distinction simply didn't develop in him.

In this chapter, we began with a disorder – synesthesia – that's been known for a century but treated mainly as a curiosity. We then showed that the phenomenon is real, pointed out what the underlying brain mechanisms might be, and spelled out its broader implications. (One day we might be able to clone the gene or genes, if we can find a large enough family. I have recently heard rumors of a whole island of synesthetes!) We then suggested that if the gene is expressed in the fusiform gyrus it results in lower synesthesia, and higher synesthesia if expressed in the angular gyrus. If it's expressed all over you get artsy types! We have learned something of the detailed perceptual psychophysics – such as the pop-out effect of 2s against 5s, which can be measured – and can perhaps begin to approach elusive phenomena such as abstract thought, metaphor, Shakespeare, even the evolution of language. All this by studying one little quirk that people call synesthesia. So I agree wholeheartedly with what Thomas Henry Huxley said in the nineteenth century: contrary to the views of Bishop Wilberforce and Benjamin Disraeli, we are not angels, we are merely sophisticated apes. Yet we don't *feel* like that – we feel like angels trapped inside the bodies of beasts, forever craving transcendence, trying to spread our wings and fly off. Really a very odd predicament to be in, if you think about it.

Our revels now are ended. These our actors,
As I foretold you, were all spirits and
Are melted into air, into thin air ...
We are such stuff
As dreams are made on,
And our little life
Is rounded with a sleep.

5

Neuroscience — The New Philosophy

He who understands baboon would do more toward metaphysics than Locke.

CHARLES DARWIN

The main theme of the book so far has been the idea that the study of patients with neurological disorders has implications far beyond the confines of medical neurology, implications even for the humanities, for philosophy, maybe even for aesthetics and art. In this final chapter I would like to continue this theme and take up the challenge of mental illness. The boundary between neurology and psychiatry is becoming increasingly blurred and it is only a matter of time before psychiatry becomes just another branch of neurology. I shall also touch on a few philosophical issues such as free will and the nature of self.

There have traditionally been two different approaches to mental illness. The first one tries to identify chemical imbalances, changes in transmitters and receptors in the brain, and attempts to correct these changes using drugs. This approach has revolutionized psychiatry and has been phenomenally successful. Patients who used to be put in straitjackets or locked up can now lead relatively normal lives. The second approach we can loosely characterize as the so-called Freudian approach. It assumes that most mental illness arises from early upbringing. I'd like to propose a third approach which is radically different from either of these but which, in a sense, complements them both.

To understand the origins of mental illness it is not enough merely to say that some transmitter has changed in the brain. We need to know how that change produces the bizarre symptoms that it does – why certain patients have certain symptoms and why those symptoms are different for different types of mental illness. I will attempt to explain the symptoms of mental illness in terms of what is known about function, anatomy and neural structures in the brain. And I will suggest that many of these symptoms and disorders seem less bizarre when viewed from an evolutionary standpoint, that is from a Darwinian perspective. I also propose to give this discipline a new name – evolutionary neuro-psychiatry.

Let us begin with the classic example of what most people consider to be a purely mental disorder or psychological disturbance – hysteria. "Hysteria" is used here in its strict medical sense, as opposed to the everyday notion of a person shouting and screaming. In the strictly medical sense, a hysteric is a patient who suddenly experiences blindness or develops a paralysis of an arm or a leg, but who has no neurological deficits that could be responsible for his or her condition: a brain MR scan reveals that the brain is apparently completely normal, there are no identifiable lesions, no apparent damage. So the symptoms are dismissed as being purely psychological in origin.

However, recent brain-imaging studies using PET scans and functional Magnetic Resonance imaging have dramatically changed our understanding of hysteria. Using PET scans and fMR, we can now find what parts of the brain are active or inactive when a patient performs a specific action or engages in a specific mental process. For example, when we do mental arithmetic, the left angular gyrus usually exhibits activity. Or if I were to prick you with a needle and cause pain, another part of your brain would light up. We can then conclude that the particular brain region that lights up is somehow involved in mediating that function.

If you were to wiggle your finger, a PET scan would reveal that two areas of your brain light up. One is called the motor cortex, which is actually sending messages to execute the appropriate sequence of muscle twitches to move your finger, but there is another area in front of it called the pre-motor cortex that *prepares* you to move your finger.

John Marshall, Chris Frith, Richard Frackowiak, Peter Halligan and others tried this experiment on a hysterically paralyzed patient. When he tried to move his leg, the motor area failed to light up even though he claimed to be genuinely intending to move his leg. The reason he was unable to is that another area was simultaneously lighting up: the anterior cingulate and the orbito-frontal lobes. It's as if this activity in the anterior cingulate and orbito-frontal cortex was inhibiting or vetoing the hysterical patient's attempt to move his leg. This makes sense, because the anterior cingulate and orbito-frontal cortex are intimately linked to the limbic emotional centers in the brain, and we know that hysteria originates from some emotional trauma that is somehow preventing him from moving his "paralyzed" leg.

Of course, all this doesn't explain exactly why hysteria occurs, but now we at least know where to look. In the future it might be possible to use a brain scan to distinguish genuine hysterics from malingerers or fraudulent insurance claimants. And it does prove that one of the oldest "psychological" disturbances – one that Freud studied – has a specific and identifiable organic cause. (Actually, an important control is missing in this experiment: no one has yet obtained a brain scan from a genuine malingerer.)

We can think of hysteria as a disorder of "free will," and free will is a topic that both psychologists and philosophers have been preoccupied with for over two thousand years.

Several decades ago the American neurosurgeon Benjamin Libet and the German physiologist Hans Kornhuber were experimenting on volunteers exercising free will, instructing subjects to, for example, wiggle a finger at any time of their own choosing within a ten-minute period. A full three-quarters of a second *before* the finger movement the researchers picked up a scalp EEG potential, which they called the "readiness potential," even though the subject's sensation of consciously willing the action coincided almost exactly with the actual onset of finger movement. This discovery caused a flurry of excitement among philosophers interested in free will. For it seemed to imply that the brain events monitored by the EEG kick in almost a second before there is any sensation of "willing" the finger movement, even though your *subjective* experience is that your will caused the finger movement! But how can your will be the cause if the brain commands begin a second earlier? It's almost as though your brain is really in charge and your "free will" is just a post-hoc rationalization – a delusion, almost – like King Canute thinking he could control the tides or an American president believing that he is in charge of the whole world.

This alone is strange enough, but what if we add another twist to the experiment. Imagine I'm monitoring your EEG while you wiggle your finger. Just as Kornhuber and Libet did, I will see a readiness potential a second before you act. But suppose I display the signal on a screen in front of you so that you can *see* your free will. Every time you are about to wiggle your finger, supposedly using your own free will, the machine will tell you a second in advance! What would you now experience? There are three logical possibilities. (1) You might experience a sudden loss of will, feeling that the machine is controlling you, that you are a mere puppet and that free will is just an illusion. You may even become paranoid as a result, like schizophrenics who think their actions are controlled by aliens or implants (I'll return to this later). (2) You might think that it does not change your sense of free will one iota, preferring to believe that the machine has some sort of spooky paranormal precognition by which it is able to predict your movements accurately. (3) You might confabulate, or rearrange the experienced sequence mentally in order to cling to your sense of freedom; you might deny the evidence of your eyes and maintain that your sensation of will preceded the machine's signal, not vice versa.

At this stage this is still a "thought experiment" – technically it is hard to get a feedback EEG signal on each trial, but we are trying to get around this obstacle. Nevertheless, it is important to note that one can do experiments that have direct relevance to broad philosophical issues such as free will – a field in which my colleagues Pat Churchland, Dan Wegner and Dan Dennett have all made valuable contributions.

Leaving aside this "thought experiment" for the moment, let's return to the original observation on the readiness potential with its curious implication that the brain events are kicking in a second or so before any actual finger movement, even though conscious intent to move the finger coincides almost exactly with the wiggle. Why might this be happening? What might the evolutionary rationale be?

The answer is, I think, that there is an inevitable neural delay before the signal arising in one part of the brain makes its way through the rest of the brain to deliver the message: "wiggle your finger." (A televisual equivalent is the sound delay experienced when conducting an interview via satellite.) Natural selection has ensured that the subjective sensation of willing is delayed deliberately to coincide not with the onset of the brain commands but with the actual execution of the command by your finger.[1]

And this in turn is important because it means that the subjective sensations which accompany brain events must have an evolutionary purpose. For if that were not the case, if they merely *accompanied* brain events, as so many philosophers believe (this is called epiphenomenalism) – in other words, if the subjective sensation of willing is like a shadow that accompanies us as we move but is not causal in making us move – then why would evolution bother delaying the signal so that it coincides with our movement?

So we have a paradox: on the one hand, the experiment shows that free will is illusory: it cannot be causing the brain events because the events kick in a second earlier. But on the other hand, the delay must have some function, otherwise why would the delay have evolved? Yet if it *does* have a function, what could it be other than moving (in this case) the finger? Perhaps our very notion of causation requires a radical revision … as happened in quantum mechanics.

Other types of "mental" illness can also be approached, perhaps, through brain imaging. Take the case of pain: when someone is jabbed with a needle, there is usually activity in many regions of the brain, but especially in the insula and in the anterior cingulate. The former structure seems to be involved in sensing the pain and the latter in giving pain its aversive quality. So when the pathways leading from the insula to the anterior cingulate are severed, the patient can feel the pain, but it doesn't hurt—a paradoxical syndrome called pain asymbolia. This leads me to wonder about the image of the brain of a masochist who derives pleasure from pain, or a patient with Lesch–Nyhan syndrome who "enjoys" mutilating himself. The insula would be activated, of course, but would the anterior cingulate also light up? Or, given, especially, the sexual overtones of masochism, a region concerned with pleasure, such as the nucleus accumbens, septum or hypothalamic nuclei? At what stage in processing do the "pain/pleasure" labels get switched? (I am reminded of the masochist from Ipswich who loved taking ice cold showers at four in the morning and therefore didn't.)

In chapter 1 I mentioned the Capgras delusion, sometimes seen in patients who have sustained a head injury, in which sufferers start claiming that someone they both recognize and know well – such as their mother – is an imposter.

The theoretical explanation for Capgras syndrome is that the connection between the visual areas and the emotional core of the brain, the limbic system and the amygdala, has been cut by the accident (Figure 1.3). So when the patient looks at his mother, since the visual areas in the brain concerned with recognizing faces is not damaged, he is able to say that she *looks* like his mother. But there is no emotion, because the wire taking that information to the emotional centers is cut, so he tries to rationalize this by believing her to be an imposter.

How can this theory be tested? Well, it is possible to measure the gut-level emotional reaction that someone has to a visual stimulus – or any stimulus – by measuring the extent to which they sweat. When any of us sees something exciting, emotionally important, the neural activation cascades from the visual centers to the emotional centers in the brain and we begin to sweat in order to dissipate the heat that we are going to generate from exercise, from action (feeding, fleeing, fighting, or sex). This effect can be measured by placing two electrodes on a person's skin to track changes in skin resistance – when skin resistance falls, we call it a galvanic skin response. Familiar or nonthreatening objects or people produce no galvanic skin response because they generate no emotional arousal. But if you look at a lion or a tiger, or – as it turns out – your mother, a huge galvanic skin response occurs. Believe it or not, every time you see your mother, you sweat! (And you don't even have to be Jewish.)

But we found that this doesn't happen in Capgras patients, supporting the idea that there has been a disconnection between vision and emotion.

There exists an even more bizarre disorder, Cotard's syndrome, in which the patient starts claiming he or she is dead. I suggest that this is similar to Capgras except that instead of vision alone being disconnected from the emotional centers in the brain, all the senses become disconnected from the emotional centers. So that nothing in the world has any emotional significance, no object or person, no tactile sensation, no sound – nothing – has emotional impact. The only way in which a patient can interpret this complete emotional desolation is to believe that he or she is dead. However bizarre, this is the only interpretation that makes sense to him; the reasoning gets distorted to accommodate the emotions. If this idea is correct we would expect no galvanic responses in a Cotard's patient whatever the stimulus.

The delusion of Cotard's is notoriously resistant to intellectual correction. For example, a man will agree that dead people cannot bleed; then, if pricked with a needle, he will express amazement and conclude that the dead *do* bleed after all, instead of giving up his delusion and inferring that he is alive. Once a delusional fixation develops, all contrary evidence is warped to accommodate it. Emotion seems to override reason rather than the other way around. (Of course, this is true of most of us to some extent. I have known many an otherwise rational and intelligent person who believes the number 13 to be unlucky or who won't walk under a ladder.)

Capgras and Cotard's are both rare syndromes, but there is another disorder, a sort of mini-Cotard's, that is much more commonly seen in clinical practice. This disorder is known as derealization and depersonalization, and is found in acute anxiety, panic attacks, depression and other dissociative states. Suddenly the world seems completely unreal – like a dream. The patient feels like a zombie.

I believe such feelings involve the same circuitry as Capgras and Cotard's. In nature, an opossum when chased by a predator will suddenly lose all muscle tone and play dead, hence the phrase "playing possum." This is a good strategy for the opossum because (a) any movement will encourage the predatory behavior of the carnivore and (b) carnivores usually avoid carrion, which might be infected. Following the lead of Martin Roth, Mauricio Sierra and German Berrios, I suggest that derealization and depersonalization, and other dissociative states, are examples of playing possum in the emotional realm and that this is an evolutionary adaptive mechanism.

There is a well-known story of the explorer David Livingstone being attacked by a lion. He saw his arm being mauled but felt no pain or even fear. He felt detached from it all, as if he were watching events from a distance. The same thing can happen to soldiers in battle or to a woman being raped. During such dire emergencies, the anterior cingulate in the brain, part of the frontal lobes, becomes extremely active. This inhibits or temporarily shuts down the amygdala and other limbic emotional centers, so temporarily suppressing potentially disabling emotions such as anxiety and fear. But at the same time, the anterior cingulate activation generates extreme alertness and vigilance in preparation for any appropriate defensive reaction that might be required.

In an emergency, this James Bond-like combination of shutting down emotions ("nerves of steel") while being hypervigilant is useful, keeping us out of harm's way. It is better to do nothing than to engage in some sort of erratic behavior. But what if the same mechanism is accidentally triggered by chemical imbalances or brain disease, when there is no emergency? A person looks at the world, is intensely alert, hypervigilant, but the world has become completely devoid of emotional meaning because the limbic system has been shut down. There are only two possible ways to interpret this strange predicament, this paradoxical state of mind. Either "the world isn't real" – derealization – or "I am not real" – depersonalization.

Epileptic seizures originating in this part of the brain can also produce these dreamlike states of derealization and depersonalization. Intriguingly, we know that during a seizure, when the patient is experiencing derealization, there is no galvanic skin response to anything. Following the seizure, skin response returns to normal. All of which supports the hypothesis that we have been considering.

Probably the disorder most commonly associated with the word "madness" is schizophrenia. Schizophrenics do indeed exhibit bizarre symptoms. They hallucinate, often hearing voices. They become delusional, thinking they're Napoleon or Ramachandran. They are convinced the government has planted devices in their brain to monitor their thoughts and actions. Or that aliens are controlling them.

Psycho-pharmacology has revolutionized our ability to treat schizophrenia, but the question remains: why do schizophrenics behave as they do? I'd like to speculate on this, based on some work my colleagues and I have done on anosognosia (denial of illness) – which results from right-hemisphere lesions – and some very clever speculations by Chris Frith, Sarah Blakemore and Tim Crow. Their idea is that, unlike normal people, schizophrenics cannot tell the difference between their own internally generated images and thoughts and perceptions that are evoked by real things outside.

If I conjure up a mental picture of a clown in front of me, I don't confuse it with reality partly because my brain has access to the internal command I gave. I am expecting to visualize a clown, and that is what I see. It is not an hallucination. But if the "expectation" mechanism in my brain that does this becomes faulty, then I would be unable to tell the difference between a clown I'm imagining and a clown I'm actually seeing there. In other words, I would believe that the clown was real. I would hallucinate, and be unable to differentiate between fantasy and reality.

Similarly, I might momentarily entertain the thought that it would be nice to be Napoleon, but in a schizophrenic this momentary thought becomes a full-blown delusion instead of being vetoed by reality.

What about the other symptoms of schizophrenia – alien control, for example? A normal person knows that he moves of his or her own free will, and can attribute the movement to the fact that the brain has sent the command "move." If the mechanism that monitors intention and compares it with performance is flawed, a more bizarre interpretation is likely to result, such as that body movements are controlled by aliens or brain implants, which is what paranoid schizophrenics claim.

How do you test a theory like this? Here is an experiment for you to try: using your right index finger, tap repeatedly your left index finger, keeping your left index finger steady and inactive.

Notice how you feel the tapping mainly on the left finger, very little on the right finger. That is because the brain has sent a command from the left hemisphere to the right hand saying "move." It has alerted the sensory areas of the brain to expect some touch signals on the right hand. Your left hand, however, is perfectly steady, so the taps upon it come as something of a surprise. This is why you feel more sensation in the immobile finger, even though the tactile input to both fingers is exactly the same. (If you change hands, you will find that the results are reversed.)

Following our theory, I predict that if a schizophrenic were to try this experiment, he would feel the sensations equally in both fingers since he is unable to differentiate between internally generated actions and externally generated sensory stimuli. It's a five-minute experiment – yet no one has ever tried it.[2]

Or imagine that you are visualizing a banana on a blank white screen in front of you. While you are doing this, if I secretly project a very low-contrast physical image of the banana on the screen, your threshold for detecting this real banana will be elevated – presumably even your normal brain tends to get confused between a very dim real banana and one which you imagine. This surprising result is called the "Perky effect" and one would predict that it would be amplified enormously in schizophrenics.

Another simple yet untried experiment: as you know, you can't tickle yourself. That is because your brain knows you're sending the command. Prediction: a schizophrenic will laugh when he tickles himself.

Even though the behavior of many patients with mental illness seems bizarre, we can now begin to make sense of the symptoms using our knowledge of basic brain mechanisms. Mental illness might be thought of as disturbances of consciousness and of self, two words that conceal great depths of ignorance. Let me try to summarize my own view of consciousness. There are really two problems here – the problem of the subjective sensations or qualia and the problem of the self. The problem of qualia is the more difficult.

The qualia question is, how does the flux of ions in little bits of jelly – the neurons – in our brains give rise to the redness of red, the flavor of Marmite or paneer tikka masala or wine?[3] Matter and mind seem so utterly unlike each other. One way out of this dilemma is to think of them really as two different ways of describing the world, each of which is complete in itself. Just as we can describe light as made up either of particles or as waves – and there's no point in asking which description is correct, because they both are, even though the two seem utterly dissimilar – the same may be true of mental and physical events in the brain.

But what about the self, the last remaining great mystery in science and something that everybody is interested in? Obviously self and qualia are two sides of the same coin. You can't have free-floating sensations or qualia with no one to experience them and you can't have a self completely devoid of sensory experiences, memories or emotions. (As we saw in Cotard's syndrome, when sensations and perceptions lose all their emotional significance and meaning, the result is a dissolution of self.)

What exactly is meant by the "self"? Its defining characteristics are fivefold. First of all, continuity: a sense of an unbroken thread running through the whole fabric of our experience with the accompanying feeling of past, present and future. Second, and closely related, is the idea of unity or coherence of self. In spite of the diversity of sensory experiences, memories, beliefs and thoughts, we each experience ourselves as one person, as a unity.

Third is a sense of embodiment or ownership – we feel ourselves anchored to our bodies. Fourth, a sense of agency, what we call free will, being in charge of our own actions and destinies. I can wiggle my finger but I can't wiggle my nose or your finger.

Fifth, and most elusive of all, the self, almost by its very nature, is capable of reflection – of being aware of itself. A self that's unaware of itself is an oxymoron.

Any or all of these different aspects of self can be differentially disturbed in brain disease, which leads me to believe that the self comprises not just one thing, but many. Like "love" or "happiness," we use one word, "self," to lump together many different phenomena. For example, if I stimulate your right parietal cortex with an electrode (while you're conscious and awake), you will momentarily feel that you are floating near the ceiling, watching your own body down below. You have an out-of-the-body experience. The embodiment of self – one of the axiomatic foundations of your self – is temporarily abandoned.[4] And this is true of all of those aspects of self I listed above. Each of them can be selectively affected in brain disease.

Keeping this in mind, I see three ways in which the problem of self might be tackled by neuroscience. First, maybe the problem of self is a straightforward empirical one. Maybe there is a single, very elegant, Archimedes-type Eureka! solution to the problem, just as DNA base-pairing was the solution to the riddle of heredity. I think this is unlikely, but I could be wrong.

Second, given my earlier remarks about the self, the notion of the self as being defined by a set of attributes – embodiment, agency, unity, continuity – maybe we will succeed in explaining each of these attributes individually in terms of what is going on in the brain. Then the problem of what the self is will vanish, or at least recede into the background, just as scientists no longer speak of vital spirits or ask what "life" is. (We recognize that life is a word loosely applied to a collection of processes – DNA replication and transcription, Krebs cycle, Lactic acid cycle, etc., etc.)

Third, maybe the solution to the problem of the self is not a straightforward empirical one. It may instead require a radical shift in perspective, the sort of thing that Einstein did when he rejected the assumption that things can move at arbitrarily high velocities. When we finally achieve such a shift in perspective, we may be in for a big surprise and find that the answer was staring at us all along. I don't want to sound like a New Age guru, but there are curious parallels between this idea and the Hindu philosophical (albeit somewhat nebulous) view that there is no essential difference between self and others, or that the self is an illusion.

Of course, I have no clue what the solution to the problem of self is, what the shift in perspective might be. If I did I would dash off a paper to *Nature* today, and overnight I'd become the most famous scientist alive. But, just for fun, I'll have a crack at describing what the solution might look like.

I will begin with qualia. It seems quite obvious that qualia must have evolved to fullfil a specific biological function – they cannot be mere by-products (an "epiphenomenon") of neural activity. In 1997 I suggested that sensory representations that are themselves devoid of qualia might acquire qualia in the process of being economically encoded or "prepared" into manageable chunks as they are delivered to a central executive structure higher up in the brain. The result is a higher order representation that serves new computational goals. Let us call this second, higher-order, representation a metarepresentation. (Though I feel a bit uncomfortable using the prefix "meta," which is often employed as a disguise for fuzzy thinking – especially among social scientists.) One could think of this metarepresentation almost as a second "parasitic" brain – or at least a set of processes – that has evolved in us humans to create a more economical description of the rather more automatic processes that are being carried out in the first brain. Ironically this idea implies that the so-called homunculus fallacy – the notion of a "little man in the brain watching a movie screen filled with qualia" – isn't really a fallacy. In fact, what I am calling a metarepresentation bears an uncanny resemblance to the homunculus that philosophers take so much delight in debunking. I suggest that the homunculus is simply either the metarepresentation itself, or another brain structure that emerged later in evolution for creating metarepresentations, and that it is either unique to us humans or considerably more sophisticated than a "chimpunculus." (Bear in mind, though, that it doesn't have to be a single new structure – it could be a set of novel functions that involves a distributed network. Ideas similar to this have also been foreshadowed by David Darling, Derek Bickerton, Marvin Minsky and many others, although usually invoked for reasons other than the ones I consider here.)

But what is the purpose of creating such a metarepresentation? Clearly it cannot be just a copy or duplicate of the first – that would serve no purpose. Just like the first representation itself, the second one serves to emphasize or highlight certain aspects of the first in order to create tokens that facilitate novel styles of subsequent computation, either for internally juggling symbols sequentially ("thought") or for communicating ideas to others through a one-dimensional sound stream ("language"). Indeed, if you combine abstraction (discussed at length in chapter 4) with sequential symbol juggling you get "thinking" – a hallmark of our species.

Once this line was crossed in evolution the brain became capable of generating what Karl Popper would call "conjectures"; it could tentatively try out novel – even absurd – juxtapositions of perceptual tokens just to see what would happen. It's a moot point whether an ape can conjure up a visual image of a horse it has just seen, but it is unlikely that it can visualize a horse with a horn – a unicorn – or imagine a cow with wings – something humans can do effortlessly.

These ideas lead to an interesting question. Are qualia and self-awareness unique to humans or are they present in other primates? And to what extent do they depend on language? Vervet monkeys in the wild have specific calls to warn their companions about different predators. A "tree snake" call will send them scurrying down to the ground and a "terrestrial leopard" call will send them climbing higher up the tree. But the caller doesn't *know* that it is warning the others; vervet monkeys have no introspective consciousness – which, as we have seen, probably requires another part of the brain (perhaps linked to aspects of language) to generate a representation of the earlier sensory representation (a metarepresentation) of the snake or leopard. We could teach a monkey that a pig is dangerous by administering a mild electric shock whenever the pig appears. But what if that monkey were put back in the treetops and a pig lifted on to an adjoining branch? I predict that the monkey would become agitated but not be capable of generating the "snake" cry to warn the other monkeys to climb down; i.e., to start using it as a verb. It is very likely that only humans are capable of the kind of consciousness of one's qualia and of the limits on one's capabilities – "will power" – that this would require.

Which parts of the brain are involved in these novel styles of computation? A tentative list would include the amygdala (that gauges emotional significance), structures like the angular gyrus and Wernicke's area that are clustered around the left temporo-parieto-occipital (TPO) junction, and the anterior cingulate, involved in "intention." As I noted in *Phantoms in the Brain*, another reason for choosing the temporal lobes – especially the left temporal lobe ... is that this is where much of language – especially semantics – is represented. If I see an apple, it is the activity in the temporal lobes that allows me to apprehend all its implications almost simultaneously. Recognition of it as a fruit of a certain type occurs in IT (inferotemporal cortex), the amygdala gauges its significance for my well-being, and Wernicke's and other areas alert me to all the subtle nuances of meaning that the mental image including the word apple evokes; I can eat the apple, I can smell it, I can bake a pie, remove its pith, plant its seeds, use it to "keep the doctor away," tempt Eve, and on and on. The choice of which implication occupies center stage ("attention") and what to do about it is partly mediated by the anterior cingulate. When it becomes damaged the patient appears fully awake but loses all desire to talk, think, choose or act; he suffers from "akinetic mutism."

An important question that emerges from all this is the extent to which the metarepresentation is linked to the emergence of language comprehension/meaning capacity.[5] One way to find out would be to see if a patient with Wernicke's aphasia – caused by damage to the language area in the left hemisphere – can lie non-verbally, even though he is incapable of understanding or engaging in meaningful conversation. For unless you have an explicit representation of your representations you cannot begin to distort them before transmitting them to others; i.e., you cannot lie.[6] (This is because if the first representation itself is distorted you deceive yourself – which defeats the whole purpose of lying. It might help you disseminate your genes to lie to a potential mate that you have a huge bank balance, but if you actually believe it – if you are delusional – you might start spending money you don't have.)

Indeed, deliberate lying is the litmus test of whether a subject – be it a chimp or an infant or a brain-damaged individual – is simultaneously capable of both modeling other people's minds *and* of reflective self-consciousness. It is true that a bird can feign a broken wing to distract a predator away from her chicks, but she doesn't realize she's doing this; she doesn't have a representation of the representation. And therefore she cannot deploy this strategy in an open-ended manner in novel situations where it might prove useful. For example, she cannot feign this injury to invite more attention and compassion from her mate (though such an ability could evolve later through natural selection).

The distinction between deliberate lying and self-deception becomes very blurred in disorders like anosognosia (chapter 2) in which a patient with a paralyzed left arm caused by right hemisphere damage will deny her paralysis. Oddly enough, when I asked one of these patients whether she could touch my nose with her left hand she said, "Sure ... but watch out – I might poke your eye!" And on another occasion when I asked a retired Army general "Can you use your left arm?" his reply was "Yes – but I won't. I am not accustomed to taking orders, doctor." Such remarks imply that somebody in there "knows" the truth and it leaks out even though "she" or "he" – the reflectively conscious person – does not. ("Methinks the lady doth protest too much." Shades of Freudian psychology here.) And, again, I would point out that the very existence of the phenomenon of self-deception implies that there must be a self to deceive. Far from being an epiphenomenon, the sense of self must have evolved through natural selection to enhance survival and, indeed, must include within it the ability to preserve its integrity and stability – even deceiving itself when necessary. I doubt very much that an ape would be capable of Freudian defense mechanisms such as a "nervous laugh," a denial or a rationalization (assuming we could even test it through signing).

All this takes us back to my earlier remark that qualia and self are really two sides of a coin – you can't have one without the other. The ability to use special brain circuits to create metarepresentations[7] of sensory and motor representations – partly to facilitate language and partly facilitated *by* language – might have been critical for the evolution of both full-fledged qualia and a sense of self. As we noted earlier, it is impossible to have free-floating qualia without a self experiencing it, nor a self existing in isolation, devoid of all feeling and sensation.

A similar distinction can be made between representations of "raw" emotions and metarepresentations of them that would allow you to reflect on the emotion and make sophisticated choices – even withholding certain actions that might otherwise follow automatically. If I sprinkle pepper near your nose you sneeze reflexively – but why is this accompanied by a distinctive sneeze quale? (Unlike the knee-jerk, which can occur without quale even in a paraplegic.) Ironically this quale may have evolved as a metarepresentation for the sole purpose of enabling you to abort the sneeze voluntarily if you need to (e.g., when pursuing game). A cat probably cannot voluntarily abort an impending sneeze, given that it doesn't – in my view – have a metarepresentation. A sneeze can hardly be described as an emotion, but the same principle probably applies to more complex human emotions. A cat simply pounces when it sees a long black moving shape but cannot contemplate a mouse or "mousiness" the way you and I do. Nor can it experience subtle emotions like humility, arrogance, mercy, desire (as opposed to need) or "tears of self-pity," all of which are based on metarepresentations of emotions requiring complex interactions with social values represented in the orbitorontal cortex. Although emotions are phylogenetically ancient and often regarded as more primitive, in humans they are probably just as sophisticated as reason.

The sense of "unity" of self also deserves comment. Why do you feel like "one" despite being immersed in a constant flux of sensory impressions, thoughts and emotions? This is a tricky question and may well turn out to be a pseudo-problem. Perhaps the self by its very nature can be experienced only as a unity. Actually experiencing two selves may be logically impossible, because it would raise the question of who or what is experiencing the two selves. True, we sometimes speak of "being in two minds," but that is nothing more than a figure of speech. Even people with so-called multiple personality disorder don't experience two personalities simultaneously – the personalities tend to rotate and are mutually amnesic: at any given instant the self occupying center stage is walled off from (or only dimly aware of) the other(s). Even in the extreme case of a split-brain patient whose two hemispheres have been surgically disconnected, the patient doesn't experience doubling subjectively; each hemisphere's "self" is aware only of itself – although it may intellectually deduce the presence of the other.[8]

Another "paradox" of sorts is that even though the self is private – almost by definition – it is very much enriched by social interactions and, indeed, may have evolved mainly in a social context, as both Nick Humphrey and Horace Barlow first pointed out in a conference that Brian Josephson and I organized in 1979.

Let me expand on this. Our brains are essentially model-making machines. We need to construct useful, virtual reality simulations of the world that we can act on. Within the simulation, we need also to construct models of other people's minds because we primates are intensely social creatures. (This is called "a theory of other minds.") We need to do this so that we can predict their behavior. For example, you need to know whether another's action in jabbing you with an umbrella was willful, and so likely to be repeated, or involuntary, in which case it's quite benign. Furthermore, for this internal simulation to be complete it needs to contain not only models of other people's minds but also a model of itself, of its own stable attributes, its personality traits and the limits of its abilities – what it can and cannot do. It is possible that one of these two modeling capacities evolved first and then set the stage for the other. Or – as often happens in evolution – the two may have co-evolved and enriched each other enormously, culminating in the reflective self-awareness that characterizes *Homo sapiens*.

At a very rudimentary level we are reminded of this reciprocity of "self" and "others" each time a newborn baby mimics an adult's behavior. Stick your tongue out at a newborn baby and the baby will stick its tongue out too, poignantly dissolving the boundary, the arbitrary barrier, between self and others. To do this it must create an internal model of your action and then re-enact it in its own brain. An astonishing ability, given that it cannot even see its own tongue, and so must match the visual appearance of your tongue with the felt position of its own. We now know that this is carried out by a specific group of neurons, in the frontal lobes, called the mirror neurons. I suspect that these neurons are at least partly involved in generating our sense of "embodied" self-awareness as well as our "empathy" for others. No wonder children with autism – who (I conjecture) have a deficient mirror-neuron system – are incapable of constructing a theory of other minds, lack empathy, and also engage in self-stimulation to enhance their sense of being a self anchored in a body. It would be interesting to see if an autistic baby (diagnosed sufficiently young) would mimic the tongue protrusion of an adult in the same way that normal infants do.

Without a "theory of other minds" an organism (or person) would also be incapable of blushing – the external marker of embarrassment. (As someone said: "Only humans blush – or need to.") Blushing is a fascinating topic that greatly intrigued

Darwin. Since it is an involuntary "flag" of violation of a social taboo, it may have evolved in humans as a "marker" or index of reliability. When courting a man, a blushing woman is saying (in effect): "I can't lie to you about my affair or cuckold you without my blush giving me away – I'm reliable, so come disseminate your genes through me." If this is correct one might expect that autistic children are incapable of blushing.

In addition to their obvious role in empathy, "mind reading" and evolution of language (chapter 4), mirror neurons may have also played a vital role in the emergence of another important capacity of our minds – namely, learning through imitation – and therefore the transmission of culture.[9] Polar bears had to go through millions of years of natural selection of genes to evolve a fur coat, but a human child can acquire the skill required to make a coat by simply watching his parent slaying a bear and skinning it. Once the mirror neuron system became sophisticated enough, this remarkable ability – imitation and mimesis – liberated humans from the constraints of a strictly gene-based evolution, allowing them to make a rapid transition to Lamarckian evolution. As noted in chapter 2, the result was a rapid horizontal spread and vertical transmission of cultural innovations of the kind that took place about 50,000 to 75,000 years ago leading to the so-called great leap forward – the relatively sudden dissemination of one-of-a-kind "accidental" cultural innovations like fire, sophisticated multi-component tools, personal adornments, rituals, art, shelter etc. Among the great apes, orangutans alone are reputed to display imitation of sophisticated skills ... often watching the keeper and picking locks or even paddling across a river in a canoe. If our species becomes extinct, they may well inherit the earth.

This type of gene-culture co-dependence suggests that the nature/nurture debate is meaningless in the context of human mental functions; it's like asking whether the wetness of water derives mainly from the H_2 or the O_2 that constitute H_2O. Our brains are inextricably bound to the cultural mileu they are immersed in and, if raised in a cave by wolves or in a culture-free environment (like Texas), we would barely be human – just as a single cell cannot exist without its symbiotic mitochondria. A Martian taxonomist watching the evolution of hominids would be struck by the observation that the behavioral difference (caused by culture) between post-twentieth-century *Homo sapiens* and early *Homo sapiens* (say 75,000 years ago – before the great leap forward) is actually much greater than the difference between *Homo erectus* and *Homo sapiens*! If he used behavioral criteria alone – rather than anatomy – he might classify the former two (late and early *sapiens*) as two different species and the latter two as a single one![10]

In chapter 2 I mentioned the "blindsight" syndrome, in which a patient with visual cortex damage cannot consciously see a spot of light shown to him but is able to use an alternative spared brain pathway to guide his hand unerringly to reach out and touch the spot. I would argue that this patient has a representation of the light spot in his spared pathway, but without his visual cortex he has no representation of the representation – and hence no qualia "to speak of." Conversely, in a bizarre syndrome called Anton's syndrome, a patient is blind owing to cortical damage but *denies* that he is blind. What he has, perhaps, is a spurious metarepresentation but no primary representation.[11] Such curious uncoupling or dissociations between sensation and conscious awareness of sensations are only possible because representations and metarepresentations occupy different brain loci and can therefore be damaged (or survive) independently of each other, at least in humans. (A monkey can develop a phantom limb but never Anton's syndrome or hysterical paralysis.) Even hypnotic induction in normal people can generate such dissociations – so-called "hidden observer" phenomena – leading to intriguing questions such as 'Can you hypnotically eliminate denial in an Anton's patient or demonstrate a form of blindsight after hypnotically inducing blindness in someone?'[12]

The flip side of this is, just as we have metarepresentations of sensory representations and percepts, we also have metarepresentations of motor skills and commands such as "waving goodbye," "hammering a nail in the wall" or "combing," which are mainly mediated by the supramarginal gyrus of the left hemisphere (near the left temple). Damage to this structure causes a disorder called ideomotor apraxia. Sufferers are not paralyzed, but, if asked to "pretend" to hammer a nail into a table, they will make a fist and flail at the tabletop. (This is not mimicry: they do not mime the action accurately by holding an imaginary hammer shaft as a normal person would.) Or when asked to mime combing her hair a patient will make a fist and bang it on her head, even though she understands the instruction and is perfectly intelligent in other respects. The left supramarginal gyrus is required for conjuring up an internal image – an explicit metarepresentation – of the intention and the complex motor–visual–proprioceptive "loop" required to carry it out. That the representation of the movement itself is not in the supramarginal gyrus is shown by the fact that if you actually give the patient a hammer and nail he will often execute the task effortlessly, presumably because with the real hammer and nail as "props" he doesn't need to conjure up the whole metarepresentation. (I have noticed that some of these patients even have difficulty with pointing, or looking at what's being pointed to, as though their sense of intentionality, of "aboutness" or "thatness," is to some extent compromised.)

For an act to be fully intentional the person has to be aware of – i.e., to anticipate – the full consequences of the act and must desire the consequences, as has been pointed out by the Oxford philosopher Anthony Kenny. (For example, if someone forces you to sign a document at gunpoint, you anticipate the signing of it but don't want to do it.) I suggest that anticipation and awareness are partly in the supramarginal gyrus and desire requires the additional involvement of the anterior cingulate and other limbic "emotional" structures. The sense of free will associated with the activity of these structures may be the proverbial carrot at the end of the stick that keeps goading the donkey in you into action.

Both a chimp and a human can reach out and grab a chocolate bar, but only a human can reflect on the long-term consequences and withhold the action because he or she is on a diet. (Intriguingly, patients with frontal lobe damage cannot do this withholding; they are not capable of "free won't," you might say. I would be very surprised if a frontal patient could go on a diet.) Patients with ideational and ideomotor apraxia have great difficulty in making judgments about whether other people's actions are intentional or not; they would probably make terrible judges or criminal lawyers. And a time may come when we may be able to do brain scans to determine whether a defendant is guilty of premeditated murder or merely of manslaughter (leading to new fields such as "neurojurisprudence" and "neurocriminology").

A crucial yet elusive aspect of self is its self-referential quality, the fact that it is aware of itself. One possibility is that as soon as other attributes of self evolved—such as coherence, continuity, embodiment, symbol juggling and planning actions—it became necessary to create a single "metarepresentation" of these representations. The resulting "awareness that you are aware," knowing that you know, or "wanting to want" (or indeed not wanting to want) is what gives the "I" its unique self-referential character.

Free will—the capacity to plan open-ended scenarios and try out even improbable scenarios entirely in the mind by juggling images and symbols—if linked with episodic memories, enables you to see yourself as an active agent doing things in the future (or past) thereby generating a full-fledged sense of self. As a bonus, this ability would also enable you to present yourself to others as a predictable human being with certain stable attributes, an important capacity for intensely social creatures like us. By combining behavioral studies on patients with brain lesions—the main theme of my work—with functional brain imaging, and viewing the results from an evolutionary perspective, we can begin to elucidate these different components of self and finally tackle the mystery of how the components interact to generate awareness and self-consciousness.

The widely used phrase "raw awareness of sensations," or "primary awareness," employed by my colleagues to designate an earlier phylogenetic stage, is an oxymoron. "Awareness" simply doesn't mean anything without a metarepresentation—an awareness of awareness and a concomitant sense of self. If you are not aware that you are aware, then by definition, you are not aware! Humans are unique in this respect. The Victorian biologists Alfred Russel Wallace and Richard Owen were correct in asserting that a huge gap separates us mentally from other beasts. But, as I have suggested in this chapter contrary to Owen, what sets us apart from other mammals, including other primates, is not any single structure—such as the hippocampus minor—but a set of circuits that includes the temporo-parieto-occipital junction (especially the angular and supramarginal gyri), the Wernickes area (concerned with semantics) and the anterior cingulate with its limbic connections ("attention," "will," "desire,") and the right parietal and insula (concerned with embodiment). These structures are for consciousness what chromosomes and DNA were for heredity. Know how they perform their individual operations, how they interact, and you will know what it means to be a conscious human being.

And there I must close. As I said in chapter 1, my goal was not a complete survey of our knowledge of the brain, but I hope I've succeeded in conveying the sense of excitement that my colleagues and I experience each time we try to tackle one of these problems, be it synesthesia, hysteria, phantom limbs, free will, blindsight, neglect or any other of these syndromes I have mentioned. Once I joked that all of philosophy consists of un*locking*, ex*hum*ing, and re*cant*ing whats been said before and then getting riled up about it. But by studying these strange cases and asking the right questions, neuroscientists can begin to answer some lofty questions that have preoccupied philosophers since the dawn of history: What is free will? What is body image? Why do we blush? What is art? What is the self? Who am I? – questions that until recently were soley the province of metaphysics.

No enterprise is more vital for the well-being and survival of the human race. This is just as true now as it was in the past. Remember that politics, colonialism, imperialism and war also originate in the human brain.

Notes

Chapter I: A Pain in the Brain

1 Hirstein and Ramachandran (1997); Ellis, Young, Quale and De Pauw (1997).

2 Ramachandran and Hirstein, 1998; Ramachandran, Rogers-Ramachandran and Stewart, 1992; Melzack, 1992. These experiments on phantom limbs were inspired, in part, by the pioneering physiological studies of Mike Merzenich, Patrick Wall, John Kaas, Tim Pons, Ed Taub and Mike Calford. Additional evidence for the remapping hypothesis comes from people who have undergone other types of sensory deprivation.

We have encountered two patients in whom after amputation of a leg, sensations were referred from the genitals to the phantom foot. One gentleman told us that even erotic sensations were referred from penis to leg so his orgasm was "much bigger than it used to be" (Ramachandran and Blakeslee, 1998). Perhaps some minor cross-wiring occurring even in normals might explain why feet are often considered erogenous zones and why we have foot fetishes. We prefer this anatomical view to Freud's far-fetched theory that foot fetishes occur because feet resemble the penis.

We predicted that after the fifth nerve (innervating the face) is cut the converse should be observed: sensations should be referred in a topographically organized manner from the face to the phantom. This has now been demonstrated by Stephanie Clarke (Clarke et al., 1996).

How massive an amputation is required to get the remapping to occur? Giovanni Berlucchi and Salvatore Aglioitti have shown that after amputation of an index finger a map of just that finger alone may be seen draped neatly across the face. And we have also seen referral from adjacent fingers. The referral was modality specific and topographically organized (Ramachandran and Hirstein, 1997).

In our first patient we had seen face-to-hand referral in four weeks and we suggested that the phenomenon was at least partly due to unmasking or activating previously silent connections that already existed between the face area and hand area rather than sprouting new axon terminals. In collaboration with David Borsook and Hans Beiter we found some degree of referral occurring less than 24 hours after deafferentation of an arm, implying that the unmasking idea must be at least partially correct. More recently there have even been intriguing reports of pressure cuff–induced anesthesia in an arm leading to sensations referred from face to hand, but this needs replication.

These referred sensations in amputees do not tell us whether the reorganization is occurring in the cortex or in the thalamus. Some years ago we suggested that the issue might be resolved by systematically looking for referred sensations in stroke patients who have zones of anaesthesia caused by lesions in touch pathways leading from thalamus to cortex. If referred sensations do occur in them (e.g., from face to arm) we can conclude that the reorganization is at least partly cortical. The existence of such referral has now been observed by several groups (Turton and Butler, 2001).

3 These observations on referred sensations also have the important implication that the sensory qualia of *location* (e.g., this sensation is coming from my hand, not my face) depends entirely on which part of the sensory cortex is activated – not on the actual source of the sensory stimulus – the face.

Yet I have found that if a person is born without arms this referral of touch from the face to the phantom does not occur, even though the subject does have a phantom; the stimulation of regions originally destined to report "hand" to higher centers must in these cases be reporting "face"! Similarly, Mriganka Sur has shown that if the visual pathways in a newborn ferret are redirected to the auditory cortex, viable connections are formed and the ferret now sees with its auditory cortex. How this reassignment of qualia labels occurs in people with congenitally missing limbs is a fascinating question (Hurley and Noe, 2003).

We are now trying to answer these questions by exploring the emergence (and precise topographic localization) of phosphenes (visual qualia) in patients with congenital vs. acquired blindness while we stimulate their visual areas artificially using transcranial magnetic stimulation.

4 These phantom movements originate because every time the motor center in the front of the brain sends a signal to the missing arm it sends a "copy" of the signal to the cerebellum and parietal lobes and these commands themselves are experienced

as movements even if there is no actual moving limb. Liz Franz, Rich Ivry and I showed this experimentally. Normal people cannot simultaneously do very dissimilar actions with their two hands – e.g., draw a circle with one hand and a triangle with the other. We found this was equally true if the patient "moved" his or her phantom to mime drawing a triangle – it interfered with the real hand's drawing, implying that commands to the phantom must be centrally monitored even if the arm is missing (Franz and Ramachandran, 1998).

5 These results obviously imply that the neural substrate of the body's image – in the parietal lobes – can be profoundly modified by experience. But there must also be an innately specified genetic template. We, and others, have found that some people born without arms experience vivid, complete phantom arms that even gesticulate and point to things.

It would be interesting to investigate male-to-female transsexuals from this point of view. A majority of patients undergoing amputation of the penis for cancer report feeling a vivid phantom penis and phantom erections. On the other hand, transsexuals report that "this appendage – this penis – doesn't feel like part of me. I have always felt like a woman trapped in a man's body," suggesting that this person's genetically specified brain-sex and corresponding body image is female rather than male. If so, one would predict a much smaller incidence of phantom penises after amputation of the organ in transsexuals than in "normal" adult men. Conversely, female to male transexuals may experience a phantom penis long before gender reassigned surgery, since their brain-sex and body image is already male.

Intriguingly, some men with intact penises also report mainly having phantom erections rather than real ones (S. M. Anstis, personal communication).

6 Perhaps even the paralysis seen after a stroke is, in part, a form of learned paralysis that can be "cured" by a mirror. Preliminary results from our group (Altschuler et al., 1999) and by others (Sathian, Greenspan and Wolf, 2000; Stevens and Stoykov, 2003) have been encouraging, but systematic double-blind placebo-controlled studies are needed. The results would have tremendous implications even if only a small proportion of patients were helped by the procedure, given that 5 percent of the world's population will eventually suffer from stroke-related paralysis of an arm or leg.

7 Our overall strategy has been the intensive investigation of neurological syndromes that have been mainly regarded as "oddities" in the past ... whether phantom limbs, synesthesia or the Capgras delusion.

One problem is that both in neurology and in psychiatry many bogus syndromes have been described that represent little more than an attempt by a clinician to have a syndrome named eponymously. It can be difficult to decide which ones are likely to be genuine and worthy of study. For example there is a syndrome called De Clerambault's syndrome which is defined as "A young woman developing a delusional fixation that a much older, successful and famous man is passionately in love with her but doesn't realise it." Ironically there is no name for the syndrome in which an old man develops the delusion that a young woman is attracted to him but is in denial about it. Surely this latter syndrome is far more common (S. M. Anstis, personal communication), yet it remains unnamed! (A feminist might argue that this is because the vast majority of psychiatrists who make up names for syndromes are male.)

On the other hand, certain syndromes such as cortical color blindness (achromotopsia), motion blindness, the commissurotomy ("split brain") patients studied by Mike Gazzaniga, Joe Bogen and Roger Sperry and anterograde amnesia (studied by Brenda Milner, Elizabeth Warrington, Alan Baddeley, Larry Squire and others) have enormously enriched our understanding of the brain even though they were originally described as single case studies. Even the cellular and biochemical mechanisms underlying the physical changes that embody the "memory trace" have now been explored in intricate detail – culminating in a Nobel prize awarded to Eric Kandel.

Chapter 2: Believing Is Seeing

1 Bear in mind, though, that this is only an analogy. One key difference is that while driving I can voluntarily switch attention to my driving and ignore the conversation, but in the case of blindsight the information processed by the "unconscious" pathway cannot be accessed even through attention.

For detailed descriptions of blindsight see Weiskrantz, 1986 and Stoerig and Cowey, 1989. Some researchers – especially Semir Zeki – regard blindsight as not being a real phenomenon.

2 Ramachandran, Altschuler and Hillyer, 1997.

3 Ramachandran, 1995; Ramachandran and Blakeslee, 1998. In 1996, I proposed in the journal *Medical Hypotheses* that the extreme mood swings of bipolar disorder and "manic-depressive illness" may result from an actual alternation between the manic, delusional left hemisphere and the "depressed" right hemisphere. J. D Pettigrew (2001, personal communication) has tested this notion and found that transcranial magnetic stimulation or caloric irrigation can shift the balance of relative activation between the two hemispheres and restore equilibrium of mood in the patient.

4 Frith and Dolan, 1997.

5 Ramachandran and Rogers-Ramachandran (1996).

6 In 1997 Eric Altschuler, Jamie Pineda and I showed that the blocking of mu waves in human EEG may provide an index of mirror neuron activity. Such blocking occurs when a normal individual voluntarily moves his hand *or* merely watches someone else's hand moving. Intriguingly, we found that in autistic children the blocking occurs for voluntary movements as in normals but not while watching someone else's hand – suggesting that they have a deficiency in their mirror neuron system that might partially explain their lack of empathy and impoverished "theory of other minds."

It's also not clear whether the mere visual appearance of someone's grasping hand is enough to set off these neurons and cause mu wave suppression or whether you need to impute intention to the hand. What if you watched a grasping dummy hand that was obviously powered mechanically or an anesthetized, paralyzed hand that was passively opened and closed by mechanical pulleys? Would mu wave suppression occur?

7 As another example of a strange neurological disorder, let me mention one that I discovered recently in collaboration with my colleague A. V. Srinivasan. We saw a middle-aged lady named Jane who started to develop a progressive dementia—probably of the Alzheimer's type. Early in her disease, even though her dementia was mild (her verbal IQ was relatively normal and she was mentally lucid in most respects), she was absolutely terrified of images in mirrors, especially of her own mirror image. What was remarkable was that except for this one bizarre symptom, she had no other delusions or neurological deficits. She had no problems recognizing people when a mirror wasn't present. We have dubbed this "enantiophobia" (fear of mirror images) or Nosferatus syndrome (after vampires, who intensely dislike being in the presence of mirrors).

Whenever she saw her own reflection she would refer to it as '"that other ugly woman who keeps haunting" her. When seeing her husband's reflection right next to her own, she would say it was her father-in-law. When we pointed to a reflection of a common object such as a cup or pen and asked her to reach out to grab the real object, she would either try groping behind the mirror or say things like, "That's an abomination . . . it doesn't exist," or "What kind of magic trick is this? What are you doing to me?" or "That thing shouldn't exist," and she would get very agitated and avert her gaze from the mirror.

She had few problems recognizing real people's faces (prosopagnosia) and also recognized herself and her husband in photographs. Yet if a mirror reflection of the photo was shown, she once again denied its existence or became agitated.

We then tried the Gallup "mirror test," which has been done on primates to purportedly demonstrate "awareness of self" but never (to our knowledge) on human patients with neurological disorders of self. When she was asleep, we put a splotch of paint on her forehead. When she woke up and saw her refection in the mirror, she reflexively raised her hand to her forehead and erased the splotch, while at the same time asserting that the woman in the mirror was an ugly old woman—not her reflection. This observation implies that her brain does have a representation of the fact that the mirror image is of her own body, but there is no metarepresentation, so she has no conscious access to this knowledge. (Which raises the question: is the Gallup mirror test really a valid test for awareness of self? Does it really prove that chimps have self-awareness in the sense you and I do? Perhaps the chimp reflexively removed the paint splotch but—just like Jane—didn't really "know" that the image in the mirror was its own. Perhaps there was a big qualitative jump between apes and us, as I have implied repeatedly in this chapter.)

Jane probably had a combination of two "lesions" in her brain. One, affecting the right parietal, created mirror agnosia or the looking-glass syndrome (which I described in chapter 2), and a second one in her temporal lobes—related to the Capgras disorder—created a mild propensity toward reduplication and paranoia. Because of the first lesion she doesn't "know" what a reflection is and assumes it's a real person. But then her brain is confronted with a dilemma: "If it's a real person, why does this other person look like me? This must be a person pretending to be me . . . an abomination . . . I refuse to accept it," and so on.

The requirement of a dual lesion of this sort may explain the rarity of some of the more exotic syndromes. I have suggested (Hirstein and Ramachandran, 1997; Ramachandran and Blakeslee, 1998) that even the Capgras delusion may require not only a disconnection between visual and emotional centers in the temporal lobes (Chapter 1), but also a second lesion in the right frontal lobe that puts the patient in a delusional frame of mind, causing him to produce absurdly far-fetched explanations for the lack of emotions when he looks at his mother.

Chapter 3: The Artful Brain

1 My book *The Artful Brain* is due to be published in 2005. Also see the website of Bruce Gooch (University of Utah) on the laws of art: http://www.cs.utah.edu/~bgooch/.

2 Experiments dating back to Francis Galton show that averaging several faces together often produces a face that is quite attractive. Does this contradict my peak shift law? Not necessarily. Averaging probably works by eliminating minor blemishes and distortions such as warts, disproportionate face parts, asymmetries, etc., which makes evolutionary sense.

Yet the peak shift principle would predict that the most attractive female face is not necessarily the "average" but usually one with the right kind of exaggeration. For example, if you subtract the average female face from the male and amplify the difference you would end up with an even more gorgeous face – a "superfemale" with neotonous features (or a male stud-muffin with exaggerated jawline and eyebrows).

3 Just for fun, let's see how far we can take this argument. Cubism involves taking the usually invisible other side of an object or face and moving it forward to the same plane as the side that is visible: two eyes and two ears visible on the profile view of a face, for example. This has the effect of liberating the observer from the tyranny of a single viewpoint: you don't have to walk around the object to see its other side. Every art student knows this is the gist of Cubism but few have raised the question of why it is appealing. Is it just shock value, or is there something else?

Let us consider the response of single neurons in the monkey brain. In the fusiform gyrus individual neurons often respond optimally to a particular face, e.g., one cell might respond to the monkey's mother, one to the big alpha male and one to a particular side-kick monkey – a "Phanka waala cell," you might say. Of course the one cell doesn't "contain" all the properties of the face; it is part of a network that responds selectively to that face, but its activity is a reasonably good way of monitoring the activation of the network as a whole. All this was shown by Charlie Gross, Ed Rolls and Dave Perrett.

Interestingly, a given neuron (say an "alpha male face neuron") will respond only to *one* view of that particular face – e.g., its profile. Another one nearby might respond to semi-profile and a third one to a full frontal of that face. Clearly, none of these neurons can by itself be signaling the concept "alpha male" because it can respond only to one view of him. If the alpha male turns slightly the neuron will stop firing.

But at the next stage in the visual processing hierarchy you encounter a new class of neurons that I'll call "master face cells" or "Picasso neurons." A given neuron will respond only to a particular face, e.g., "alpha male" or "mother," but unlike the neurons in the fusiform the neuron will fire in response to *any* view of that particular face (but not to any other face). And that, of course, is what you need for signaling: "Hey – it's the alpha male: watch out."

How do you construct a master face cell? We don't know, but one possibility is to take the outgoing wires – axons – of all the "single viewpoint" cells in the fusiform that correspond to a single face (e.g., alpha male) and make the axons converge on to a single master face cell – in this case the alpha male cell. As a result of this pooling of information you can present any view of the alpha male and it will make at least one of the individual view cells in the fusiform fire, and that signal will in turn activate the master cell. So the master cell will respond to any view of that face.

But now what would happen if you were simultaneously to present two ordinarily incompatible views of the face in a single part of the visual field in a single plane? You would activate two individual face cells in parallel in the fusiform and hence the master cell downstream will get a double dose of activation. If the cell simply adds these two inputs (at least until the cell's response is saturated), the master cell will generate a huge jolt, as if it were seeing a "super face." The net result is a heightened aesthetic appeal to a Cubist representation of a face – to a Picasso!

Now the advantage of this idea – however far-fetched – is

that it can be tested directly by recording from face cells at different stages in the monkey brain and confronting them with Picasso-like faces. I may be proved wrong, but that is its strength – it can at least be *proved* wrong. As Darwin said, when you close one path to ignorance, you often simultaneously open a new one toward the truth. This cannot be said for most philosophical theories of aesthetics.

4 If these arguments about "aesthetic universals" are correct then an obvious question arises: why doesn't everyone like a Picasso? The surprising answer to this question is that everyone *does*, but most people are in denial about it. Learning to appreciate Picasso may consist largely in overcoming denial! (Just as the Victorians initially denied the beauty of Chola bronzes until they overcame their prudishness.) Now I know this sounds a bit frivolous, so let me explain. We have known for some time now that the mind isn't one "thing" – it involves the parallel activity of many quasi-independent modules. Even our visual response to an object isn't a simple one-step process – it involves multiple stages or levels of processing. And this is especially true when we talk about something as complex as aesthetic response … it is sure to involve many stages of processing and many layers of belief. In the case of Picasso I would argue that the "gut level" reaction – the "a-ha" jolt – may indeed exist in everyone's brain, caused, perhaps, by early limbic activation. But then in most of us higher brain centers kick in telling us, in effect, "Oops! That thing looks so distorted and anatomically incorrect that I had better not admit to liking it." Likewise, a combination of prudery and ignorance might have vetoed the Victorian art critics' reaction to voluptuous bronzes – even though neurons at an earlier stage are firing away, signaling peak shifts. Only when these subsequent layers of denial are peeled off can we begin to enjoy a Picasso or a Chola. Ironically, Picasso himself derived much of his inspiration from "primitive" African art.

5 In my book *Phantoms in the Brain* I suggested that many of these laws of aesthetics – especially peak shift – may have powerfully influenced the actual course of evolution in animals, an idea that I call the "perceptual theory of evolution." A species needs to be able to identify its own species in order to mate and reproduce, and to do so it uses certain conspicuous perceptual "signatures" – not unlike the gull chick pecking the stick with three stripes. But because of the peak shift effect (and ultranormal stimuli) a mate might be preferred that doesn't "resemble" the original. In this view the giraffe's neck grew longer not merely to reach tall acacia trees but because giraffes' brains are wired to automatically show greater propensity to mate with more "giraffe-like" mates, i.e., mates with the giraffe trait of longer necks. This would lead to a progressive caricaturization of descendants in phylogeny. It also predicts less variation in the externally visible morphology and colors in creatures which don't have well-developed sensory systems. (e.g., cave dwellers) and less florid variations of internal organs which cannot be seen.

This notion is similar to Darwin's idea of sexual selection – i.e., peahens preferring peacocks with larger and larger tails. But it is different in three respects.

My argument, unlike Darwin's, doesn't apply only to secondary *sexual* characteristics. It argues that many morphological features and labels identifying species (rather than sexual) differences might propel evolutionary trends in certain directions.

Although Darwin invokes "liking larger tails" as a principle in sexual selection he doesn't explain *why* this happens. I suggest that it results from the deployment of an even more basic psychological law wired into our brains that initially evolved for other reasons, such as facilitating discrimination learning.

Given our seagull chick principle, i.e., the notion that the optimally attractive stimulus need not bear any obvious surface resemblance to the original (because of idiosyncratic aspects of neural codes for perception), it is possible that new trends in morphology will start that have no immediate functional significance and may seem quite bizarre. This is different from the currently popular view that sexual selection of absurdly large tails occurs because they are a "marker" for the absence of parasites. For example, certain fish are attracted to a bright blue spot applied by the experimenter on a potential mate, even though there is nothing remotely resembling it in the fish. I predict the future emergence of a race of blue spotted fish even though the blue spot is not a marker of sex or of species, or an advertisement for good genes that promote survival. Or perhaps a race of seagulls with striped beaks!

Note that this principle sets up a positive feedback between the observer and the observed. Once a "species label" is wired into the brain's visual circuits, then offspring who accidentally have more salient labels will survive and reproduce more, causing an amplification of the trait. That in turn will make the trait an even more reliable species label, thereby enhancing the survival of those whose brains are wired up to detect the label more efficiently. This sets up a progressive gain amplification.

6 Another way to test these ideas would be to obtain a skin conductance response (SCR) which measures your gut-level emotional reaction to something by measuring increases in skin conductance caused by sweating. We know that familiar faces usually evoke a bigger response than unfamiliar ones – because of the emotional jolt of recognition. The counterintuitive prediction would be that an even bigger response would be shown to a caricature or Rembrandt-like rendering of a familiar face than to a realistic photo of the same face. (One could control for the effects of novelty caused by the exaggeration by comparing this response to that elicited by a randomly distorted familiar face or an "anticaricature" that reduced rather than amplified the difference.)

I am not suggesting that an SCR is a complete measure of a person's aesthetic response to art. What it really measures is "arousal," and arousal doesn't always correlate with beauty – it only implies "disturbing." Yet few would deny that "disturbing" is also part of the aesthetic response: just think of a Dalí or Damien Hirst's pickled cows. This is no more surprising than the fact that we seem to, paradoxically, "enjoy" horror movies or white-knuckle rides. Such activity may represent a playful rehearsal of brain circuits for future genuine threats and the same could be said of visual aesthetic responses to disturbing, attention-grabbing, visual images. It's as if anything salient and attention-grabbing – almost by its very nature – encourages you to look at it more to process it further, thereby fulfilling at least the first requirement of art. But the "attention-grabbing" component would be the same for the randomly distorted face and the caricature, whereas only the latter will have an additional component added by the peak shift. These different "components" of the aesthetic response will eventually be dissected more as we develop a clearer understanding of the connections between the visual areas and limbic structures and of the logic that drives them (the "laws" we have been discussing). So a randomly distorted nude might excite only the amygdala ("interest + fear") whereas the peak-shifted Chola bronze would excite the amygdala (interest) *and* the septum and nuclus accumbens (adding "pleasure" to the cocktail, so you end up with "interest + pleasure").

An analogy with IQ tests might be illuminating. Most people would agree that it is ludicrous to measure something as multidimensional and complex as human intelligence using a single scale such as IQ. Yet it's better than nothing if you are in a hurry: trying to recruit sailors, for example. An individual with an IQ of 70 is unlikely to be bright by any standard and one of IQ 130 is unlikely to be stupid.

In a similar vein I would suggest that even though the SCRs can provide only a very crude measure of aesthetic response, better a crude measure than none. And it can be especially useful if combined with other measures such as brain imaging and single neuron responses. For example, a caricature or a Rembrandt might activate face cells in the fusiform more effectively than a realistic photo.

It may be helpful, also, to make a further distinction between "aesthetic universals" vs. "art" – which is in some ways a more loaded term. Aesthetic universals would include so-called design, but wouldn't include pickled cows.

7 It isn't clear what "kitsch" is, but unless we tackle this we cannot really claim to have completely understood art. After all, kitsch art also sometimes deploys the same "laws" I am talking about – e.g., grouping or peak shifts. So one way of finding out what neural connections are involved in "mature aesthetic appreciation" would be to do brain imaging experiments in which you subtract the subject's reaction to kitsch from her reaction to high art.

One possibility is that the difference is entirely arbitrary and idiosyncratic, so one man's high art might be another's kitsch. This seems unlikely, since we all know that you can evolve from appreciating kitsch to appreciating the genuine thing, but you can't slide backwards. I would suggest instead that kitsch involves merely *going through the motions* of applying the laws we have talked about, without a genuine understanding of them. The result is "pseudo art" of the kind found in hotel lobbies in North America.

As an analogy we can compare kitsch to junk food. A strong solution of sugar elicits a gustatory jolt, as every child knows, and powerfully activates certain taste neurons. This makes sense from an evolutionary standpoint: our ancestors (as Steve Pinker points out) often had to go on carbohydrate binges in preparation for enduring frequent famines. But such junk food cannot begin to compete with gourmet food in producing a complex multidimensional titillation of the palate (partly because of reasons divorced from the original evolutionary functions, e.g., peak shift and contrast, etc. applied to taste responses and partly to provide a more balanced meal that's more nutritious in the long run). Kitsch, in this view, is visual junk food.

8 Do animals have art? Some of these universal laws of aesthetics (e.g., symmetry, grouping, peak shift) not only may hold across different human cultures but may even cross the species barrier. The male bower bird is quite a drab fellow but an accomplished architect and artist, often building enormous colorful bowers – the avian equivalent of a bachelor pad; a sort of Freudian compensation for his personal appearance, you might say. He makes elaborate entryways, groups berries and pebbles according to color similarity and contrast, and even collects shiny bits of cigarette-foil "jewelery." Any of these bowers could probably fetch a handsome price if displayed in a Fifth Avenue gallery in Manhattan and falsely advertised as a work of contemporary art.

The existence of aesthetic universals is also suggested by the fact that we humans find flowers beautiful, even though they evolved to be beautiful to bees and butterflies, which diverged from our ancestors in Cambrian times. Also, principles such as symmetry, grouping, contrast and peak shift used by birds (e.g., birds of paradise) evolved to attract other birds, but we are similarly moved by them.

In response to this chapter, Richard Gregory and Aaron Schloman have pointed out to me that if such laws exist it might be possible to program at least some of them into a computer and thereby generate visually appealing pictures. Something along these lines was in fact attempted by Harold Cohen many years ago at UCSD and his algorithms do indeed produce attractive pictures that fetch handsome prices.

9 Not all Western art critics were as obtuse as Sir George. Listen to the French scholar Renée Grousset describing the Shiva Nataraja (Figure 3.4):

"Whether he be surrounded or not by the flaming aureole of the Tiruvasi – the circle of the world, which he both fills and oversteps – the king of the dance is all rhythm and exaltation. The tambourine, which he holds with one of his right hands, calls all creatures into this rhythmic motion, and they dance in his company. The conventionalized locks of flying hair and the blown scarf tell of the speed of this universal movement, which crystallizes matter and reduces it to powder in turn. One of his left hands holds the fire which animates and devours the world in this cosmic whirl. One of the god's feet is crushing a titan, for 'this dance is upon the bodies of the dead,' yet one of the right hands is making the gesture of reassurance (*abhayamudra*), so true it is that, seen from the cosmic point of view..., the very cruelty of this universal determinism is kindly, and the generative principle of the future. And, in more than one of our bronzes, the king of the dance wears a broad smile. He smiles at death and life, at pain and at joy alike, or rather, if we may be allowed so to express it, his smile *is* both death and life, both joy and pain ... From this lofty point of view, in fact, all things fall into their place, finding their explanation and logical compulsion ... The very multiplicity of arms, puzzling as it may seem at first sight, is subject in turn to an inward law, each pair remaining a model of elegance in itself, so that the whole being of the Nataraja thrills with a magnificent harmony in his terrible joy. And as though to stress the point that the dance of the divine actor is indeed a force (*lila*) – the force of life and death, the force of creation and destruction, at once infinite and purposeless – the first of the left hands hangs limply from the arm in a careless gesture of the *gajahasta* (hand as the elephant trunk). And lastly as we look at the back view of the statue, are not the steadiness of these shoulders that support the world, and the majesty of the Jove-like torso, as it were a symbol of the stability and immutability of substance, while the gyration of the legs in their dizzy speed would seem to symbolize the vortex of phenomena."

Chapter 4: Purple Numbers and Sharp Cheese

1 In many "lower synesthetes" not only numbers but also letters of the alphabet – what we call graphemes – evoke specific colors. It is very likely that the visual appearance of a letter is also represented in the fusiform, so the "cross-activation" hypothesis can explain this as well.

In others it is the sound of the letter – the phoneme – that determines the color, and this may be based on cross-activation at a higher stage, near the TPO junction and angular gyrus (Ramachandran and Hubbard, 2001a, b).

2 This raises the possibility of a novel therapeutic approach to at least some forms of dyslexia or inherited reading disability. It has been suggested by Jerome Lettvin, Gad Geiger and

Janet Atkinson that at least one form of dyslexia might be caused by an attentional defect that causes excess crowding. An individual letter is easily recognized, but when embedded in a word it becomes indiscernible because of increased crowding or distractibility. Given our observation that synesthetes show less crowding because of the evoked color differences between adjacent letters, one wonders whether dyslexia can be overcome by coloring adjacent letters (or words) differently. We have had some promising preliminary results with this but additional experiments are needed.

3 This theory should not be taken to imply that learning in early childhood plays no role in synesthesia. In a sense it *must* play a role, given that we are not born with number-signaling neurons in the brain. So the cross-activation merely provides a substrate – it confers a propensity to link numbers with colors – but doesn't determine which number evokes which color.

Hence it is not surprising that the same number can evoke different colors in different synesthetes. Yet the distribution isn't completely random across synesthetes – "o" tends to be white far more than green, for example. Similarly, the correspondence between specific phonemes and colors may initially seem arbitrary, but if we classify phonemes into bilabial, dental alveolar, palatal, velar, labio dental (and voiceless vs. voiced, nasal, etc.) depending on their mode of representation in motor maps, certain correlations and patterns may emerge. Let us not forget the lesson from the periodic table of elements. The elements seemed to form clusters (e.g., halogens or alkaline metals) but no clear pattern could be discerned until Mendeleev discovered the "atomic number rule" that later resulted in the periodic table.

4 Additional evidence for this view comes from looking at the effects of changing the contrast of the numbers. In lower synesthetes, as the contrast is reduced the color becomes less and less saturated until below 8 percent contrast it vanishes completely, even though the number itself is still clearly visible (Ramachandran and Hubbard, 2002). This high level of sensitivity to physical stimulus parameters such as contrast points to cross-wiring at early stages in neural processing. What happens when the subject visualizes or imagines the number in front of him? Oddly enough we found that many subjects report the color to be more vivid. To understand this we have to bear in mind that when you imagine an object there is partial activation of the same sensory pathways in the brain as when that object is actually seen. This "top down" internally generated activation may be sufficient to cross-activate the color nodes. But when you actually look at a black number there is simultaneous activation of neurons in the brain that signal black, and they partially veto the synesthetic color. For a number representation evoked internally through imagery this vetoing doesn't occur, hence the color becomes more vivid.

5 Another relatively common type of synesthesia, described by Francis Galton, is the "number line." When asked to visualize numbers, such synesthetes see each number always in a particular location with different numbers (sometimes up to 30 or even 100) arranged sequentially along a line. Usually the line is highly convoluted – sometimes even doubling back on itself so that (say) 9 might be nearer to 2 than to 8 in Cartesian space.

We recently devised a technique for showing this objectively (Ramachandran and Hubbard, 2001b). When normal people are asked to say which of two numbers is larger, their reaction increases linearly with the numerical "distance" between the numbers – as if they were reading them off a perfectly straight number line – so that numbers closer together become harder to tell apart. (This was shown by Stanislas Dehaene.) But when we tested our synesthetes who had convoluted number lines that doubled back, we found this wasn't true. The reaction time no longer varied with the numerical distance alone – there seemed to be some sort of compromise between the Cartesian distance and numerical distance (Ramachandran and Hubbard, 2002).

There is another curious twist to this number form. We asked another subject who experienced a convoluted—but mainly horizontal—number line to lean rightward and tilt her head in the coronal plane to the right so that her ear touched her right shoulder. Remarkably, her number line also rotated 90 degrees along with her head, so it was now vertical with respect to gravity (and the individual numbers also rotated correspondingly). This observation implies that at least in this subject, the Galtonian number line is computed and represented in retinocentric coordinates—prior to vestibular correction for head tilt, rather than in an abstract world-centered coordinate system. We have previously shown that this is also true for other visual phenomena, such as shape from shading. Intriguingly, if she stood next to a wall the portion of the number line that went "through" the wall or obstacle became "fuzzy"—almost invisible. Putting a mirror instead of a wall next to her produced acute anxiety: "Those numbers are on my right but that means I ought to see them on my left . . .," reminiscent of mirror agnosia described in chapter 2 and mirror phobia, in chapter 5. These experiments were done with my students Shai Azoulai and Ed Hubbard.

We then asked her to "project" her number line on to a screen in front of her that was either one meter away or three meters way, and she had to point to the numbers' locations using a laser pointer. Despite the threefold increase in distance, unlike a retinal afterimage, the perceived length of the number line on the screen remained largely unchanged, becoming only 10 percent bigger. This shows a strong influence of size constancy, the effect of distance cues on perceived size of the number line.

It remains to be seen if these results hold for all number-line synesthetes; we have come across one, at least, in whom the number line seemed anchored in world-centered rather than retinocentric coordinates. We are presently exploring these effects quantitatively using reaction time measures to obtain what psychologists refer to as "psychometric functions," that is, graphs.

6 The *directionality* of synesthesia also requires comment. Many have noted that numbers evoke colors but colors rarely evoke numbers (though see below). Perhaps the manner in which "color space" is represented in maps in the brain vs. the manner of representation of graphemes confers an automatic bias toward unidirectional cross-activation (Ramachandran and Hubbard, 2001b).

7 Contrary to popular wisdom, *metaphors* in ordinary language

are not arbitrary either – certain directions are preferred (Lakoff and Johnson, 1999) and this supports our claim about the analogy between metaphors and synesthesia. For example, we speak of a "loud shirt" but not of a "red sound"; a "soft" or "rough" sound but not a "loud" texture. And we say "sharp taste" but never "sour touch." All of which, we suggest, may reflect anatomical constraints.

We have encountered only one synesthete who sees numbers when she sees colors but not the other way around. Indeed when she sees (say) a polka-dotted or check shirt composed of two colors, she instantly sees the *sum* of the two numbers and then decomposes it to realize that she has unconsciously added them. Such examples remind us that we are not dealing with physics here but with biology, where exceptions abound.

Synesthesia can also be used as a mnemonic aid. Many have told us how their color associations helped them learn to type (or learn musical scales) more rapidly than their peers because the letters (or notes) were "color coded" (Ramachandran and Hubbard, 2001a).

What about more exotic forms of synesthesia such as touch to taste (as in Cytowick's (2002) celebrated "Man who tasted shapes" or in our subject Matt Blakeslee). We have suggested this probably reflects the close proximity of the insular cortex which receives taste and the somatosensory hand area in the Penfield map.

Brain maps that are already partially connected are more likely to become involved in synesthetic cross-activation. Such maps are often close together anatomically (like the color and number areas in the fusiform, the color and hearing areas near the temporo-parieto-occipital (TPO) region or the touch/taste maps in the insula). But they don't have to be. Jamie Ward has recently studied synesthetes in whom phonemes evoke tastes and he implicates connections between the insula and Broca's area.

Do normal people also experience synesthesia? We all speak of certain smells – such as nail polish – being sweet, even though we have never tasted it; this might involve the close neural links and cross-activations between smell and taste – you can think of this as a form of synesthesia that exists in all our brains. This would make sense not only functionally – e.g., fruits are sweet and also smell "sweet," like acetone – but also structurally: the brain pathways for smell and taste are closely intermingled and project to the same parts of the frontal cortex.

Lastly, consider the fact that even as infants we scrunch up our noses and raise our hands when we encounter disgusting smells and tastes. Why is it that all cultures use the same word, "disgusting," and make the same facial expression for a person who is morally offensive? Why the same word as to describe a horrible taste? (Why not describe an offensive person as "painful," for example?) I suggest that once again this is because of evolutionary and anatomical constraints. In lower vertebrates certain regions of the frontal lobes have maps for smell and taste, but as mammals became more social the same maps become usurped for social functions such as territorial marking, aggression and sexuality, eventually culminating in mapping a whole new social dimension – morality. Hence the interchangeable words and facial expressions for olfactory/gustatory and moral disgust (Ramachandran and Hubbard, 2001a, b).

8 Are there neurological disorders that disturb metaphor and synesthesia? This has not been studied in detail but, as noted in the lecture, we have seen disturbances in the booba/kiki effect as well as with proverbs in patients with angular gyrus lesions. I tested one patient with anomia caused by a left angular gyrus lesion recently and found that he got fourteen out of fifteen proverbs wrong – usually interpreting them literally rather than metaphorically.

Howard Gardner has shown that patients with right-hemisphere lesions show problems with metaphor. It would be interesting to see if their deficits are mainly with spatial metaphors – such as "He stepped down as director" or "He has moved on in life" – or arbitrary ones. I have noticed that, paradoxically, they are actually good at generating puns (which may be mainly a left-hemisphere tendency that's disinhibited by the right-hemisphere lesion).

Schizophrenics are also bad at interpreting proverbs but curiously they often unwittingly generate puns and "clang associations." I am stuck by the similarities between schizophrenics and patients with right fronto-parietal lesions. Delusions ("I am Napoleon," "I am not paralyzed") and hallucinations ("My left hand is touching your nose") occur in both and, like schizophrenics, patients with right-hemisphere lesions come up with puns and facetious humor. This suggests that they may have right fronto-parietal hypofunction and some abnormal excess of activity in the left.

Matha Farrah and Steve Kosslyn have shown that the generation and control of internally generated imagery is mainly a left-frontal function and I have suggested that "checking" it against reality is done in the right (Ramachandran and Blakeslee, 1998). So the lesions we are postulating here in schizophrenia would lead to uncontrolled, unchecked imagery (hallucinations) and beliefs (delusions).

9 A word is much more than a mere label. This was brought home to me vividly when I recently examined an Indian patient with a language disorder called anomia, or "tip of the tongue" phenomenon, caused by damage to his left angular gyrus. In addition to his anomia (difficulty naming objects shown to him and in finding the right words during spontaneous speech) he had other symptoms of Gerstmann's syndrome: finger agnosia (inability to name fingers, neither his own nor the physician's) and left/right confusion (intriguingly he couldn't point to which shoe went with which foot even when actually seeing the shoes and feet simultaneously – a form of "chirality blindness").

When shown an object he would often produce semantically related words – e.g., when shown glasses he said "Eyesight medicine" – supporting the standard view that he *knew* what it was but the name label eluded him. However, there were many categories of objects for which this wasn't true. When shown a statue of the Indian god Krishna (whom any Indian child can identify instantly) he got it wrong, saying, "Oh, he's the god who helps Rama cross the ocean" (meaning the monkey god, Hanuman). When I then hinted "The name starts with Kr …" he said, "Oh, of course, it's Krishna … he *doesn't* help Rama." The same was true for a number of other objects that he initially miscategorized; correcting the name enabled him also to evoke correctly the appropriate semantic associations. The observation implies that, contrary to popular wisdom, a name is not merely a label; it's a magic key that opens up a treasury of meanings associated with what you are looking at.

Given that my patient couldn't name fingers, I wondered what he would do if I made the rude sign of "giving him the finger" (extended middle finger). He said I was pointing to the ceiling ... suggesting, again, that what is lost is not only a word name but even highly salient associations.

Lastly, he was also terrible with metaphors: this was the man who interpreted fourteen out of fifteen proverbs literally instead of metaphorically (see note 8), despite being perfectly normal in other intellectually demanding tasks. (This supports my idea that the left TPO – especially the angular gyrus – may have played a pivotal role in the emergence of metaphor in humans.)

10 If there is anything to this view then why don't all languages use the same word for the same object? After all we say "dog" in English, "chien" in French and "nai" in Tamil.

The answer is that our principle applies only to the ancestral proto-language when things were just getting off the ground. Once the basic framework was in place arbitrary differences emerged as languages diverged: a switch to the Sassurian mode. It's getting things started that's often a problem in evolution.

Support for this view also comes from studies in comparative linguistics (Berlin, 1994). A certain South American tribe has several dozen words for different fish species and an equally large number of bird names. If an English speaker is asked to classify these words – incomprehensible to him – into birds vs. fish, he scores well above chance. This points to a non-arbitrary link between object appearance and the sound used to denote it.

11 The syntactical "nesting" of clauses bears a striking resemblance to the execution of arm movements. If I say "touch your nose" you effortlessly move your hand along the shortest trajectory, adjusting the elbow angle, fingers, etc., using the appropriate sequence of muscle twitches. But you could also, if you wanted to, move your hand behind your neck, curving it forward to touch your nose, even though you have never done this before. So it's the *goal* (nose touching) and overall strategy (contract proximal muscles first and progressively "nest" other subroutines for more and more distal joints) alone that are specified – not the precise sequence of muscle twitches. This goal-directed "nesting" of motor subroutines is not unlike the embedding of clauses within larger sentences.

We also have to distinguish the question of the functional autonomy of syntax and semantics in the modern brain from the question of evolutionary origins. Syntax is almost certainly modular in a modern human because we know that patients with damage to the Wernicke's area can have one without the other. They can produce grammatically flawless but meaningless sentences (such as Chomsky's fictitious example "Colorless green ideas sleep furiously"), implying that the isolated Broca's area can generate syntactic structure on its own. But it does not follow from this that syntax didn't evolve from some preceding ability.

As an analogy, consider the three little bones in your middle ear used for amplifying sounds. It's a defining characteristic of mammals – our reptilian ancestors didn't have them. It turns out, though, that reptiles have three bones on each side of their multihinged lower jaw – suitable for swallowing large prey but not for chewing – whereas mammals have only one, the mandible. From studies in comparative anatomy we now know that because of their fortuitous anatomical location the two posterior bones in the jaws of reptiles became assimilated into the mammalian ear for hearing.

In modern mammals hearing and chewing are "modular" – independent of each other both structurally and functionally (i.e., you can lose your jawbone without becoming deaf). Yet once the evolutionary sequence is spelled out it becomes crystal clear that one function evolved from the other. And in my view the same sort of thing could have happened time and again in the emergence of syntax and other language capacities as outlined here, an idea that is abhorrent to many linguists.

One reason for tension between "pure" linguists and neuroscientists is that the former group is interested *only* in rules intrinsic to the system – not how and why the rules came to be, or how they are enshrined in neural architecture, and how they are related to other brain functions. To an orthodox linguist such questions are as meaningless as they would be for a number theorist interested in prime numbers, Fermat's theorem or Goldbach's conjecture. (And any talk of evolution or of neurons or the role of the angular gyrus in numerical skills would seem remote from his interests!) The key difference is that syntax evolved over 200,000 years or longer, through natural selection, whereas number theory is less than 2,000 years old and its intrinsic rules were neither selected for nor adaptive in any sense. Indeed, it's their very uselessness that makes them so seductive to many pure mathematicians!

Chapter 5: Neuroscience – The New Philosophy

1 Another possibility is that this "delay" has no function and occurs because of an inevitable smearing of neural events in space-time. Since there is no "cinema screen" in the brain being watched by a little person (homunculus) in real time, there's no reason to expect a precise synchrony between one's sense of volition and the neural cascades that generate the corresponding movements. This view has been eloquently championed by the eminent American philosopher Dan Dennett, and it has the advantage of parsimony. (Although the parsimony rule can be misleading in biology, given the way evolution works; as Crick once said: God is a hacker – not an engineer.) Wegner (2002) and Churchland (1996, 2002) have made important contributions to the problem of free will. Thanks to them, and to Francis Crick, Cristoff Koch and Gerald Edelman, the study of consciousness is now considered respectable.

One difficulty I have with the time-smearing notion is

why the error in judging synchrony between brain events and the feeling of will is systematic and always in one direction; if it is really just "error" you might expect to feel the willing at random points of time clustered around the brain events.

In general, it's fair to say that philosophers have made very little progress in understanding consciousness, but there are a few exceptions, notably Pat and Paul Churchland, John Searle, Dan Dennett, Jerry Fodor, David Chalmers, Bill Hirstein, Ned Block, Rick Grush, Alva Noe and Susan Hurley (although even these enlightened few tend to take perverse delight in disagreeing among themselves ad nauseam – an occupational hazard "that goes with the territory").

2 If these ideas are right, then we can also make another prediction. A normal person sending a command to move an arm receives feedback from vision and proprioception (joint and muscle sense) that the arm is obeying the command. But using a system of mirrors and a hidden assistant wearing a glove it is possible to make someone see his arm as perfectly stationary – i.e., *not* moving. Even though the motor *commands* are being monitored by his brain and the arm is *felt* to be moving, it is *seen* as stationary (Ramachandran and Blakeslee, 1998). Normal people experience a powerful jolt when confronted with this incongruity, saying things like, "My God, what's going on! Why isn't the arm moving?" But when we tried this on the non-paralyzed right hand of a patient with anosognosia caused by right-hemisphere damage, she calmly reported that she could see the arm moving perfectly well – she ignored the mismatch. Pursuing the analogy between anosognosia and schizophrenia further,

I predict that schizophrenics will do the same thing when confronted with this type of mirror box – they will hallucinate their arm moving.

3 There are actually two different versions of the qualia problem (Ramachandran and Blakeslee, 1998). The first one – which in my view is intimately linked to one's sense of self – is the puzzle of why there should be any subjective sensations at all. Why can't everyone, including me, be zombies going about their business? Why are there two parallel descriptions or "stories" about the world – the subjective "I" story and the objective "It" story? The second problem is why the sensations take the particular form that they do. This second is, in my view, more tractable by the methods of science and its solution may take us closer to also answering the first question.

The first question can be illustrated by the following paradox. Imagine I show you two completely identical human beings – one (without his knowledge) condemned from this moment on to live in a cave and be tortured and the other outside, perpetually enjoying himself. If I ask you, "Would it be OK if I swapped the two people while they were asleep?" you would say fine, or at least you wouldn't see any particular reason why they shouldn't be. But if I now modify the question and say, "Assume one of them (the one outside the cave) is *you* … would it be OK if I did the swap?" you would say, "No … don't." Yet how can you logically justify this if you believe that only an "objective world" – a third person account – exists? A question similar to this was raised in Sankhya philosophy in ancient India (as quoted by Erwin Shrödinger in "Mind and matter").

As an example of the second question – about qualia – consider the manner in which we experience two different physical dimensions: wavelength (in vision) and pitch (in sound). Even though wavelength is a continuous dimension we experience colors as four qualitatively distinct sensations – red, yellow, green and blue. These four seem subjectively "pure" – they do not seem to be made of other colors or intermediate between other colors. Adjacent colors in this set of four are "miscible," i.e., we can see orange as being a blend of red and yellow and purple as comprising red and blue. But non-adjacent ones are immiscible, like oil and water – it is hard even to imagine a bluish yellow or a reddish green. Thus color sensations seem "chopped up" into four immiscible bits. But this isn't true of the frequency of sound waves: we hear the full range from very low to very high pitch as a continuum, with no breaks in qualia.

All this is obvious, but the question is why it should be so. To say that it's because of the way colors are coded (using three receptors in the eye, for red, green and blue, and four neural channels) doesn't explain why the *qualia* should also be chopped up into four more elementary subjective sensations. After all, once the wavelength information has been extracted (by computing ratios of activity of the three classes of cones), it could, in theory, have been represented in the brain and experienced subjectively as a continuum – just as we do with pitch. The fact that different modes of experience apply to wavelength and pitch suggests that qualia cannot be epiphenomenal; they must have an evolutionary function – such as serving as a mnemonic aid for labeling and talking about things as edible fruits (red), inedible fruits (green) vs. edible leaves (green) or sexually receptive female primate rumps (reds and blues), etc. Pitch isn't used to label things in the same way. This is, admittedly, a far-fetched argument for the distinctiveness of color qualia but it is hard to avoid being far-fetched when speculating on this topic (see also Crick, 1994; Ramachandran and Hirstein, 1997; Crick and Koch, 1998). I was asked by Richard Dawkins whether bats, which "see" objects and their surface textures using echo location, might use color labels when experiencing and designating textural qualities in hearing: a not unreasonable suggestion.

Another view on introspective consciousness is that it originally emerged in order to help simulate other people's minds – to help develop a sophisticated theory of other minds. Nick Humphrey originally proposed this at a conference I organized in Cambridge (Josephson and Ramachandran, 1979). Similar ideas have also been proposed by David Premack and Marc Hauser. At the same meeting Horace Barlow suggested an intimate link between language and consciousness.

Qualia may require a sense of self but I have difficulty accepting that it requires full-fledged language in the sense that we usually understand that term. As I noted in that same Cambridge meeting, qualia in general and colors in particular are vastly more "fine grained" than the words used to describe them.

4 It would be interesting to see how the patient would react to being poked with a needle while having an out-of-body experience. Would there be a galvanic skin response? Would the patient feel pain or simply feel detached from it all as if his *body* was experiencing pain but he was merely a spectator? The same question – about galvanic responses – could be asked of patients on Ketamine who also have out-of-body experiences.

5 Another ability closely linked to semantic aspects of language is symbol manipulation: the ability to juggle visual images of objects in your brain "off line."

To illustrate this I'll invent a thought experiment. (Unlike philosophers' thought experiments, this one can actually be done!) Imagine I show you three boxes of three different sizes on the floor and a desirable object dangling from a high ceiling. You will instantly stack the three boxes, with the largest one at the bottom and the smallest at the top, then climb up to retrieve the reward.

A chimp can also solve this problem, but presumably requires trial-and-error physical exploration of the boxes.

But now I modify the experiment. I add three colored luminous spots – one on each of the boxes – say red (big box), blue (intermediate box) and green (small box) and have them lying separately on the floor. I bring you into the room for the first time and expose you to the boxes long enough for you to realize which box has which spot. Then I switch the room lights off so that only the symbols of the boxes – the luminous colored dots – are visible. Finally, I bring a luminous reward into the dark room and dangle it from the ceiling. If you have a normal brain you will without hesitation put the red dot at the bottom, blue in the middle and green on the top and then climb to the top. In other words, as a human being you can create arbitrary symbols (loosely analogous to words) and then juggle them entirely in your brain, doing a virtual reality simulation to discover the solution. You could even do this if during the first phase you were shown only the red and green dot-labeled boxes and then, separately, the green and blue, and if then in the test phase I showed you the red and green dot-labeled boxes alone. (Assume that stacking even two boxes gives you better access to the reward.) I bet you could now juggle the symbols entirely in your head to establish the transitivity using conditionals – if red is bigger than blue and blue is bigger than green then red must be bigger than green – and then proceed to stack the green box on the red box in the dark to reach the reward, even though the relative size of the boxes was not currently visible.

An ape would almost certainly fail this task, which requires off-line manipulation of Sassurian (arbitrary) signs – the basis of language. But to what extent is language a *requirement* for "if/then" conditional statements done off line – especially in novel situations? What if you tried the experiment on patients with Wernicke's aphasia who have no comprehension of language? Or a Broca's aphasic who has difficulty with grammatical-function concepts such as if/then? Such experiments may go a long way in helping us explore the elusive interface between language and thought.

And what of abilities like playing chess (which requires "if/then" conditionals), doing both formal and informal algebra (John and Mary together have nine apples: John has twice as many as Mary. How many does each have?) or computer programming? Can Wernicke's and Broca's aphasics perform such tasks, assuming they were skilled chess players, mathematicians or programmers prior to the stroke? After all, formal algebra has its own "syntax" of sorts and programming, too, is a "language," but to what extent do they tap into the same brain machinery as natural language?

But isn't all this an overkill? After all, most of us can "visualize" juggling visual images without explicity using (silent) internal words like if/then – so why even invoke language? But here we have to be careful not to be deceived by introspection; it is quite possible that even what *feels* like visual symbol juggling may be tacitly using the same neural machinery as certain aspects of language, without your being aware of it.

6 But doesn't this notion of a "representation of a representation" lead to an endless regress? Wouldn't you need a third representation of the second representation as well?

Not necessarily. Imagine the sentence "I know he knows that I know he stole my car." This entails a representation of his representation of my representation. But if I take it any further, I can no longer hold the representations in my head simultaneously; they start fading like an echo. (Although I can figure it out intellectually by counting.) A single metarepresentation is already a major advance and there may have simply been no selection pressure in evolution to develop this ability to unwieldy limits, given our already existing limitations of memory and attention span. Consciousness is a much more limited capacity than we usually realize.

7 What exactly does this "preparing the input into manageable chunks" for qualia, language and thought (see page 98) really entail? Here we enter the twilight zone: that magic step in evolution that transformed ape-like mentation into human consciousness and self-awareness. In *Phantoms in the Brain* I suggested that there are four functional characteristics associated with neural events that become linked to qualia as opposed to those (like blindsight) which don't. I call these the four laws of qualia: (1) indubitability/irrevocability; (2) evocation of explicit meaning or semantic implications; (3) short-term memory; (4) attention.

I have described the first "law" in detail in *Phantoms in the Brain*, so I will just briefly summarize the other three here. A paraplegic will have an intact knee-jerk reflex – the tapping of his tendon invariably evokes a knee jerk; but he experiences no qualia. The reason is that the sensation processed by the spinal cord is hooked up only to *one* output: the muscle contraction – it can't be used for anything else. A qualia-laden percept (such as seeing a yellow splotch of paint), on the other hand, has an enormous number of implications – a penumbra of associations such as "banana," "yellow teeth," "lemon," the word "yellow" – which give you the luxury of choosing which "implication" to make explicit for current needs dictated by the cingulate and other frontal structures. A choice that, in its turn, requires you to hold the information long enough in "working" memory (law 3) to allow you to deploy attention using the cingulate (law 4). That completes our examination of my "four laws of qualia."

The advantage of spelling out these criteria is that you can apply them to any system to determine whether it enjoys qualia and reflective self-awareness. (For example, would a sleepwalker have them?) It also eliminates silly questions such as "Does a Venus fly trap experience sensory 'insect quale' as it closes shut?"' (it doesn't); "Does a thermostat have temperature quale?," etc. Such questions have as little meaning to a neuroscientist as the question "Is a virus really alive?" does to a post-Watson/Crick molecular biologist. Borderline cases should be used to illuminate rather than blur distinctions.

It is fashionable for both neuroscientists and Indian mystics to assert that the self is an illusion, but if this is so then the burden of proof is on us to show *how* the illusion comes about. The clearest exposition of this problem comes from Zoltan Torey, who has made the ingenious suggestion that both the sense of self and qualia may be based on attention switching between the two cerebral hemispheres. Both he and David Darling also explore the link between language and reflective self-consciousness (Torey, 1999; Darling, 1993), just as I have done in this chapter. Torey's book, in particular, is full of dazzling new insights on many of these issues.

8 The key role played by hemispheric specialization in human consciousness has been emphasised by Marcel Kinsbourne, Jack Pettigrew, Mike Gazzaniga, Joe Bogen and Roger Sperry.

Some years ago William Hirstein and I published a study showing that the non-verbal right hemisphere of a split-brain patient can lie (e.g., by non-verbally signing the wrong answer to experimenter B after receiving instructions from experimenter A to lie to B), which shows that lying doesn't require language. Bear in mind, though, that even though the right hemisphere doesn't have syntax and cannot talk, it does have some protolanguage – a rudimentary semantics and a reasonable lexicon of words that "refer" to things.

The only way finally to resolve this may be to test the left hemisphere of a split-brain patient who then has a stroke which damages the Wernicke's language area in the left hemisphere! Would his left hemisphere be capable of "off line" symbol manipulation and introspective self-consciousness? And can it lie?

We also tried testing the personality and aesthetic preferences of the two hemispheres independently using the same procedure – namely by training the right hemisphere to communicate "yes," "no" or "I don't know" non-verbally to us by picking one of three abstract shapes with the left hand. Imagine our surprise when we noticed that in patient LB the left hemisphere said it believed in God whereas the right hemisphere signaled that it was an atheist. The inter-trial consistency of this needs to be verified but at the very least it shows that the two hemispheres can simultaneously hold contradictory views on God, an observation that should send shock waves through the theological community. When a patient like this eventually dies, will one hemisphere end up in hell and the other in heaven?

9 See V. S. Ramachandran, Mirror neurons and the great leap forward (1999), http://www.edge.org/3rd_culture/ramachandran/ramachandran_index.html.

10 A similar distinction between metarepresentions and representations can be made for memory. When the hippocampus is damaged on both sides of the brain, the patient develops anterograde amnesia; he can remember events that took place in the past—prior to the damage—but he is unable to form new memories. For instance, if an attending physician introduces himself to the patient, walks out to go to the restroom, and returns after five minutes, the patient has no recollection of ever having seen him. Almost a hundred years ago, the French psychologist Edouard Claparede performed an ingenious experiment. He walked in, introduced himself, and shook the patient's hand. Concealed in Claparede's palm was a pin, so the patient shouted "Ouch!" and withdrew his hand. The next day when Claparede returned, the patient had no glimmer of recognition, no inkling that he had seen Claparede before, yet when offered a handshake, he instinctively withdrew! As Endel Tulving has emphasized, such dissociations imply different memory acquisition mechanisms in the brain, only one of which depends on the hippocampus. In our terminology his brain has a rich enough representation to create vague emotional propensities such as avoidance, but it doesn't have a metarepresentation linked to his sense of self, so there is no conscious recollection of ever having seen the physician, and hence no semantic associations such as: How much money do I owe him? What's his name? Is he an M.D. or a mere Ph.D.? Can I sue him? And so on. Which raises an intriguing question: How rich is the representation that evokes reflex avoidance? (Assuming it's a memory trace that's evoked in parallel.) Can it elicit only crude emotions or behavioral propensities like avoidance, or can it also evoke—when appropriate—pity, filial love and jealousy? Will a face that *resembles* the original but is obviously perceived as different by the explicit (conscious) memory system nevertheless "fool" the tacit memory system and evoke avoidance? How sophisticated is the explicit/conscious recollection memory system in monkeys and apes, if it exists in them at all? After all, the ability to link a long string of salient "episodes" in our lives in the correct sequential order is vital for the sense of coherence and continuity of self, and it is a moot point whether the great apes have this ability.

About eight years ago, I tried waking up an amnesic patient during REM (rapid eye movement) sleep. Intriguingly, her dream reports contained specific episodic memories from the previous day even though she had no conscious recollection of these episodes when awake. This implies that the patient continues to acquire at least some new memories but without the hypocampus, she cannot index them in a consciously retrievable manner (Ramachandran 1996). Similar observations have recently been made by R. Stickgold and colleagues (2000).

Glossary

I would like to thank the Society for Neuroscience for their permission to reproduce this glossary. Some emendations have been made.

Acetylcholine A neurotransmitter both in the brain, where it may help regulate memory, and in the peripheral nervous system, where it controls the actions of skeletal and smooth muscle.
Action potential This occurs when a neuron is activated and temporarily reverses the electrical state of its interior membrane from negative to positive. This electrical charge travels along the axon to the neuron's terminal where it triggers or inhibits the release of a neurotransmitter and then disappears.
Adrenal cortex An endocrine organ that secretes corticosteroids for metabolic functions: aldosterone for sodium retention in the kidneys, androgens for male sexual development, and estrogens for female sexual development.
Adrenal medulla An endocrine organ that secretes epinephrine and norepinephrine for the activation of the sympathetic nervous system.
Adrenaline *see* **Epinephrine**
Affective psychosis A psychiatric disease relating to mood states. It is generally characterized by depression unrelated to events in the life of the patient, which alternates with periods of normal mood or with periods of excessive, inappropriate euphoria and mania.
Agonist A neurotransmitter, a drug or other molecule that stimulates receptors to produce a desired reaction.

Amino acid transmitters The most prevalent neurotransmitters in the brain, these include glutamate and aspartate, which have excitatory actions, and glycine and gamma-amino butyric acid (GABA), which have inhibitory actions.
Amygdala A structure in the forebrain that is an important component of the limbic system.
Androgens Sex steroid hormones, including testosterone, found in higher levels in males than in females. They are responsible for male sexual maturation.
Anosognosia A syndrome in which a person with a paralyzed limb claims it is still functioning. (Anosognosia means denial of illness.) An explanation may involve close analysis of the different roles of the left and right hemispheres of the brain.
Antagonist A drug or other molecule that blocks receptors. Antagonists inhibit the effects of agonists.
Aphasia Disturbance in language comprehension or production, often as a result of a stroke.
Auditory nerve A bundle of nerve fibers extending from the cochlea of the ear to the brain, which contains two branches: the cochlear nerve that transmits sound information and the vestibular nerve that relays information related to balance.
Autonomic nervous system A part of the peripheral nervous system responsible for regulating the activity of internal organs. It includes the sympathetic and parasympathetic nervous systems.
Axon The fiber-like extension of a neuron by which the cell sends information to target cells.
Basal ganglia Clusters of neurons, which include the caudate nucleus, putamen, globus pallidus and substantia nigra, that are located deep in the brain and play an important role in movement. Cell death in the substantia nigra contributes to Parkinsonian signs.

Blindsight Some patients who are effectively blind because of brain damage can carry out tasks which appear to be impossible unless they can see the objects. For instance they can reach out and grasp an object, accurately describe whether a stick is vertical or horizontal, or post a letter through a narrow slot. The explanation appears to be that visual information travels along two pathways in the brain. If only one is damaged, a patient may lose the ability to see an object but still be aware of its location and orientation.

Blindspots Blindspots can be produced by a variety of factors. In fact everyone has a small blindspot in each eye caused by the area of the retina which connects to the optic nerve. These blindspots are often filled in by the brain using information based on the surrounding visual image. In some cases, patients report seeing unrelated images in their blindspots. One reported seeing cartoon characters. This phenomenon may involve other pathways in the brain.

Brainstem The major route by which the forebrain sends information to and receives information from the spinal cord and peripheral nerves. It controls, among other things, respiration and regulation of heart rhythms.

Broca's area The brain region located in the frontal lobe of the left hemisphere that is important for the production of speech.

Capgras delusion A rare syndrome in which the patient is convinced that close relatives, usually parents, spouse, children or siblings, are imposters. It may be caused by damage to the connections between the areas of the brain dealing with face recognition and emotional response. A sufferer might recognize the faces of his loved ones but not feel the emotional reaction normally associated with the experience.

Catecholamines The neurotransmitters dopamine, epinephrine and norepinephrine that are active both in the brain and in the peripheral sympathetic nervous system. These three molecules have certain structural similarities and are part of a larger class of neurotransmitters known as monoamines.

Cerebral cortex The outermost layer of the cerebral hemispheres of the brain. It is responsible for all forms of conscious experience, including perception, emotion, thought and planning.

Cerebral hemispheres The two specialized halves of the brain. The left hemisphere is specialized for speech, writing, language and calculation; the right hemisphere is specialized for spatial abilities, face recognition in vision and some aspects of music perception and production.

Cerebrospinal fluid A liquid found within the ventricles of the brain and the central canal of the spinal cord.

Cholecystokinin A hormone released from the lining of the stomach during the early stages of digestion which acts as a powerful suppressant of normal eating. It also is found in the brain.

Circadian rhythm A cycle of behavior or physiological change lasting approximately 24 hours.

Classical conditioning Learning in which a stimulus that naturally produces a specific response (unconditioned stimulus) is repeatedly paired with a neutral stimulus (conditioned stimulus). As a result, the conditioned stimulus can become able to evoke a response similar to that of the unconditioned stimulus.

Cochlea A snail-shaped, fluid-filled organ of the inner ear responsible for transducing motion into neurotransmission to produce an auditory sensation.

Cognition The process or processes by which an organism gains knowledge of or becomes aware of events or objects in its environment and uses that knowledge for comprehension and problem solving.

Cone A primary receptor cell for vision located in the retina. It is sensitive to color and used primarily for daytime vision.

Cornea A thin, curved transparent membrane on the surface of the front of the eye. It begins the focusing process for vision.

Corpus callosum The large bundle of nerve fibers linking the left and right cerebral hemispheres.

Cortisol A hormone manufactured by the adrenal cortex. In humans, it is secreted in greatest quantities before dawn, readying the body for the activities of the coming day.

Cotard's syndrome A disorder in which a patient asserts that he is dead, claiming to smell rotting flesh or worms crawling over his skin. It may be an exaggerated form of the Capgras delusion, in which not just one sensory area (i.e., face recognition) but all of them are cut off from the limbic system. This would lead to a complete lack of emotional contact with the world.

Dendrite A tree-like extension of the neuron cell body. Along with the cell body, it receives information from other neurons.

Dopamine A catecholamine neurotransmitter known to have multiple functions depending on where it acts. Dopamine-containing neurons in the substantia nigra of the brainstem project to the caudate nucleus and are destroyed in Parkinson's victims. Dopamine is thought to regulate emotional responses, and to play a role in schizophrenia and cocaine abuse.

Dorsal horn An area of the spinal cord where many nerve fibers from peripheral pain receptors meet other ascending nerve fibers.

Endocrine organ An organ that secretes a hormone directly into the bloodstream to regulate cellular activity of certain other organs.

Endorphins Neurotransmitters produced in the brain that generate cellular and behavioral effects similar to those generated by morphine.

Epinephrine A hormone, released by the adrenal medulla and the brain, that acts with norepinephrine to activate the sympathetic division of the autonomic nervous system. Sometimes called adrenaline.

Estrogens A group of sex hormones found more abundantly in females than males. They are responsible for female sexual maturation and other functions.

Evoked potentials A measure of the brain's electrical activity in response to sensory stimuli. This is obtained by placing electrodes on the surface of the scalp (or, more rarely, inside the head), repeatedly administering a stimulus, and then using a computer to average the results.

Excitation A change in the electrical state of a neuron that is associated with an enhanced probability of action potentials.

Follicle-stimulating hormone A hormone released by the pituitary gland. It stimulates the production of sperm in the male and growth of the follicle (which produces the egg) in the female.

Forebrain The largest division of the brain, which includes the cerebral cortex and basal ganglia. It is credited with the highest intellectual functions.

Frontal lobe One of the four divisions (the others are parietal, temporal and occipital) of each hemisphere of the cerebral cortex. It has a role in controlling movement and associating the functions of other cortical areas.

Gamma-amino butyric acid (GABA) An amino acid transmitter in the brain whose primary function is to inhibit the firing of neurons.

Glia Specialized cells that nourish and support neurons.

Glutamate An amino acid neurotransmitter that acts to excite neurons. Glutamate probably stimulates N-methyl-D-aspartate (NMDA) receptors that have been implicated in activities ranging from learning and memory to development and specification of nerve contacts in a developing animal. Stimulation of NMDA receptors may promote beneficial changes, while overstimulation may be the cause of nerve cell damage or death in neurological trauma and stroke.

Gonad Primary sex gland: testis in the male and ovary in the female.

Growth cone A distinctive structure at the growing end of most axons. It is the site where new material is added to the axon.

Hippocampus A seahorse-shaped structure located within the brain and considered an important part of the limbic system. It functions in learning, memory and emotion.

Hormones Chemical messengers secreted by endocrine glands to regulate the activity of target cells. They play a role in sexual development, calcium and bone metabolism, growth and many other activities.

Hypothalamus A complex brain structure composed of many nuclei with various functions. These include regulating the activities of internal organs, monitoring information from the autonomic nervous system and controlling the pituitary gland.

Immediate memory A phase of memory that is extremely short-lived, with information stored only for a few seconds. It also is known as short-term and working memory.

Inhibition In reference to neurons, a synaptic message that prevents the recipient cell from firing.

Ions Electrically charged atoms or molecules.

Iris A circular diaphragm that contains the muscles which alter the amount of light that enters the eye by dilating or constricting the pupil. It has an opening in its center.

Korsakoff's syndrome A disease associated with chronic alcoholism, resulting from a deficiency of vitamin B-1. Patients sustain damage to part of the thalamus and cerebellum. Symptoms include inflammation of nerves, muttering delirium, insomnia, illusions and hallucinations and a lasting amnesia.

Limbic system A group of brain structures – including the amygdala, hippocampus, septum and basal ganglia – that work to help regulate emotion, memory and certain aspects of movement.

Long-term memory The final phase of memory in which information storage may last from hours to a lifetime.

Mania A mental disorder characterized by excessive excitement. A form of psychosis with exalted feelings, delusions of grandeur, elevated mood, psychomotor overactivity and overproduction of ideas.

Melatonin Produced from serotonin, melatonin is released by the pineal gland into the bloodstream. It affects physiological changes related to time and lighting cycles.

Memory consolidation The physical and psychological changes that take place as the brain organizes and restructures information in order to make it a permanent part of memory.

Metabolism The sum of all physical and chemical changes that take place within an organism and all energy transformations that occur within living cells.

Mitochondria Small cylindrical particles inside cells that provide energy for the cell by converting sugar and oxygen into special energy molecules.

Monoamine oxidase (MAO) The brain and liver enzyme that normally breaks down the catecholamines norepinephrine, serotonin and dopamine.

Motor neuron A neuron that carries information from the central nervous system to the muscle.

Myasthenia gravis A disease in which acetylcholine receptors on the muscle cells are destroyed, so that muscles can no longer respond to the acetylcholine signal in order to contract. Symptoms include muscular weakness and progressively more common bouts of fatigue. Its cause is unknown but is more common in females than in males and usually strikes between the ages of 20 and 50.

Myelin Compact fatty material that surrounds and insulates axons of some neurons.

Nerve growth factor A substance whose role is to guide neuronal growth during embryonic development, especially in the peripheral nervous system.

Neuron Nerve cell. It is specialized for the transmission of information and characterized by long fibrous projections called axons, and shorter, branch-like projections called dendrites.
Neurotransmitter A chemical released by neurons at a synapse for the purpose of relaying information via receptors.
Nociceptors In animals, nerve endings that signal the sensation of pain. In humans, they are called pain receptors.
Norepinephrine A catecholamine neurotransmitter, produced both in the brain and in the peripheral nervous system. It seems to be involved in arousal, reward and regulation of sleep and mood, and the regulation of blood pressure.
Occipital lobe One of the four subdivisions (the others are frontal, temporal and parietal) of each hemisphere of the cerebral cortex. It plays a role in vision.
Organelles Small structures within a cell that maintain the cells and do the cells' work.
Pain asymbolia People with this condition do not feel pain when, for example, stabbed in the finger with a sharp needle. Sometimes patients say they can feel the pain, but it doesn't hurt. They know they have been stabbed, but they do not experience the usual emotional reaction. The syndrome is often the result of damage to a part of the brain called the insular cortex. The stabbing sensation is received by one part of the brain, but the information is not passed on to another area, the one which normally classifies the experience as threatening and triggers – through the feeling of pain – an avoidance reaction.
Parasympathetic nervous system A branch of the autonomic nervous system concerned with the conservation of the body's energy and resources during relaxed states.
Parietal lobe One of the four subdivisions (the others are frontal, temporal and occipital) of each hemisphere of the cerebral cortex. It plays a role in sensory processes, attention and language.

Peptides Chains of amino acids that can function as neurotransmitters or hormones.

Periaqueductal grey area A cluster of neurons lying in the thalamus and pons. It contains endorphin-producing neurons and opiate receptor sites and thus can affect the sensation of pain.

Peripheral nervous system A division of the nervous system consisting of all nerves not part of the brain or spinal cord.

Phantom limbs People who lose a limb through an accident or amputation sometimes continue to feel that it is still there. These sensations may be the result of the brain forming new connections.

Phosphorylation A process that modifies the properties of neurons by acting on an ion channel, neurotransmitter receptor or other regulatory molecule. During phosphorylation, a phosphate molecule is placed on another molecule, resulting in the activation or inactivation of the receiving molecule. It may lead to a change in the functional activity of the receiving molecule. Phosphorylation is believed to be a necessary step in allowing some neurotransmitters to act and is often the result of second messenger activity.

Pineal gland An endocrine organ found in the brain. In some animals, it seems to serve as a light-influenced biological clock.

Pituitary gland An endocrine organ closely linked with the hypothalamus. In humans, it is composed of two lobes and secretes a number of hormones that regulate the activity of other endocrine organs in the body.

Pons A part of the hindbrain that, with other brain structures, controls respiration and regulates heart rhythms. The pons is a major route by which the forebrain sends information to and receives information from the spinal cord and peripheral nervous system.

Qualia A term for subjective sensations.

Receptor cell Specialized sensory cells designed to pick up and transmit sensory information.

Receptor molecule A specific molecule on the surface or inside of a cell with a characteristic chemical and physical structure. Many neurotransmitters and hormones exert their effects by binding to receptors on cells.

Reuptake A process by which released neurotransmitters are absorbed for subsequent re-use.

Rod A sensory neuron located in the periphery of the retina. It is sensitive to light of low intensity and specialized for nighttime vision.

Second messengers Recently recognized substances that trigger communications between different parts of a neuron. These chemicals are thought to play a role in the manufacture and release of neurotransmitters, intracellular movements, carbohydrate metabolism and, possibly, even processes of growth and development. Their direct effects on the genetic material of cells may lead to long-term alterations of behavior, such as memory.

Sensitization A change in behavior or biological response by an organism that is produced by delivering a strong, generally noxious, stimulus.

Serotonin A monoamine neurotransmitter believed to play many roles including, but not limited to, temperature regulation, sensory perception and the onset of sleep. Neurons using serotonin as a transmitter are found in the brain and in the gut. A number of antidepressant drugs are targeted to brain serotonin systems.

Short-term memory A phase of memory in which a limited amount of information may be held for several seconds to minutes.

Stimulus An environmental event capable of being detected by sensory receptors.

Stroke A major cause of death in the West, a stroke is an impeded blood supply to the brain. It can be caused by a blood clot forming in a blood vessel, a rupture of the blood vessel wall, an obstruction of flow caused by a clot or other material, or by pressure on a blood vessel (as by a tumor). Deprived of oxygen, which is carried by blood, nerve cells in the affected area cannot function and die. Thus, the part of the body controlled by those cells cannot function either. Stroke can result in loss of consciousness and brain function, and death.

Sympathetic nervous system A branch of the autonomic nervous system responsible for mobilizing the body's energy and resources during times of stress and arousal.

Synapse A gap between two neurons that functions as the site of information transfer from one neuron to another.

Synesthesia A condition in which a person quite literally tastes a shape or sees a color in a sound or number. This is not just a way of describing experiences as a poet might use metaphors. Synesthetes actually experience the sensations.

Temporal lobe One of the four major subdivisions (the others are frontal, parietal and occipital) of each hemisphere of the cerebral cortex. It functions in auditory perception, speech and complex visual perceptions.

Temporal lobe epilepsy A condition which may produce a heightened sense of self and which has been linked to religious or spiritual experiences. Some people may undergo striking personality changes and may also become obsessed with abstract thoughts. One possible explanation is that repeated seizures may cause a strengthening of the connections between two areas of the brain – the temporal cortex and the amygdala. Patients have been observed to have a tendency to ascribe deep significance to everything around them, including themselves.

Thalamus A structure consisting of two egg-shaped masses of nerve tissue, each about the size of a walnut, deep within the brain. It is the key relay station for sensory information flowing into the brain, filtering out only information of particular importance from the mass of signals entering the brain.

Ventricles Of the four ventricles, comparatively large spaces filled with cerebrospinal fluid, three are located in the brain and one in the brainstem. The lateral ventricles, the two largest, are symmetrically placed above the brainstem, one in each hemisphere.

Wernicke's area A brain region responsible for the comprehension of language and the production of meaningful speech.

Bibliography

Sources

Altschuler E., Wisdom, S., Stone, L., Foster, C. and Ramachandran, V. S. (1999). Rehabilitation of Hemiparesis After Stroke with a Mirror, *Lancet* 353: 2035–2036

Armel, K. C. and Ramachandran, V. S. (1999). Acquired Synesthesia in Retinitis Pigmentosa, *Neurocase* 5(4): 293–296

Baron-Cohen, S., Burt, L., Smith-Laittan, F., Harrison, J. and Bolton, P. (1996). Synaesthesia: Prevalence and Familiarity, *Perception* 25(9): 1073–1080

Berlin, B. (1994). Evidence for Pervasive Synthetic Sound Symbolism in Ethnozoological Nomenclature, in L. Hinton, J. Nichols and J. J. Ohala (eds.), *Sound Symbolism*, New York: Cambridge University Press, chapter 6

Churchland, P. (1996). *Neurophilosophy*, Cambridge, MA: MIT Press

(2002). *Brain Wise: Studies in Neurophilosophy*, Cambridge, MA: MIT Press

Clarke, S., Regali, L., Janser, R. C., Assal, G. and De Tribolet, N. (1996). Phantom Face, *Neuroreport* 7: 2853–2857

Crick, F. (1994). *The Astonishing Hypothesis: The Scientific Search for the Soul*, New York: Scribner

Crick, F. and Koch, C. (1998). Consciousness and Neuroscience, *Cerebral Cortex* 8(2): 97–107

Darling, D. (1993). *Equations of Eternity*, New York: MJF Books

Deacon, T. (1997). *The Symbolic Species*, Harmondsworth: Penguin

Domino, G. (1989). Synesthesia and Creativity in Fine Arts Students: an Empirical Look, *Creativity Research Journal* 2(1–2): 17–29

Ellis, H., Young, A. W., Quale, A. H. and De Pauw, K. W. (1997). Reduced Autonomic Responses to Faces in Capgras Syndrome, *Proceedings of the Royal Society of London* B 264: 1085–1092

Franz, E. and Ramachandran, V. S. (1998). Bimanual Coupling in Amputees with Phantom Limbs, *Nature Neuroscience* 1: 443–444

Frith, C. and Dolan, R. (1997). Abnormal Beliefs, Delusions and Memory. Conference presentation, Harvard conference on memory and belief

Galton, F. ([1880]1997). Colour Associations, in S. Baron-Cohen and J. E. Harrison (eds.), *Synaesthesia: Classic and Contemporary Readings*, Oxford: Blackwell, pp. 43–48

Greenfield, S. (2002). *Private Life of the Brain*, Harmondsworth: Penguin

Harris, A. J. (1999). Cortical Origin of Pathological Pain, *Lancet* 354: 1464–1466

Hirstein, W. and Ramachandran, W. S. (1997). Capgras Syndrome, *Proceedings of the Royal Society of London* B 264: 437–444

Humphrey, N. (1983). *Consciousness Regained*, Oxford: Oxford University Press

Hurley, S. and Noe, A. (2003). Neural Plasticity and Consciousness, *Biology and Philosophy* 18: 131–168

Josephson, B. and Ramachandran, V. S. (1979). *Consciousness and the Physical World*, Oxford: Pergamon Press

La Cerra, P. and Bingham, R. (2002). *The Origin of Minds: Evolution, Uniqueness, and the New Science of the Self*, New York: Harmony Books

Lakoff, G. and Johnson, M. (1999). *Philosophy in the Flesh: The Embodied Mind and Its Challenge to Western Thought*, New York: Basic Books

Lueck, C. J., Zeki, S., Friston, K. J., Deiber, M. P., Cope, P., Cunningham, V. J., Lammertsma, A. A., Kennard, C. and Frackowiak, R. S. (1989). The Colour Centre in the Cerebral Cortex of Man, *Nature* 340: 386–389

McCabe, C. S., Haigh, R. C., Ring, E. F., Halligan, P., Wall, P. D. and Blake, D. R. (2003). A Controlled Pilot Study of the Utility of Mirror Visual Feedback in the Treatment of Complex Regional Pain Syndrome (type 1), *Rehematology* 42: 97–101

Melzack, R. (1992). Phantom Limbs, *Scientific American* 266: 120–126

Merikle, P., Dixon, M. J. and Smilek, D. (2002). The Role of Synaesthetic Photisms on Perception, Conception and Memory. Speech Delivered at the 12th Annual Meeting of the Cognitive Neuroscience Society, San Francisco, CA, 14–16 April

Merzenich, M. and Kaas, J. (1980). Reorganization of Mammalian Somatosensory Cortex Following Peripheral Nerve Injury, *Trends in Neuroscience* 5: 434–436

Miller, S. and Pettigrew, J. D. (2000). Interhemispheric Switching Mediates Binocular Rivalry, *Current Biology* 10: 383–392

Nielsen, T. L. (1963). Volition: A New Experimental Approach, *Scandinavian Journal of Psychology* 4: 215–230

Nunn, J. A., Gregory, L. J., Brammer, M., Williams, S. C. R., Parslow, D. M., Morgan, M. J., Morris, R. G., Bullmore, E. T., Baron-Cohen, S. and Gray, J. A. (2002). Functional

Magnetic Resonance Imaging of Synesthesia: Activation of V4/V8 by Spoken Words, *Nature Neuroscience* 5(4): 371–375

Pons, T. P., Garraghty, P. E., Ommaya, A. K., Kaas, J., Taub, E. and Mischkin, M. (1991). Massive Cortical Reorganization After Sensory Deafferentation in Adult Macaques, *Science* 252: 1857–1860

Ramachandran, V. S. (1995). Anosognosia, *Consciousness and Cognition* 1: 22–46

(2001). Sharpening Up "the Science of Art," *Journal of Consciousness Studies* 8(1): 9–29

(1996). Illusions of Body Image, in *The Mind Brain Continuum*, ed. R. Llinas and P. F. Churchland, Boston: MIT Press.

Ramachandran, V. S., Altschuler, E. and Hillyer, S. (1997). Mirror Agnosia, *Proceedings of the Royal Society of London* B 264: 645–647

Ramachandran, V. S. and Blakeslee, S. (1998). *Phantoms in the Brain*, New York: William Morrow

Ramachandran, V. S. and Hirstein, W. (1997). Three Laws of Qualia: What Neurology Tells Us about the Biological Functions of Consciousness, *Journal of Consciousness Studies* 4(5–6): 429–457

(1998). The Perception of Phantom Limbs; the D. O. Hebb Lecture, *Brain* 121: 1603–1630

(1999). The Science of Art: A Neurological Theory of Aesthetic Experience, *Journal of Consciousness Studies* 6(6–7): 15–51

Ramachandran, V. S. and Hubbard, E. M. (2001a). Psychophysical Investigations into the Neural Basis of Synaesthesia, *Proceedings of the Royal Society of London* B 268: 979–983

(2001b). Synaesthesia – a Window into Perception, Thought and Language, *Journal of Consciousness Studies* 8(12): 3–34

(2002). Synesthetic Colors Support Symmetry Perception,

Apparent Motion, and Aambiguous Crowding. Speech Delivered at the 43rd Annual Meeting of the Psychonomics Society, 21–24 November

(2003). Hearing Colors and Tasting Shapes, *Scientific American*, May: 52–59

Ramachandran, V. S. and Rogers-Ramachandran, D. (1996). Denial of Disabilities in Anosognosia, *Nature* 377: 489–490

Ramachandran, V. S., Rogers-Ramachandran, D. and Stewart, M. (1992). Perceptual Correlates of Massive Cortical Reorganization, *Science* 258: 1159–1160

Sathian, K., Greenspan, A. I. and Wolf, S. L. (2000). Doing It with Mirrors: A Case Study of a Novel Approach to Rehabilitation, *Neurorehabilitation and Neural Repair* 14: 73–76

Shödinger, Erwin (1992). Mind and Matter, in *What Is Life?*, New York: Cambridge University Press

Stevens, J. and Stoykov, M. E. (2003). Using Motor Imagery in the Rehabilitation of Hemiparesis, *Archives of Physical and Medical Rehabilitation* 84: 1090–1092

Stickgold, R., A. Malia, D. Maguire, D. Roddenberry, M. O'Connor (2000). "Hypnagogic Images in Normals and Amnesiacs." *Science* 290: 350–353.

Stoerig, P. and Cowey, A. (1989). Wavelength Sensitivity in Blindsight, *Nature* 342: 916–918

Torey, Z. (1999). *The Crucible of Consciousness*, Oxford: Oxford University Press

Treisman, A. M. and Gelade, G. (1980). A Feature-Integration Theory of Attention, *Cognitive Psychology* 12(1): 97–136

Turton, A. J. and Butler, S. R. (2001). Referred Sensations Following Stroke, *Neurocase* 7(5): 397–405

Wegner. D. (2002). *The Illusion of Conscious Will*, Cambridge, MA: MIT Press

Weiskrantz, L. (1986). *Blindsight*, Oxford: Oxford University Press

Whiten, A. (1998). Imitation of Sequential Structure of Actions in Chimpanzees, *Journal of Comparative Psychology* 112: 270–281

Young, A. W., Ellis, H. D., Quayle, A. H. and De Pauw, K. W. (1993). Face Processing Impairments and the Capgras Delusion, *British Journal of Psychiatry* 162: 695–698

Zeki, S. and Marini, L. (1998). Three Cortical Stages of Colour Processing in the Human Brain, *Brain* 121(9): 1669–1685

General reading

Baddeley, A. D. (1986). *Working Memory*, Oxford: Churchill Livingstone

Barlow, H. B. (1987). *The Biological Role of Consciousness in Mindwaves 361–381*, Oxford: Basil Blackwell

Baron-Cohen, S. (1995). *Mindblindness*, Cambridge, MA: MIT Press

Bickerton, D. (1994). *Language and Human Behaviour*, Seattle: University of Washington Press

Blackmore, Susan. (2003). *Consciousness: An Introduction*, New York: Oxford University Press

Blakemore, C. (1997). *Mechanics of Mind*, Cambridge: Cambridge University Press

Carter, R. (2003). *Exploring Consciousness*, Berkeley: University of California Press

Chalmers, D. (1996). *The Conscious Mind*, New York: Oxford University Press

Corballis, M. C. (2002). *From Hand to Mouth: The Origins of Language*, Princeton: Princeton University Press

Crick, F. (1993). *The Astonishing Hypothesis*, New York: Scribner

Cytowick, R. E. (2002). *Synaesthesia: A Union of the Senses*, 2nd Edition (originally published 1989), New York: Springer-Verlag

Damasio, A. (1994). *Descartes' Error*, New York: G. P. Putnam

Dehaene, S. (1997). *The Number Sense: How the Mind Creates Mathematics*, New York: Oxford University Press

Dennett, D. C. (1991). *Consciousness Explained*, New York: Little, Brown, and Co.

Edelman, G. M. (1989). *The Remembered Present: A Biological Theory of Consciousness*, New York: Basic Books

Ehrlich, P. (2000). *Human Natures*, Harmondsworth: Penguin Books

Gazzaniga, M. (1992). *Nature's Mind*, New York: Basic Books

Glynn, I. (1999). *An Anatomy of Thought*, London: Weidenfeld and Nicolson

Greenfield, S. (2000). *The Human Brain: A Guided Tour*, London: Weidenfeld and Nicolson

Gregory, R. L. (1966). *Eye and Brain*, London: Weidenfeld and Nicolson

Hubel, D. (1988). *Eye, Brain and Vision*, New York: W. H. Freeman

Humphrey, N. (1992). *A History of the Mind*, New York: Simon and Schuster

Kandel er Schwartz, J. and Jessel, T. M. (1991). *Principles of Neural Science*, New York: Elsevier

Kinsbourne, M. (1982). Hemispheric Specialization, *American Psychologist* 37: 222–231

Milner, D. and Goodale, M. (1995). *The Visual Brain in Action*, New York: Oxford University Press

Mithen, Steven. (1999). *The Prehistory of the Mind*, London: Thames & Hudson

Pinker, S. (1997). *How the Mind Works*, New York: W. W. Norton

Posner, M. and Raichle, M. (1997). *Images of Mind*, New York: W. H. Freeman

Premack, D. and Premack, A. (2003). *Original Intelligence*, New York: McGraw-Hill

Quartz, S. and Sejnowski, T. (2002). *Liars, Lovers and Heroes*, New York: William Morrow

Robertson, I. (2000). *Mind Sculpture*, New York: Bantam

Sacks, O. (1985). *The Man Who Mistook His Wife for a Hat*, New York: HarperCollins

(1995). *An Anthropologist on Mars*, New York: Alfred Knopf

Schacter, D. L. (1996). *Searching for Memory*, New York: Basic Books

Wolpert, L. (2001). *Malignant Sadness: The Anatomy of Depression*, Faber and Faber

Zeki, S. (1993). *A Vision of the Brain*, Oxford: Oxford University Press

Acknowledgments

First, I must thank my parents, who always nurtured my curiosity and interest in science. My father bought me a Zeiss microscope when I was eleven and my mother helped me set up a chemistry lab under the staircase in our house in Bangkok, Thailand. Many of my teachers at the Bangkok British (Patana) school, especially Mrs. Vanit and Mrs. Panachura, gave me chemicals to take home and "experiment" with.

My brother V. S. Ravi played an important role in my early upbringing: he would often recite Shakespeare and the *Rubaiyat* to me. Poetry and literature have a great deal more in common with science than people realize; they both involve unusual juxtapositions of ideas and a certain "romantic" view of the world.

My thanks to Semmangudi Sreenivasa Iyer, whose divine music was always a tremendous catalyst to all my endeavors.

Gratitude to Jayakrishna, Chandramani and Diane for being a constant source of delight and inspiration; to the BBC Reith lecture staff Gwyneth Williams and Charles Siegler for the excellent job they did in editing the lectures and to Sue Lawley for hosting the events; to the staff of Profile Books, Andrew Franklin and Penny Daniel, who helped transform the lectures into a readable book for the United Kingdom. And for the North

American edition, I thank Stephen Morrow and Pi Press for their good creative work.

Science flourishes best in an atmosphere of complete freedom and financial independence. No wonder it reached its zenith during times of great prosperity and patronage of learning—in ancient Greece, where the science of logic and geometry first emerged; in the golden age of the Guptas in India (around the fifth century AD), when the number system, trigonometry and much of algebra as we now know it were born; and during the Victorian era—the era of gentleman scientists like Humphry Davy, Darwin and Cavendish. The closest thing to this that we now have in the United States is the tenure system and federally funded grants, and I am especially grateful to the NIH for having provided uninterrupted support for my research over the years. (But, as many of my students have learned, the system isn't perfect, often unwittingly rewarding the conformist and punishing the visionary. As Sherlock Holmes told Watson, "Mediocrity knows nothing higher than itself; it requires talent to recognize genius.")

My career as a medical student was strongly influenced by six prominent physicians: K. V. Thiruvengadam, P. Krishnan Kutti, M. K. Mani, Sharada Menon, Krishnamurti Sreenivasan and Rama Mani. Later, when I went up to Trinity College, Cambridge, I found myself in a very intellectually stimulating environment. I remember the many conversations with other research students and colleagues: Sudarshan Iyengar, Ranjit Kumar Nair, Mushirul Hasan, Hemal Jayasurya, Hari Vasdudevan, Arfaei Hessam, and Vidya and Prakash Virkar.

Among the teachers and colleagues who influenced me most

I should mention Jack Pettigrew, Richard Gregory, John Allman, Oliver Sacks, Horace Barlow, Sir Alan Gilchrist, Dave Peterzell, Edie Munk, P. C. Anand Kumar, Sheshagiri Rao, T. R. Vidyasagar, V. Madhusudhan Rao, Vivian Barron, Oliver Braddick, Fergus Campbell, C. C. D. Shute, Colin Blakemore, David Whitteridge, Donald McKay, Don McLeod, David Presti, Alladi Venkatesh, Carrie Armell, Ed Hubbard, Eric Altschuler, Ingrid Olson, Pavithra Krishnan, David Hubel, Ken Nakayama, Marge Livingstone, Nick Humphrey, Brian Josephson, Pat Cavanagh, Bill Hurburt and Bill Hirstein. I have also maintained strong links with Oxford over the years: Ed Rolls, Anne Treisman, Larry Weiskrantz, John Marshall and Peter Halligan. I am grateful to All Souls College for electing me to a fellowship in 1998, an affiliation that is unique in that it carries no formal responsibilities of any kind (indeed, too much hard work is frowned upon). It gave me the leisure to think and write about the neurology of aesthetics—the topic of my third Reith lecture. My interest in art was also fueled by Julia Kindy, a brilliant art historian at UCSD. Her inspiring courses on Rodin and Picasso got me thinking about the science of art.

Thanks to the Athenaeum Club, which provided excellent library facilities and a safe haven whenever I wanted to escape the hustle and bustle of the city during my visits to London; and to Esmeralda Jahan, the eternal muse to all aspiring scientists and artists.

I also had the good fortune of having many uncles and cousins who are distinguished scientists and engineers. I thank my uncle, Alladi Ramakrishnan, an eminent physicist who encouraged my early interest in science; when I was nineteen years old he had his secretary Ganapathy type my manuscript on stereopsis for *Nature* and, to my amazement (and his!), it was accepted and published without revision. The physicist P. Hariharan—one of the inventors of white light holography—had a major influence on my early intellectual development. I have also enjoyed many stimulating conversations with Alladi Prabhakar, an outstanding scholar in the field of telecommunications who has inspired many generations of students; Krishnaswami Alladi, a wizard in number theory; Ishwar (Isha) Hariharan, whose rapid rise to preeminence in the highly competitive field of molecular biology I have watched with avuncular pride (and who has now joined the UC system, I'm happy to say); and Kumpati Narendra, the first of the Alladi cousins to defy Brahminical prohibitions against crossing the oceans to the United States—where he became a leading expert in the field of control theory and systems engineering.

Other friends, relatives and colleagues: V. R. Arjun, Shai Azoulai, Liz Bates, Roger Bingham, Jeremy Brockes, Steve Cobb, Nikki De Saint Phalle, Gerry Edelman, Rosetta Ellis, Jeff Ellman, K. Ganapathy, Lakshmi Hariharan, Bela Julesz, Kristof Koch, Dorothy Kleffner, P. C. Anand Kumar, S. Lakshmanan, Steve Link, Kumpati Narendra, Malini Parathasarathy, Hal Pashler, Dan Plummer, R. K. Raghavan, K. Ramesh, Ravi (editor of *The Hindu*), Bill Rosar, Krish Sathian, Spencer Seetaram, Terry Sejnowski, Chetan Shah, Gordon Shaw, Lindsey Shenk, Alan Snyder, A. V. Sreenivasan, Subramanian Sriram, K. Sriram, V. R. Suresh, Claude Valenti, Ajit Varki, Alladi Venkatesh, Nairobi Venkatraman and Ben Williams, many of whom hosted my visits to Madras.

Special thanks to Francis Crick, who at eighty-six continues to be more ebullient and passionate about science than most of my junior colleagues. Also to Stuart Anstis, a distinguished vision scientist who has been my friend and collaborator for over two decades. And to Pat and Paul Churchland, Leah Levi and Lance Stone, my colleagues here at UCSD. It also helps to have enlightened administrators and chairpersons such as Paul Drake, Jim Kulik, John Wixted, Jeff Ellman, Robert Dynes and Marsha Chandler. Financial support for research has come from Richard Geckler and Charlie Robins, who have, over the years, taken a keen interest in the work being done at our center.

Index

O

P

Q

R

S

About the Author

V.S. Ramachandran M.D., Ph.D., is director of the Center for Brain and Cognition and professor of psychology and neuroscience at the University of California, San Diego, and adjunct professor of biology at the Salk Institute. He trained as a physician and later obtained a Ph.D. from Trinity College at the University of Cambridge. He has received many honors and awards, including a fellowship from All Soul's College, Oxford, an honorary doctorate from Connecticut College, the Ariens Kappers Gold Medal from the Royal Nederlands Academy of Sciences, for landmark contributions in neuroscience, a gold medal from the Australian National University, a fellowship from the Society for Experimental Psychology (USA), the presidential lecture award from the American Academy of Neurology and the Ramon y Cajal award from the International Neuropsychiatry Society. He gave the Decade of the Brain lecture at the twenty-fifth annual (Silver Jubilee) meeting of the Society for Neuroscience (1995), the inaugural keynote lecture at the Decade of the Brain conference held by NIMH at the Library of Congress, the Dorcas Cumming Lecture at Cold Spring Harbor, the Raymond Adams Lecture at Massachusetts General Hospital, Harvard, and the Jonas Salk memorial lecture at the Salk Institute. Dr. Ramachandran has published over 120 papers in scientific journals (including four invited review articles in *Scientific American*), and is author of the critically acclaimed book *Phantoms in the Brain*, which has been translated into eight languages and formed the basis for a two-part series on Channel 4 TV in the UK and a one-hour PBS special in the USA. *Newsweek* has named him a member of "the century club": one of the hundred most prominent people to watch in the twenty-first century.

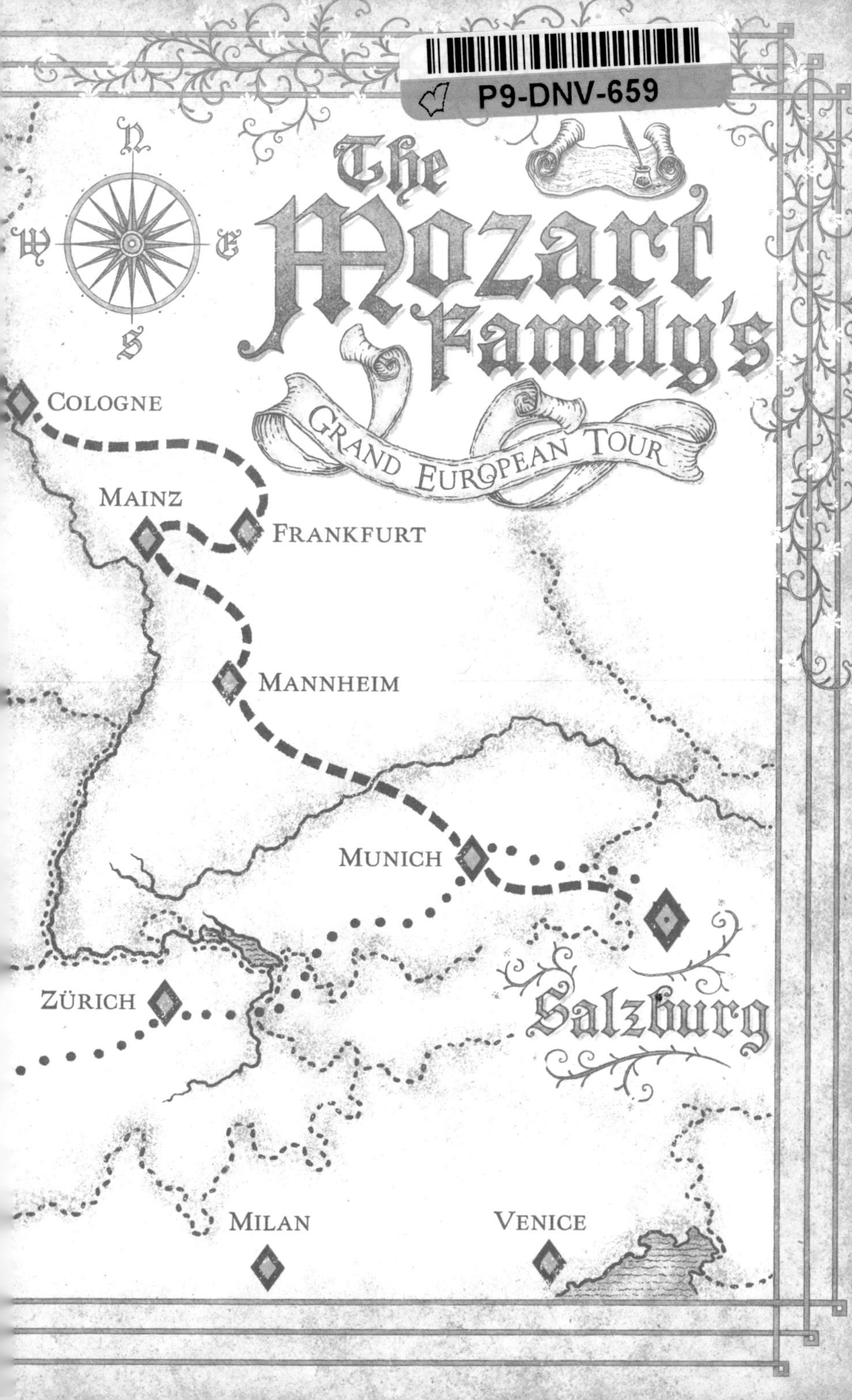

The Mozart Family's Grand European Tour

N
W
E
S

COLOGNE

MAINZ

FRANKFURT

MANNHEIM

MUNICH

Salzburg

ZÜRICH

MILAN

VENICE

THE KINGDOM OF BACK

ALSO BY MARIE LU

Warcross

Wildcard

Batman: Nightwalker

The Young Elites

The Rose Society

The Midnight Star

Legend

Prodigy

Champion

Rebel

THE KINGDOM OF BACK

MARIE LU

G. P. PUTNAM'S SONS

G. P. Putnam's Sons
An imprint of Penguin Random House LLC, New York

Visit us online at penguinrandomhouse.com

Library of Congress Cataloging-in-Publication Data
Names: Lu, Marie, 1984– author.
Title: The Kingdom of Back / Marie Lu.
Description: New York: G. P. Putnam's Sons, [2020]
Summary: "Desperate to be forever remembered for her music, Nannerl Mozart makes a dangerous pact with a mysterious stranger from a magical land, which may cost her everything"—Provided by publisher.
Identifiers: LCCN 2019002623 |
ISBN 9781524739010 (hardcover) | ISBN 9781524739027 (ebook)
Subjects: LCSH: Berchtold zu Sonnenburg, Maria Anna Mozart, Reichsfreiin von, 1751–1829—Juvenile fiction. | Mozart, Wolfgang Amadeus, 1756–1791—Juvenile fiction. | CYAC: Berchtold zu Sonnenburg, Maria Anna Mozart, Reichsfreiin von, 1751–1829—Fiction. | Mozart, Wolfgang Amadeus, 1756–1791—Fiction. | Musicians—Fiction. | Brothers and sisters—Fiction. | Mozart, Leopold, 1719–1787—Fiction. | Fantasy.
Classification: LCC PZ7.L96768 Kin 2020 | DDC [Fic]—dc23
LC record available at https://lccn.loc.gov/2019002623
Printed in the United States of America
ISBN 9781524739010

1 3 5 7 9 10 8 6 4 2

Design by Kristie Radwilowicz
Text set in FreightText Pro

For Kristin, who believed first.

This is the book that started it all.

I'm forever grateful.

THE KINGDOM OF BACK

I AM GOING TO TELL YOU A STORY YOU ALREADY KNOW. But listen carefully, because within it is one you have never heard before.

The story you know is about a boy named Wolfgang Amadeus Mozart.

You recognize his name. Even if you do not, you know him well, because you have heard his music all your life.

He was here and then gone, a brief, brilliant shard of life, a flash of stardust that ignited the sky. I knew his mind better than anyone else, understood its every winding path and quiet corner as deeply as my own. I remember everything about the way his tiny hand fit into mine, the sweep of his long lashes against his baby cheeks, the expression he would turn on me in the darkness of our shared bedchamber, his wide, fragile eyes glittering, always dreaming of some faraway place. I will tell you how the space in his small chest held so much joy and beauty that, if he wasn't careful, it might all spill out into the streets, drenching the world in too much light. He knew this, and so he held back, made rigid symmetry of the unimaginable so that the world could understand it, and for that his music became all the more sublime.

The story you have never heard is about the sister who composed beside him. In a way, you know her too, for you have also heard her music all your life. She is not the stardust but the steady wick, the one who burns low and quiet. You do not see her by the way she lights up the sky but by the way she steadies herself against the darkness, alone, at night, beside a window while the world sleeps around her. She writes when others do not see. By morning, none would know that her flame had ever been there. Her music is the ghost in the air. You know it because it reminds you of something you cannot quite grasp. You wonder where you have heard it before.

The story you already know is set in a real land, full of real kings and castles and courts. There are long carriage rides and summer concerts and a little boy in a royal coat.

The story you have never heard is set in a dream of fog and stars, faery princelings and queens of the night. It is about the Kingdom of Back, and the girl who found it.

I am the sister, the other Mozart. And her story is mine.

Salzburg, Austria
1759

Mozart by the Ocean

Sometimes, a day comes along that seems possessed by a certain shade of magic. You know those moments. There is a peculiar pattern to the silhouettes of leaves quivering against the sunbeam on the floor. The dust in the air glows white, charmed. Your voice is a note suspended in the breeze. The sounds outside your window seem very far away, songs of another world, and you imagine that this is the moment just before something unusual happens. Perhaps it is happening right now.

My day of magic arrived on a bright autumn morning, when the poplar trees swayed against a golden city. I had recently turned eight years old. My brother, Wolfgang, was not yet four.

I was still playing through my exercises when Papa came through the door with Herr Schachtner at his side, the two of them discussing some matter or other about the archbishop, their hair blown wispy from the bustle of the Getreidegasse, the city's main thoroughfare, on which our home stood.

I paused in the middle of my arpeggios and folded my hands in my lap. Even now, I can remember the uneven stitching of my blue petticoat, my white hands against black clavier keys, the skeleton leaves clinging to Herr Schachtner's shoulders. His voice had been steeped in something rich and baritone. The scent of the street—wind and smoke and baked bread—lingered like a perfume on his coat.

My lips were rosy and dry. My hair stayed neatly curled behind my neck in loose dark waves, held back with pins. I was still too young to fuss over my appearance, so my mother had left me in a simple state.

"Herr Schachtner!" My mother's voice sweetened with surprise at the sound of men in the room. She said this as if we were not expecting him at all, the esteemed court trumpeter of Salzburg, as if we had not planned everything in advance for his visit. "Such talk of the archbishop and the orchestra, it's no wonder you and my husband are always tired. Sebastian," she added, nodding at our manservant. "The Herr's coat and hat."

Sebastian hung up the court trumpeter's belongings. They were finely made, velvet encrusted with gold lining, his hat made of beaver pelt and trimmed with lace. Beside them, my father's coat appeared worn, the threads thin at the elbows. My eyes wandered to the hem of my mother's dress—it was fraying, the color dull. We were the look of a family forever on the edge of respectable.

My father was too busy with our guest to pay me any mind, but Mama noticed the stiffness in my posture and the paleness of my cheeks. She gave me an encouraging glance as she passed me.

Steady, little one, she had said to me earlier in the morning. *You have practiced hard for this. Do not be nervous.*

I kept her words in mind and tried to loosen my shoulders. But Papa had timed their arrival a bit too early this morning. I had only played my scales so far. My fingers had not yet shaken the cold from their tips, and when I pressed down on the keys, they still felt as if they were somewhere far away.

My brother stayed mercifully out of sight today, hiding somewhere in our parents' bedroom, no doubt up to some mischief. I hoped he would remain quiet until Herr Schachtner left, or at least until I finished playing.

The Herr gave Mama a warm smile that crinkled the edges of his mouth and molded his face into a pleasant sight. "Ah, Frau Mozart," he replied, winking at her as he kissed her hand. "I always tell Leopold how lucky he is to have found the rare woman with a good ear."

My mother blushed and thanked him for his kind words. Her skirts glided against the floor as she curtsied. "I can only claim my gifted ear from my father," she told him. "He was a talented musician, you know."

As she moved, I memorized the polite tilt of her head and the way she tucked a stray hair behind her ear. Somewhere in those movements must have been her true reaction to his statement, but her face remained as it always was, serene and secretive, sweet and mild. It was clear she pleased the Herr, because his grin broadened.

"Yes, God has blessed me in many ways," my father said. His smile was coiled as tightly as my nerves. His eyes flashed in my direction, hard and glittering. "Nannerl inherits her good ear from her mother, as you'll soon see."

It was my unspoken cue. At my father's words, I rose obediently from my bench to greet our guest. Papa disliked it when I curtsied without stepping away from the clavier or let my gaze wander anywhere that was not the floor. He said it made visitors think me a distracted and careless young lady.

I could not give Herr Schachtner any reason to find me rude.

Serene and sweet. I thought of Mama and tried to imitate the way she had lowered her head just so, the demure way she'd swept her skirts across the floor. Still, my curiosity stirred, and my eyes darted immediately to the court trumpeter's hands, searching for proof of musical talent in the way his fingers moved.

Mama called for Sebastian to bring some coffee and tea, but Papa waved her off. "Later," he said. It was best, perhaps, if the Herr did not see our porcelain set. I pictured the old saucers with their small chips, the teapot's fading paint. Mama had begged him for a new one for proper company, but it had been ages since we had a reason to entertain such guests. Until today.

Herr Schachtner brushed the leaves from the velvet of his justaucorps. "Thank you, Frau Mozart, but I will not stay long. I am here to listen to your lovely daughter's progress on the clavier."

"Johann followed me home after I mentioned Nannerl's talents." My father patted Herr Schachtner's shoulder. "He could not help himself."

"What luck," Mama said. She arched a brow at me. "You are just in time, then. Nannerl happens to be in the middle of practicing."

My hands trembled and I pressed them together harder, trying to warm them. Today would be the first time I ever performed for an audience. My father had sat at the clavier with me for weeks as we prepared, studying my technique, slapping my wrists when I erred.

Music is the sound of God, Nannerl, he would say. *If given the talent, it means God has chosen you as an ambassador for His voice. Your music will be as if God has given you eternal life.*

My father, God . . . there was little difference between them to me. A frown from Papa might as well have been a frown from Heaven, for what it did to my mood. Every night, I'd go to bed hearing the way my hands would move across the keys, the notes crisp in their perfection. I'd dreamed of how the Herr would stand, clapping heartily, and how my father would sit back in his chair with a satisfied smile. I'd imagined the Herr demanding I play before a wider audience. My father making the arrangements. Coins filling our family's coffers and the strain easing from his eyes.

That was the reason behind everything this morning. Children my age, Papa said, could not play the clavier with the skill that I did. I was the miracle. Chosen by a divine hand. Destined to be noticed. If I could demonstrate this to Herr Schachtner, he could extend an invitation for me to perform before Herr Haydn, the most acclaimed composer in Austria. He would be my gateway to the royal courts of Europe, to the kings and queens.

From my hands could sing the voice of God, worth its weight in gold.

"Nannerl, is it?" Herr Schachtner's voice addressed me.

I nodded in his direction. My chest fluttered as if it were brimming with moths. My fingers twitched, eager to dance. "Yes, Herr," I said. The last time he visited our home, he had not noticed me. But, then, he had no reason to.

"How long has your papa instructed you on the clavier, Fräulein?"

"Six months, Herr."

"And do you think you play well?"

I hesitated. It was a tricky question. I did not want to speak too proudly, so that he may think me arrogant, or too meekly, so thus a poor player. "I don't know, Herr," I finally said. "But I believe you will know best when you hear me."

He laughed, pleased, and I allowed myself a small smile in relief. Men, my mother had always advised, were incapable of resisting praise. If you needed something from them, you first told them about all the ways they impressed.

When I chanced a look up at him, his smile widened and he tugged at both sides of his justaucorps' collar. "Well, what a charming girl, Leopold," he said to my father. "Delightfully reserved for her age. She'll marry well, I'll say."

I turned my eyes down again, forcing a smile at his compliment, even as my hands tensed against the fabric of my dress. I'd once heard a coachman call his mare delightfully reserved as he tightened her bridle.

Papa turned to me. "We learned a new menuett yesterday," he said. "Let's start with that, Nannerl."

It was not a new menuett, truthfully, but one that Papa had written for me weeks ago and that I'd practiced for ten days. But Herr Schachtner did not need to know that. So I said, "Yes, Papa," then sat back down at the clavier and reached for my notebook.

In my nervousness, I started to play before I had counted to three in my head. *Careful,* I scolded myself. Herr Schachtner would notice every mistake I made. I took a deep breath and let the world still around me. The slant of light in the air, the sound of my father's voice, the weight of a stranger's presence in the room. They faded now, leaving me with only my hands and the keys.

Here, I was alone. This was my world. I began to play and my fingers steadied against the music. A major scale, a shift, a drawn-out A, another scale, a trill. I closed my eyes. In the darkness, with only myself, I searched for the pulse of the music and let my hands find it.

It was like coming upon a web in the woods so fragile that a single puff of air would blow it away. I thought of the clouds right before they shifted, a butterfly on the underside of a leaf, velvet-white edelweiss on a lonely rock, rain at midnight against the windowpanes. When I played, it was as if I were discovering the harmony of everything I already knew, but in a way that revealed itself only to me.

My entire heart pulled with yearning at the music. I leaned into the web and let it encase me.

Then—

A bubble of laughter came from somewhere in my parents' bedchamber. The web around me wavered, threads of it starting to burn away to reveal the room again, the light and the stranger and my father.

I furrowed my brows and tried to concentrate again. But from the corner of my eye, a blur of motion emerged from behind the bedroom door and ran over to where Papa sat. I caught sight of a head of warm brown curls. Small, stout limbs. A bright smile that beckoned to everything around him.

My brother, Wolfgang.

"Ah, Woferl!" I heard the familiar affection in Papa's voice as he used my brother's pet name. Of course, he did not scold him for the interruption. "What are you so eager about? It will have to wait. Do you see? Your sister is playing for us."

My brother simply smiled and lifted himself up onto the tips of his toes to whisper something into Herr Schachtner's ear. In spite of myself, I strained to hear what he said. My focus elsewhere, I felt my fingers speed up, disturbing the web in the woods and the flower on the rock. I bit my lip and forced myself back into rhythm.

Herr Schachtner laughed loudly. He shared the joke with my father, who chuckled, and then said something in return to my brother.

The music that filled my head began to fragment, and in the slots between the notes grew guesses at what they could be discussing. *Look at the funny faces she makes. See how stiffly she sits. Her tempo is uneven.*

Or, perhaps worse, they weren't talking about me at all.

My hands stumbled over each other—I managed to catch this mistake before it ruined the piece, but one of my fingers still slipped off its key.

The note came out silent, an ugly gap between rising arpeggios.

Heat rushed to my cheeks. I cast a glance toward my audience to see my father's eyes dart at me, surprise and disapproval sharp on his face. Herr Schachtner tucked one hand under each of Woferl's underarms and picked him up to sit upon his lap. My brother's legs swung idly.

"Thank you, Nannerl," Papa said.

His voice startled me. I hadn't even realized that the menuett was done, that my hands had already retreated to my lap. The web in the woods was gone. The clouds and butterflies and rain vanished from my mind. No one was listening to me anymore.

I straightened and rose, trembling, from the bench to curtsy. The floor beneath me swayed in the sudden silence of the room. My father's smile swayed on artificial hinges.

From where he sat on the Herr's lap, Woferl met my eyes with all the innocence of a little boy. His cheeks were round, still flushed from the remnants of a lingering fever that had struck just days ago. His eyes shone as brightly as pebbles winking in a stream. I softened at the angelic face of the brother I loved, even though I did not want to linger on his gaze.

Do not blame Woferl, Papa would say later. *He could not have distracted Herr Schachtner if you had played well.*

Herr Schachtner put his hands together and clapped. "Ah! Splendid, child!" he exclaimed. "You are a true talent." He turned to my father. "You are absolutely right, Leopold. She plays such smooth measures, and with such control. I've no doubt she will perform for royalty when she is older."

My father thanked him politely at those words, but I could see the strain in his pride, the disappointment in his expression.

Herr Schachtner was supposed to say more. He was supposed to be astounded. He should have extended an invitation to us, arranged for me to perform before Herr Haydn and Austria's other masters of music, offered to introduce me to his friends at court. Suggested a grand tour, to showcase me across all Europe. *Just think of the Italians!* he should have said. *A prodigy hailing from the Rome of the North, worthy of Rome herself!*

But instead, what he said was: *When she is older.*

I was not the miracle, destined to be noticed. Already, the Herr had moved on to telling my father a story about an argument between the orchestra's horns, my brother still bouncing on his knee. My performance was thoroughly forgotten.

Six weeks, I'd prepared for this. I felt the numb tingle return to my fingertips, and the shame of the note I'd let slip spilling out onto my cheeks.

I never let notes slip.

Later that night, after Papa had already retired to his chamber, I sat up in bed with my music notebook in my lap, the pages still open to the measures I'd played earlier in the day. As usual, Woferl lay curled lengthwise at my side. I thought about pushing him away, but instead I watched his chest rise and fall in a gentle rhythm, weighing my mood against the incessant complaints that I'd hear if I shook him out of sleep.

I ran my fingers across the dried ink, replaying my performance in my mind. Finally, I closed it and placed it on my shelf, reaching instead for a round pendant I always kept nearby, its glass surface painted bright blue and black. Faint oil streaks lingered on its surface where my thumbs had rubbed away its glossy sheen.

Mama noted my silence from where she was gathering up a few of Woferl's toys on the floor. She sighed. "Remember, Nannerl, your brother is only a child," she said to me. The skin under her eyes was soft and wrinkled, her hair a mixture of mahogany and silver. "He does not know any better."

"He knows what a performance means." My eyes went to hers. "He distracted Herr Schachtner today. You saw him."

Mama smiled in sympathy, her eyes warm with understanding. "Ah, *mein Liebling*. He means no harm. You played very well today."

I looked back down at Woferl. His face was flushed, his waves of brown curls in complete disarray. Mama was right, of course, and out of guilt, I reached over to smooth my brother's hair. He stirred, yawning like a pause between measures, his tongue tiny and pink.

"Can you tell me a story?" he murmured, and pressed himself closer to me. Before I could answer, his breathing evened again into sleep.

It was a request he made almost every day. Sharing stories with Woferl was our constant game—we spun myths of elves and dwarves, chimera that emerged from the dark woods, gnomes guarding the sleeping emperor in the Untersberg Mountain. But we told them to each other in secret, for Papa disapproved of them. At worst, they were stories about the Devil's creatures, here to torment and tempt us. At best, they were faery-tale nonsense.

Mama, however, indulged us with them. When I was very small, she used to gather me in her arms at night and whisper such stories to me in a hushed voice. After Woferl came along and Papa complained about our mother filling our heads with fables, I became the one to tell them. They soon turned into something that belonged wholly to us.

In this moment, his dreaming voice sounded so small, his question to me so true, that I felt my heart soften, as it always did, to him.

Mama came over to sit with us on the edge of the bed. She glanced at the pendant in my hands that I kept rubbing. It had been my birthday present from her, a trinket acquired when she'd visited our uncle Franz in Augsburg. *To give you luck,* she'd told me with a kiss on each cheek. Now she looked on as I ran my fingers idly across its smooth surface.

"Do you need good fortune so desperately?" Mama finally asked, taking my hand in hers.

My hand tightened against the pendant. "Yes," I said.

"And what for, my little love?"

I paused for a moment and turned my eyes up to her. *A silver wolf,* Papa had once called her, for although my mother was as steady and graceful as the snow, she was also warm, her eyes alight with intelligence for those attentive enough to notice. It was the gaze of a survivor, a woman who had fought through poverty and debt and somehow carried on after the deaths of the five children who Woferl and I had outlived.

My own insecurities embarrassed me. How could I explain to her the feelings that pressed against my chest? My mother, who glided through every moment in her life with serenity and grace. Who seemed to have faced every misfortune without fear.

"Mama," I finally said. "What are you afraid of?"

She laughed and leaned over to tap my nose. Her voice was full of vibrato, the music of a fine cello. "I am afraid of the cold, little one, because it makes my bones ache. I am afraid when I hear stories of plague and war." A graveness flickered in her gaze, as it often did when she thought of her childhood. "I am afraid for you and Woferl, as mothers always are." She raised an eyebrow at me, and I felt myself drawn into her gaze. "And you?"

My hands returned to the pendant, its black eye staring silently back up at me. I wondered if it could see into all the drawers and pockets of my father's mind, if it could tell me if I was still kept carefully in there. If I played poorly again, perhaps my father would lose interest altogether in teaching me. I thought of how the men had looked away from me after my performance today, how little the Herr seemed to have heard of what I played.

"I am afraid of being forgotten," I said. The truth emerged fully formed, empowered somehow by being named.

"Forgotten?" She laughed, a rich, throaty sound. "What a fear for a little girl."

"Someday I won't be little anymore," I replied.

Mama sobered at the words of an old soul emerging from her daughter's lips. "Everyone is forgotten, *mein Liebling*," she said gently. "Except the kings and queens."

And the talented, I added in silence, studying my brother's dark curls. They were words my father had once said. *Only the worthy are made immortal.*

With a sigh, Mama leaned toward me and kissed me gently on my cheek. "You will have plenty of years to weigh yourself down with such thoughts. Tonight, love, let yourself sleep." She turned her back and closed the door behind her, leaving us alone.

I stared at the door that Mama had just stepped through, then turned to look out at the dark city through our window. In that moment, I made a wish.

Help me be worthy. Worthy of praise, of being loved and remembered. Worthy of attention when I bared my heart at the clavier. Worthy enough for my music to linger long after I was gone. Worthy of my father. Make them remember me.

The thought trailed through my mind in a circle. I saw myself seated at the bench again, this time with the Herr never turning away in distraction, my father looking on with pride, the web in the woods unbroken and perfect. I let the image linger so long that when I finally went to sleep, I could still see it imprinted behind my closed eyes.

I thought no one heard my secret prayer, not even God, who seemed to have little interest in the wants of small girls.

But someone *was* listening.

That night, I dreamed of a shore lit by twin moons, each bright as a diamond, both suspended low at the water's edge. Their images were mirrored perfectly against a still ocean. The line of a dark forest curved along the horizon. The shore's sand was very white, the seashells very blue, and through the curling sea foam walked a boy. He looked like a wild child, clad in nothing more than black bark and silver leaves, twigs entangled in his hair, a flash of pearly white teeth brightening his smile, and although he was too far away for me to make out his features, his eyes glowed, the blue of them reflected against his cheeks. The air around him rippled with a melody so perfect, so unlike anything I'd ever heard, that I woke with my hand outstretched before me, aching to grasp it.

That was the first time I ever saw the Kingdom of Back.

The Waking Dream

I spent days sitting before the clavier after that first dream, trying in vain to find the perfect melody I'd heard. But no matter what I did, I couldn't get it to sound quite right.

"What is it that you keep playing over and over?" Woferl asked me whenever he came to watch me practice.

"Just something I heard in a dream," I told him.

He looked thoughtfully at me, his eyes wide as if searching for the melody too. "But the notes are not the same, are they?" he said.

I still don't know how he knew, except that he must have guessed by the frown on my face. "No, not the same," I replied. "Because what I heard in my dream wasn't real."

Weeks passed, then months, and soon my memory of it blurred. My attempts turned scattered, the tune shifting until it became unrecognizable. Eventually, I let myself believe that maybe it hadn't been such a perfect melody after all.

The seasons drifted from ice to rain to sun to wind. The hills that hemmed in Salzburg became white with snow, then green with new buds, then orange and gold, then white once more. My mother fixed my dresses as I grew. I began to hear murmured conversation between my parents at night, about how soon I would no longer be a child, about marriage and what prospects I had, how they would fill my dowry chest. Outside, the New Year's rifles fired and the star singers visited our door, slapping their arms against the Christmas cold, their voices warm with good cheer. Here and there, I'd catch a snippet of music in the streets that would just barely touch the edges of my memory, reminding me of something from a faraway dream.

Papa continued my lessons as I aged, filling the notebook he had bought me with menuetts, and I continued to practice the pieces. No more guests came to listen to me. Most days I was glad for it. The clavier was my cocoon of a world, my haven. In here, I could listen to my secrets in peace. But at night I lay awake and replayed the music in my mind, my thoughts circling the wish I'd spoken from my heart.

In my dreams, I was haunted by the way my father leaned away from me after a lesson, the weight of his disappointment that I couldn't grasp what he was offering me. I wondered what it might feel like to fade into the air one day. Whether my father would notice it. There was only so much time before I would leave childhood behind and he would stop teaching me entirely.

One morning, when Papa finished his lessons with me and I closed my notebook carefully, Woferl climbed onto the clavier bench beside me and reached his hands toward the keys. He had grown too, although perhaps not as much as a boy his age should. His eyes still looked enormous in the small, plump set of his face, and when he turned toward the music stand, I could see his long lashes against his cheeks, haloed in the light. He was a fragile child, both in body and health. It made me want to curl my arm protectively around his shoulders.

"Woferl," I chided gently. "Papa does not want you to play yet." My father said he was too young, his fingers too small and tender to press the keys properly. He did not want him to damage his hands. For now, selfishly, I was glad to keep music lessons something between only my father and me.

Woferl seemed to stare through my notebook, his eyes yearning for somewhere far away. His lashes turned up for a moment as he looked at me. "Please, Nannerl," he said, scooting closer to me so that he pressed against my side. "Can't you teach me a little? You are the best player in the world."

He had been asking me this for weeks, climbing onto my bench after Papa had left for the day, and each time I had turned him away. But this morning, his expression was particularly coaxing, and my mood was light, my hands warm and sure against the keys.

I laughed at him. "Surely you don't think I'm better than Papa," I replied.

When I looked at him again, he seemed serious. "I promise I won't tell."

Whatever a promise meant to a small boy. Still, the sweetness of his face made me surrender.

"You are too far away," I said at last. "Let's move the bench closer, *ja*?"

Everything about him illuminated. His eyes, his smile, his

posture. He let out a soft squeak under his breath as I drew him close to the clavier, then helped him position his fingers against the keys. His hands looked so tiny against mine that I held them in my palms a beat longer, as if to protect them. Only when he made a sound, pushing me to move aside, did I release him.

"This is a chord," I said, stretching my own hand out beside his. I played a harmonious trio of notes for him, each key spaced one out from the next, at first all together, then one after the other.

He watched me in fascination. He was still small enough that he had to use two hands to play it properly, the thumb of his left hand holding down the lowest of the three notes while two fingers of his right hand tapped out the middle and highest notes. E, G#, B. He listened curiously to it, tilting his head this way and that at the sound.

I smiled and played another chord. He followed my example.

This was when the first sign appeared. I don't think that anyone else could have noticed it, not even Papa, who never had the patience to see these things.

When Woferl pressed down on the keys, one of the notes that he struck sounded very slightly out of tune.

He frowned, then played it again. Again, the note came out at the wrong pitch.

I leaned toward him, about to tell him that the string must need tightening. But the frustration that clouded his gaze made me pause. He pressed the key a third time, thinking that it might fix itself, and when it didn't, he hummed the right pitch in the back of his throat, as if he couldn't understand how the same note could be correct in his mind and incorrect outside of it.

I knew, in that moment, that he had a remarkable ear. Sharper than our father's, sharper than Herr Schachtner's. Perhaps even sharper than mine, at least at that age. Already he understood the sound of perfection.

I now think this was how he first learned that the world was an imperfect place.

"Very good, Woferl," I said to him.

He paused to give me a relieved smile. "You hear it too," he said, and in that moment, I felt the warmth of his presence in my world, a second soul who understood.

We played a few more sets of chords before Woferl finally leaned away, looked from the clavier to the window's golden light, then back to me. "Can you tell me a story?" he asked absently.

So, he was in a whimsical mood. I glanced toward our parents' bedroom, as if Papa could still hear us even though he had left hours ago. Mama had gone with Sebastian to the clothier. No one else was home.

"All right," I said, and closed my eyes to think of something.

I still don't know why it returned to me then. Perhaps it was the chords we'd played together, which still seemed to hang in the air. But there, in the darkness, I found myself hearing the achingly pristine music from my dream years ago. The memory resurfaced of a beautiful young face that I couldn't quite recall. Of waking with my hand outstretched before me, yearning to stay longer.

I opened my eyes. The sun was slanting against the floors just so, and a new haze hung about the light in the room. We were bathed in its glow. "There is a forest," I said, looking down at my brother. "That surrounds a kingdom."

Woferl grinned at that. He clapped his hands. "What kingdom?" he asked. "What forest?" This was the game between us. He would ask me questions. I would invent answers for him, and slowly, our story would grow.

"It is a place where moss and flowers coat the floor," I said in a hushed voice. "Trees grow in thick bundles. But, Woferl, they are not trees like what we know."

"What are they like?"

Now my dream returned in glittering pieces: the moon, the sea, the black line of woods, and the strange shapes of the trees. The boy walking through the sea foam. I lowered my voice and gestured him closer. My imagination wandered free, constructing the rest of what this fantasy of a land might be. "They stand upside down, with their roots pointing up to the sky and their leaves curling against the ground, forming deep pools of rainwater along the lone path. You must be careful, for they feed on those who slip and fall in."

Woferl's eyes turned round as coins. "Do you think ghosts live there?"

"All manner of creatures do." I pondered on what to tell him next. "They are not what they seem. Some are good and kind. Others will tell you they are one thing when they are another. You must follow the good ones, Woferl, and if you do, they will lead you to a shore with sand white as snow."

Woferl had forgotten everything else around him now. He stared up at me with such an intent face that I laughed at his attention. My fingers danced across the clavier's keys as I played a few light notes for him. To my pleasure, every note drew his admiration, as if he could not get enough of this world I'd chosen to share with him.

"Come here," I suddenly said, putting my arm around him. "I know a piece that sounds just like this forest, if you want to hear it."

Woferl giggled as I turned to a blank page in my notebook, careful not to crinkle the edges of the paper. I took a deep breath, then attempted yet again to reconstruct the music I'd heard in my sleep. I thought of the snippets of sounds from the streets that would awaken my memories, and added them to the melody.

Note by note, a strange song emerged from another world.

Woferl's fingers danced in the air. He hummed the tune under his breath, his pitch perfect, and a part of me knew that he must be the only other person in the world who could hear the same beauty I could. "Can I play it like you, do you think?"

"When your fingers grow a little." I gripped the bottom of our bench, then stood and pulled it toward the clavier. Woferl's hands scooted closer to the keys. "Would you like to try?" I asked him.

Woferl did. He mimicked my notes. And again, I found myself pausing to notice that he could remember everything I'd played, that even with his small hands, he could follow along almost as if he'd been practicing with me for days.

I watched him in wonder, and within that wonder, a small twinge of something—envy, fear—took root. The feeling sat cold against my chest. The wish I'd made so long ago came back to me in a sudden wave. *Make them remember me.*

That was when it first happened.

Woferl saw it before I did. He sucked in his breath and cooed in delight, and then stretched his little arms toward the open pages of my notebook. I looked at what had captured his attention.

There, right on the first page, was a tiny cluster of grass blades and three beautiful white blooms of flowers, all growing from the parchment at a straight angle. I blinked, hardly believing what I was seeing. They were edelweiss flowers, treasures of the Alps.

"Don't touch them, Woferl," I whispered, pulling his arm back.

"Are they real?" he asked.

I leaned closer to inspect the strange sight. Edelweiss did not grow at such low altitudes, and certainly not out of music paper. They were flowers of the mountains, plants that men sometimes died seeking out for their beloveds. Mama once told us that the Virgin Mary herself had blessed our land with edelweiss by dusting the mountains with stars.

And yet, there they were—snow white, their petals thick and velvet, their edges hazy in the glow of the afternoon. A clean, fragile scent hung in the air. The light in the room seemed very strange now, as if perhaps we were part of a waking dream.

"They must have come from the forest," I said. I reached one finger out.

My brother made an irritated sound. "You said not to touch them."

"Well, I'm older than you." I let my finger skim the surface of one flower. The petal felt like the collar of my winter coat, fuzz against my fingertips. I drew my hand back. Part of the color came away when I did, leaving a streak of white across my skin like paint.

"I'm going to tell Papa," he said.

I grabbed his hand. "No, don't. Please, Woferl? Papa will think I've been filling your head with silly stories."

He looked at me for a moment, his expression wavering between emotions. I patted his cheeks gently in the way that our mother did. It was this that finally won him over. I saw the resistance go out of him, the sway of his body toward me as he savored the affection. He scooted back beside me. I rubbed the streak on my skin between two fingers, watching as it smeared and faded away into the air. Perhaps it had never been there at all. When we glanced back to the notebook's open pages, the edelweiss had disappeared. Beside me, Woferl held his breath, waiting for the dream to return. My hands trembled.

But that was not all. When my finger had touched the flower petal, I'd heard a distinct musical note. No, something more than that. A sound too perfect to be from this world. A secret. I could tell by my brother's expression that he had not heard it. I played it over in my mind until I realized the note was not a note at all, but a sweet and beautiful voice that bubbled with bright laughter. I knew, immediately, that it belonged to the boy by the ocean. It spoke only one sentence.

I can help you, Nannerl, if you help me.

The Boy from Another World

I think it all strange now, of course—a boy from another world, born from somewhere in my dreams. But the voice was very real then. I thought about it late into the night, turning it this way and that in my mind in an attempt to make sense of it, aching to hear its perfection one more time.

Woferl lay next to me in our shared bed and watched me with bright, sleepless eyes. Finally, he propped himself up on one elbow. "Do you think we'll see the edelweiss flowers again?" He leaned toward me. He was still so small that his arms sank almost entirely into the folds of the bed. "Did they come from the forest in your story?"

I sighed and rolled over to look at him in a knowing manner. "Perhaps," I said, to appease his curiosity. "I don't know. But I do know I'm quite tired. Aren't you?"

Woferl stared back innocently at me. "Yes. But you know

everything, Nannerl. Don't you also know what the forest is like?"

His chatter distracted me. All I wanted was to close my eyes and drift off to sleep with that musical note again in my mind. I sighed. "If I tell you a little more of the story, will you go to bed?"

"Yes," he promised in a rush.

I couldn't help smiling at his eagerness. "All right." I snuggled closer and wrapped my arms around him. "The forest is very large," I went on. My imagination loosened again. The world from my dream reappeared in my mind, parts of it blank and waiting for me to fill them in. "Larger than anything we've ever seen."

"Larger than Salzburg?" Woferl asked.

"Yes, much larger than Salzburg. Or Vienna. Or all of Austria. It is an endless place."

Woferl shifted in bed so that he could look at me. "Nothing is bigger than *all* of Austria," he declared.

I laughed. "Well, this place is. And while edelweiss is only in the Alps here, in the forest they grow everywhere, because it is their birthplace, where all such flowers come from."

Woferl made an impressed sound at that. "It must be a special place."

"Well, a special forest needs a guardian, doesn't it?"

He nodded without hesitation. "Of course it does."

A memory glimmered in my mind of an outfit stitched together from black bark and silver leaves. A smile of white teeth. "Because you have said so," I replied formally, "the forest has a guardian now."

Woferl leaned eagerly toward me. "Who is it?"

"Well, who do you think it is?"

"An imp?" He was picturing the ones from old German tales, wicked pranksters who could shift into the shape of a rabbit or snatch children from their cradles.

"Surely not just any imp, Woferl?" I insisted. "They aren't clever enough on their own to guard an entire forest. They need someone to help them with their plans."

Woferl considered this with a serious face. "A faery princeling, then, of the forest."

A princeling. The memory in my mind sharpened further. A pair of glowing blue eyes, twigs tangled in hair. A voice too beautiful for this world. I yearned toward the thought. "A princeling," I agreed. "Someone unafraid to play pranks on trespassers to drive them away. Someone clever and lovely enough to lure in whomever he wants, someone capable of conducting the forest's symphony. Someone"—I thought for a moment, then winked at my brother—"wild."

A crash sounded out from the other side of the wall.

I bolted upright in bed. Woferl's eyes turned wide, illuminated by an edge of moonlight slipping into our room. The living room had fallen silent again, but we did not dare move. I tried to keep my breathing even, but I could feel Woferl trembling at my side, and his fright stirred my own. Where was Mama's voice or Papa's steps, someone who should check on the noise? We heard nothing. I glanced toward our closed bedroom door. Even though I heard no footsteps, I did see a faint light wander back and forth under the door.

I tucked my feet into my nightgown. It suddenly seemed very cold.

After a long silence, I finally loosened my knot of legs and swung them over the side of the bed. Perhaps Mama or Papa had tripped over something and needed help. I couldn't hear their voices, though.

Woferl stared at me. "Are you going out there?" he whispered.

I turned my eyes back toward our bedroom entrance. Lights still reflected from its bottom slit, hovering. It did not look like candlelight or the light of a fireplace or sunlight. I motioned for Woferl to stay in bed, then crept over and peered out into the living room.

There, on the other side of our door, drifted a world of fireflies.

It did not occur to me that I might be dreaming. The air seemed too alive. The fireflies were everywhere, too bright to be an illusion.

I'd never seen so many, certainly none in the winter. They clustered the most brightly near the music room. One flew so close to my face that I stepped back and blinked, afraid it would land on me. But perhaps they were not fireflies at all—for in that moment, I glimpsed a tiny figure behind the light and caught sight of slender arms, legs as fine and delicate as flower stems. It made a bell-like sound before darting away.

I wandered out of our room, awed into silence. Moonlight spilled through the windows to paint patterns on the floor. Outside, I could see the dark outlines of the Getreidegasse's buildings asleep under the stars. The tiny creatures' glow gave our flat a strange color, somewhere between this world and another. I wanted to say it looked yellow, or blue, but I could not. It was like describing the color of glass.

The shadows stirred near the music room's door. I turned toward it. My feet moved forward on their own, and my brother followed close behind. The dots of light drifted aside for us, letting us carve a dark blue trail through their golden mist.

Someone was humming near our clavier. When I saw him, I gasped and lifted a hand to point in his direction.

The boy swiveled to face us. He flashed me a smile that revealed pearl-white canines.

He was taller than I, his frame as young and willowy as a dancer at the ballet. His skin glinted pale in the moonlight, and his fingers were long and lithe, his nails sharp. Sapphire hair tumbled shining down his back, and among the strands hung twisted trails of black ivy, shimmers of moss and forest, night and jewels. His eyes were large, luminous, and wondrously blue. They glowed in the darkness and lit up his lashes. His lips were full and amused. When I looked closer, I noticed the catlike slant of his pupils. His cheekbones were high and elegant in his youthful face, and he looked so unbearably beautiful that I blushed at the sight of him.

Of all my memories, this first meeting remains the sharpest.

"Who are you?" I asked.

Beside me, Woferl's eyes were round with awe. "Are you the guardian from the forest?" he added.

The boy—the creature—tilted his head at me. "You don't know?" he replied. There was a wildness about his voice, like wind that made the leaves dance, and I recognized it immediately as the sound I'd yearned toward in my dream. *This is who whispered to me at the clavier, the same boy I'd seen walking beside the ocean in my dream.*

It was him, and he was here. The breath in my chest tightened in fear and excitement.

Was he an imp, as Woferl had first suggested? I'd seen black-and-white ink drawings of those gnarled little creatures in collections of faery tales, legends, and myths, but this beautiful boy bore hardly any resemblance to them. It was as if he were the original mold and the drawings merely his crooked shadows.

When I said nothing, he smiled and beckoned to me. Several of the fireflies danced close to him now, tugging affectionately at his hair and kissing his cheeks. He brushed them away and they scattered, only to return to hover around him.

"You are the Mozart girl," he answered. "Maria Anna."

"Yes," I whispered back. "But I'm called Nannerl, for short."

"Little Nannerl," he said, his grin tilting playfully up at one side. "Of course." The way he said my name sent a shiver down my spine. He turned to look at the clavier, and the gesture set the jewels in his hair clinking. "The girl with the glass pendant. I heard your wish."

How could he have heard something I only held in my heart? A wave of fear rose in me that he might say it aloud. "You were the one in my dream," I replied.

"Was it *your* dream, Nannerl?" His fanged smile gleamed in the darkness. "Or are you in mine?"

The lights hovering about his face twinkled. *Hyacinth!* they cried at him with their tiny bell voices, and he cocked his head at their calls. "Go back to bed," he said. "We will talk again soon."

Then, he reached out to the clavier's music stand, grabbed my notebook, and tucked it under his arm.

Woferl cried out before I did, his baby hands stretching out toward the boy. "He's stealing your notebook!"

The boy shot me one last glance. "There is a trinket shop at the end of the Getreidegasse," he said. "Come tomorrow, and I will return your music to you." He didn't wait for me to reply. Instead, he turned his back to us and threw himself at the window. My cry choked in my throat.

The glass shattered and the boy blended in with a thousand

glittering shards that spilled from the frame. His figure vanished as he fell to the street below. Woferl and I both darted to the windowsill. There, the scene made me step back in shock.

The Getreidegasse, its shops and carriages and silent iron posts, had disappeared. In its place lay a forest thick with upside-down trees, their roots reaching up to the stars, their leaves spreading out on the ground like velvet pools. Twin moons washed the scene into ivory and blue. A faint hum lingered on the night breeze, that same perfect, enticing melody from my dream, whispering for us to come closer. A trodden path wound its way from our building far into the forest's belly, deep into somewhere we could no longer see, where it faded away into the darkness.

A crooked wooden signpost stood right at the forest's entrance, pointing to the path. I squinted to read what it said, but couldn't make out the letters.

The music hanging in the air made my hands tremble, and a sudden urge surged in me. I tugged on Woferl's hand. "Let's follow him!" I whispered.

Woferl obeyed without hesitation. Our feet took flight. I unlocked our flat's front door, swung it open, and hurried out with my brother. My nightgown hugged my thighs as I ran, and the winter floors numbed my bare feet. I ran down the stairs and past the archways, down the third and second stories, down, down, all the way down until I stumbled to a halt at the arched entrance leading to the main street.

I blinked.

The forest, the moons, the upside-down trees, the trodden path, the sign. The music. They were gone. The Getreidegasse had returned to normal, the bakery and winery and the pubs, their wrought-iron signs dangling quietly over their doors, shutters closed and flags pointed up toward the sky. In the distance loomed the familiar, black silhouette of Hohensalzburg Fortress, below which curved the silver ribbon of the Salzach River. I simply stood there, trembling from the cold, clutching the edge of my gown, starving to hear more of the melody from that other world.

Woferl came panting behind me. I caught him right as he ran toward the street, and pressed him close to my side. He looked as surprised as I did.

"Where did he go?" he asked. His breath rose in cloudy wisps.

A sick feeling crept into my stomach. I did not look forward to seeing Papa's face when he found out that the notebook had gone missing. He would think I'd lost it and shake his head in disappointment. Beside me, my brother noticed my crestfallen expression and sobered immediately, slumping his shoulders and lowering his eyes.

"Woferl. *Nannerl!*"

The familiar voice startled me. Both of us whirled around in unison. It was Mama, her hair tucked underneath a nightcap, racing down the steps toward us. Her hands clutched at her coat. The image of our mother looked so real, the lines of her face so defined in contrast to the halo of light that had surrounded the boy. Suddenly, I felt how solid the ground was beneath my feet, how biting the chill was in the air. She frowned at me. *I am afraid of the cold,* Mama had told me before, and I turned my eyes down in shame for forcing her out here on an autumn night.

"Nannerl, what in the world are you both doing down here?" She shivered, her breath rising in a cloud. "Have you lost your senses?"

I started to explain what we had seen. But when I pointed up to where our windows were, where the boy had thrown himself down to the streets—I saw that the glass panes had returned to their normal state. Nothing was broken.

My words died on my lips. Even Woferl stayed quiet.

"I'm sorry, Mama," I finally said. "We were dreaming."

Our mother looked from me to my brother, then back again. The hint of a smile danced on her lips before it disappeared again behind her frown. There was a question in her eyes, something curious beyond her stern gaze that wondered what could really have brought us out here.

After a pause, Mama shook her head and held out a hand to each of us. We took them, and she began to lead us back up the stairs. "The very idea," she murmured, frowning at how cold our hands felt in her warm ones. "I'd not thought you capable of such mischief, Nannerl. Rushing out here with your brother in the darkest hour of night. And in this cold! Thank goodness your father sleeps so heavily, otherwise he'd never let you hear the end of this."

I looked up at her. "Didn't you hear the crash in the music room, Mama?" I asked.

Our mother raised a slender eyebrow. "Nothing of the sort."

I fell silent again. As we stepped back inside our building, I saw the trinket shop at the end of the Getreidegasse from the corner of my eye. The boy's final words lingered. I wondered what would happen if I met him there.

When I looked at Woferl, he looked ready to say something to Mama—but after a while, his mouth relaxed into a line and he turned his face down. The matter was dropped.

The Princeling in the Grotto

PAPA DISCOVERED THAT MY NOTEBOOK WAS missing the next morning.

He did not shout when he became upset. Instead, his voice would turn quiet like a storm on the horizon, so soft that I'd have to strain to hear what he was saying.

Careless. You are so careless, Nannerl.

Each of his words lashed at me. I bore it and kept my head turned down, my eyes focused on the embroidery of our rug. It was a hunting scene of three brothers riding in the sun-dappled clearing of a forest, their hounds forever frozen in the throes of tearing a doe to pieces.

"Well?" my father asked. "What do you have to say for yourself, now that we must buy you a new notebook?"

I counted the number of hounds and horses as I tried to still my thoughts. "I'm sorry, Papa," I replied.

"Sorry," he echoed me in disbelief, then shook his head and looked away.

Beside him, Mama glanced quickly at us and cleared her throat. "They are still children, Leopold," she said, putting a comforting hand on our father's arm. "You are a grown man, and yet how many times have I scolded you about your misplaced quills and your lost spectacles?"

Papa just scowled. "Young ladies should be more responsible," he said, looking back at me again. "How will you care for a husband if you cannot even care for your belongings?"

The word burrowed into my mind. *A husband, a husband,* it repeated in a whisper that quickly evolved into a roar. *You will be forgotten,* it said. I watched as my mother smoothed my father's sleeve. *One day, you will disappear.*

I did not know how to defend myself. How does a daughter explain such a thing to her father? Even I could not be sure anymore what had happened. Sleep had already fogged the memory of last night. Could someone really have been in our home, standing by the clavier? Who had drawn us out into the street?

No, my father must have been right. I simply misplaced it. Last night was a dream, nothing more. And yet I kept staring at the rug, studying the doe's wide eyes as my mother coaxed Papa with soft words.

Then, as my father resumed his scolding of me, Woferl rose from the dining table. He went up to the clavier, pulled himself onto the bench, and placed his hands on the keys.

"Don't be angry, Papa," he said over his little shoulder. "I can remember the pages. Then we can write them down again."

Of course he could not. Of course this was just another one of his whims. I stood there and almost wanted to smile at his strange attempt to defend me, for trying to turn our father's shadow away from where I stood.

Papa's eyes softened in amusement. "Can you, now?" he said.

Woferl's expression stayed serious. He turned back around on the bench and started to play.

At first, he struck the wrong note, and hit a few more strays before he shook his head and paused. The piece was supposed to be a menuett in C. I saw him frown, knew that the same thought had just crossed his mind, and watched him start over.

This time, Woferl hit the right note. Then another and another and another. One of his fingers slipped, but that was the last mistake. He managed to make it through sixteen measures, all correct, of the menuett, and though his rhythm was off because he had to think about each measure, he remembered all of it.

My father stared at him, all signs of his earlier tirade completely vanished. I looked at my brother in disbelief. None of us dared move a muscle, as if what we'd witnessed was only a figment of our minds, and that if we disturbed this moment, Woferl's playing would have never happened at all.

My brother was barely old enough to read. What he just did was impossible.

I looked at our father. His smile had disappeared, but his eyes had turned very bright. He said nothing. He needed not to, for even then, I could see in his mind the thought that lit his face.

This was the song of God he yearned for, emerging from the small hands of his son.

My affection for Woferl wavered then, and suddenly I felt that cold twinge return to my chest. The same one I'd felt as I'd let him play on the clavier beside me, when he'd remembered what I played so easily. It had taken me a week to remember the same piece! Surely, he could not have memorized so much in such a short amount of time. I wondered, suddenly, if Woferl could have been the one who hid away my notebook.

My brother climbed off the bench and looked at me. There was only curiosity in his gaze, that perpetually innocent smile on his face. He was waiting for me to compliment his playing. I hesitated, unsure of what I might say.

Several minutes later, Papa hurried out of our flat on his way to Herr Schachtner's home. He was in such an eager mood that he had to return to grab his hat, which he'd completely forgotten.

I stayed quiet as Woferl and I prepared to accompany Sebastian down to the Getreidegasse for bread and meat. My brother hummed the tune under his breath while I helped him into his coat. When I listened closely, I could tell that he knew far more of the piece than he'd played.

"When did you learn the first page?" I finally asked him as we stepped out of our building and into the street. It was a brisk, busy morning, full of the music of carriages and conversations.

Woferl made me lean over to hear his reply, so that I walked awkwardly with my body tilted sideways. "When we saw the flowers." He kept his eyes on Sebastian's back. "When they were growing on the first page. Did you like it?" he added in a hopeful voice.

This couldn't be the answer, and I was so humored that I laughed. The edelweiss in my notebook had been a daydream. "You mean, you remembered the notes from yesterday?"

"Yes."

"Just from the few moments I kept the page open?"

Woferl seemed puzzled by my shock. "Yes," he said again.

I looked at him again. "Woferl," I said, "you could not possibly have remembered the entire piece from our session. How could you? It was too long. Now, tell me the truth, Woferl—I won't tell anyone if you don't want me to. Did you take my notebook and hide it somewhere?"

He shook his head, sending his curls bouncing. "It was not too long," he insisted. His eyes turned up to me in frustration. "I don't need to take your notebook to remember the music."

What Woferl said could not possibly be true. He must have practiced at another time, when no one else was around. Even if he hadn't taken my notebook, he must have stolen peeks at it when I wasn't looking. But his words were so sincere, so absent of his usual mischief, that I knew he wasn't lying.

He huffed. His breath floated up in the air and faded away. "Besides," he said, "we both know who stole it."

I thought again of the fireflies that had floated in the darkness of our apartment, then the midnight dream of the boy in the music room. He had spoken so clearly to me. I'd seen him tuck my notebook under his arm and throw himself from our window against a silhouette forest. Even Woferl remembered.

The skin on my neck prickled. Last night was, of course, nonsense. But this time I did not laugh at the thought.

"You are very talented, Woferl," I said to him after a long pause.

It was what he had been waiting all morning for me to say, and he brightened right away, forgetting all his frustration with our talk. His hand tightened in mine. My other hand rubbed at the glass pendant in the pocket of my petticoat. Acknowledging my brother's playing frightened me less than the thought of last night being anything more than a dream.

The Getreidegasse was still wet today from a cleaning, and the air hung heavy with the smell of soups, carriages, horses, and smoke. Hohensalzburg Fortress towered over the city's baroque roofs, a faded vision today behind a veil of fog. Farther down, where the streets met the banks of the Salzach, we could hear the splash of water from the butchers hunched behind their shops, cleaning freshly culled livestock in the river. Everything bustled with the familiar and the ordinary. Woferl and I blew our warm breaths up toward the sky and watched them turn into puffs of steam. The clouds looked gray, warnings of snow. Several ladies passed us with their faces partially obscured behind bonnets and sashes. One of them carried at her hip a fine, pink-cheeked boy swaddled in cloth.

I watched her and tried to picture myself doing the same, hoisting a child in my arms and following a faceless husband down these uneven sidewalks. Perhaps the weight of carrying a child would damage my delicate fingers, turning my music coarse and unrefined.

We reached the bakery. Sebastian ducked his head under the wrought-iron sign, greeted the baker affectionately, and disappeared inside. While he did, I turned my attention to the end of the street, squinting through the morning haze to where the trinket shop stood. I half expected to see a shadowy figure standing there already, a tall, willowy creature with his glowing blue eyes, my notebook tucked under his arm.

"Let's go," I whispered to Woferl, tugging his hand. He needed no encouragement, and slipped out of my grasp to go skipping toward the shop, his shoes squeaking against the street.

The trinket shop was a familiar sight. Woferl and I liked to stop here often and admire the strange collections of figurines behind its windowpanes. Sometimes we would make up stories about each one, how happy or sad they might feel, how old they were. Herr Colas, the elderly glassmaker who owned the shop, would humor us by playing along. Some of the trinkets were thousands of years old, he'd say, and once belonged to the faeries.

Woferl blew air at the window and left a circle of fog on it. The circle began shrinking right away.

"Woferl," I scolded, frowning at him. He stared back with big eyes.

"Do you see the boy in there?" he whispered, as if afraid to be overheard.

I bent down to study the trinkets. Some had colors painted on, deep-red dishes and gold-trimmed butterflies, blue glass pendants like my own, crosses, the Virgin Mary. Others had no color at all. They were simply glass, reflecting the colors around them, reminiscent of the faery lights we had seen in our flat. My gaze shifted from them to the shop beyond.

"I don't see anyone," I replied, looking back to the trinkets.

Then something scarlet caught my eye. I turned to look toward it and noticed a tiny sculpture I'd never seen before.

"Woferl," I breathed, pulling him closer to me. I pointed through the windowpane. "Look."

The trinket was of three perfect, white edelweiss, frozen in porcelain, their centers golden, their velvet petals gleaming in the light. One of the flowers had a missing streak of white paint.

My memory flooded with the image of the flower from my music notebook.

"Do you like this one, Fräulein?"

We both jumped, startled. Herr Colas stood near the shop's door, squinting down at Woferl and me. Thin white bandages wrapped around his hand that clutched the doorframe. As he peered at us, I could see the deep pockmarks on his face crinkled up into slants. I sometimes wondered how he looked before the smallpox ravaged his skin, if he had ever been young and smooth-faced.

I pinched Woferl's arm before he could say anything, and then I curtsied to Herr Colas. "Good morning, Herr," I said. "The trinket is very pretty."

He smiled and waved us forward with his bandaged hand. "Come, come, children," he said. "Come out of the cold and have a look around." He glanced at me. "You've grown taller since I last saw you, Fräulein. Your father has nothing but praise for your musical talents. I hear all the gossip, you know. The young girl with an ear like a court musician!" He gestured to my fingers, now covered in my gloves.

The Herr might know the gossip in the streets, but he'd never heard the barbed words my father said in our home, or seen the disappointment in his eyes. Papa would never belittle my skills in public. After all, that would embarrass him. Still, the Herr's words warmed me, and I found myself blushing, murmuring my thanks.

"Where is this trinket from, Herr Colas?" Woferl piped up as we stepped inside, his eyes locked on where the porcelain edelweiss sculpture sat in the window display.

The old shopkeeper scratched the loose skin under his chin. "Vienna, I believe." He leaned down to give us both a conspiratorial grin. In the light, one of his eyes flashed and I thought I caught a glint of blue. He wagged a finger not at the trinket, but at the windowpane, and I thought again of the strange boy shattering our window into a thousand pieces. "Who knows, though, really? Perhaps it's not from our world at all."

My skin prickled at his words. I wanted to ask him what he meant, but he had already left us alone to our wandering and returned muttering to his little desk in the shop's corner.

The shop looked hazy, the light filtering in from the windows illuminating the dust in the air. Shelves of trinkets were everywhere, music boxes in painted porcelain and strange creatures frozen in yellowing ivory, their lips twisted into humanlike grins. The stale scent of age permeated the room. While Woferl wandered off to a corner decorated with wind chimes, my eyes shifted to a dark corner of the shop hidden behind shelves and boxes. A thin ribbon of light cut through the shadows there. A door.

"Herr Colas," I called out politely. "Are there more trinkets in your back room?"

He didn't answer. All I could hear was the faint sound of humming.

My attention returned to the door. The humming seemed familiar now, a voice so perfectly tuned that it pulled at my chest, inviting me closer. My feet started moving of their own accord. I knew I shouldn't have been back there without Herr Colas's permission, and a small part of me wanted to step away—but as I drew closer, my fear faded away into nothing until I found myself standing right in front of the door.

The humming voice came from within, beautiful and coaxing.

I pushed the door with slow, steady hands and stepped inside.

At first, I saw nothing. Darkness. The door edged open without a sound, and I felt a touch of cool air. It smelled different from the air outside, not of winter and spices or of stale antiques, but of something green and alive.

I stepped onto moss, the dampness of it soaking the bottom hem of my petticoat. A faint glow gathered at my feet, a quivering mist of faery lights, skittish in their movements. The darkness crept away as I continued forward, until I could see the ground clearly without bending over, and I realized for the first time that I was walking inside a tunnel—the walls dripped with moss and green ivy, baby ferns and tiny rivulets of water. Strange fruits hung from the ivy trails, wet and bright blue and as plump as bird eggs, their shapes like musical notes. Eating one was surely a quick invitation to be poisoned, but in that moment, I felt such a surge of want tingling on my tongue that I reached out, unable to stop myself, and plucked a single fruit free of its stem. My movement jerked the ivy forward and then quickly back. Drops of water rained down from the vines in a shower.

I popped the fruit into my mouth and bit down until its skin burst. Sugar and citrus and some otherworldly spice flooded my mouth. I closed my eyes, savoring the flavor of it.

I reached to take another and my fingers sank into the soft vegetation. One of them brushed past something familiar—a soft, velvet surface. I looked at where my hand had been.

A patch of edelweiss was growing against the wall, their velvet petals glistening with dew, and when I blinked, several more popped out from the wall's moss to hang sideways, their buds drooping toward the floor.

It was impossible, truly, to see a flower of the mountains in a place like this. But nothing about this place seemed real at all.

A few notes of music caught my attention. I turned instinctively toward the sound, seeking it out. It came from farther down, where the tunnel ended in a circle of light, playing like a secret insulated from the rest of the world. My heart ached for it. *Music from my notebook?* I picked up my skirts and quickened my steps. Ahead of me, the tunnel began to widen, sloping higher until it opened abruptly into a circular cove.

The ceiling appeared to be formed from a lattice of leaves and fruit. Patches of silver moonlight filtered through to the ground, where edelweiss carpeted the floor in a white blanket. Moss and foliage enveloped every wall. And sitting there, in the center of this strange space, was the most beautiful clavier I'd ever seen, covered with baroque art and wrapped in lengths of ivy.

No one sat at the bench, even though the velvet cushion upon it had an indent as if someone had just left. When my eyes went up to the clavier's music stand, I saw with a surge of joy that my notebook was sitting there, waiting for me.

"It's here!" I called out into the tunnel, hoping that Woferl would hear me from the shop. I stared in wonder at the clavier. The keys had rounded tips that glowed under the light like polished gulden, and the entire instrument looked carved not from wood, but from marble. I ran my fingers across its surface, searching for the gaps where the body of the instrument should meet the legs, where the hinges of the lid should be screwed into the belly. But there were no gaps. The entire clavier was carved from a continuous slate of marble, as if it had always been molded in this form.

My hand drifted across the clavier, afraid to touch it and yet unable to bear not doing so. How could something so lovely be real? What would it sound like? I hesitated there for a moment, torn in two directions, before I finally pulled the bench forward so that I could sit. The legs scraped against the moss on the ground.

My notebook was already flipped open to a menuett in C, the latest piece my father had composed and the same one that Woferl and I had been playing when we first saw the edelweiss against the parchment. The very piece that Woferl had committed to memory from a single session. Even glancing at the written notes filled my mind with its music. I could distinctly hear the measures of the menuett as if I were practicing them during my lessons.

I lifted my fingers to the keys and touched their glowing surface. The keys were cold as ice. Instantly I drew my hands back, but the burn of it tingled like snow on my tongue, dangerous and enticing. I placed my fingers in position again, savoring the strange chill of the instrument. This time I tried a few notes. The sound hovered in the air, surrounding me, richer in tone than any clavier I'd ever played. My eyes closed. I realized I was humming now, trying instinctively to match that perfect melody around me. My heart fluttered with the thrill of the music.

A carefree laugh echoed from behind me. "You can have it back, Fräulein."

I stopped playing and whirled around to see the speaker.

There, underneath the shadows and the dripping moss, emerged a figure. Immediately I recognized him as the boy from the music room, the same silhouette who had walked along the shore in my very first dream. Under this new light, his pale skin took on a hint of blue. His grin was quick and lighthearted, his expression as much like a human boy's as it could be.

In that moment, I realized that perhaps that first dream was not a dream at all. Nor were the edelweiss growing against my notebook, or the sight of this boy in my music room. Perhaps even this moment was real. The world around me felt so sharp and alive that I couldn't possibly think otherwise.

"You can have it back," he repeated in his perfect voice. "I'm done with it."

"Who are you? Where do you come from?" I whispered.

The boy walked over to the clavier and performed a little jump. He settled comfortably on top of the instrument, then peered down at me with his head propped thoughtfully against one hand. His fingernails clicked against the clavier's marble surface.

"From somewhere far away," he said, "and very near."

I tilted my head at him. "That's not helpful at all."

"Isn't it? You know where it is. You've seen it before. You've been there."

The twin moons hanging silver in the sky. The blue seashells dotting a white beach. A feeling of wistfulness crept over me then, as if I were thinking back on a place I'd once known. I looked at his feet, expecting to see sand between his toes.

"Where did it come from, then?" I went on. "This place both near and far away?"

"It's been around since long before you or me. Everyone has seen it in some way, you know, although most will not remember it."

A deep longing lodged in my throat. "Will I get to go there again?"

"Perhaps. I heard your wish," he said, repeating what he had told me in the music room. "You want to be worthy of being remembered. By your father, by those your father regards highly. By the world. You're afraid of being forgotten." He studied me curiously. "That's a large wish for a Fräulein to make. Why are you so afraid?"

My thoughts snapped to my father, how he would look away from me in disinterest if I did not play well. His talks with Mama, the whisper that followed me down the Getreidegasse. *A husband, a husband.* I thought of fading into the light so quietly that my father might never notice. If I could fill our family's coffers . . . If I could create with the voice of God given to me, my father would not forget that I was here.

The voice of God. I thought of this boy's beautiful words, the music of his voice that trembled on the air of my dream, in that strange and vibrant place. That was it, the perfect sound.

At last, I met the boy's eyes. "Papa once told me that if nobody remembers you after you're gone, it's as if you never lived at all."

His smile widened at that. He looked like he had heard every thought unspoken in my mind. "It's immortality you seek, then," he said. "You burn with the ambition to leave your voice in the world. You fear your father will forget about you if you cannot do this. All your life, you have ached to be seen." He leapt off the clavier, then came to sit beside me on the bench. There, he leaned over, reached out his arm, and touched my chin with his cool, slender fingers. A sigh emerged from his lips. "Oh, Nannerl! You are an interesting one."

"Interesting? How?"

"Your need to leave a memory of yourself long after you have gone. Desire is your lifeblood, and talent is the flower it feeds." He gave me a sideways look as his hands sought out the clavier's keys. He began to play a soft melody I did not recognize. It was so lovely that I found myself touching my hand to my chest, steadying myself against the sound. "I can help you . . . but first, we must play a little game." His grin widened, childlike in its delight.

My heart lurched in excitement and fear at his words. "What kind of game?"

"You have your desires, and I have mine." He leaned his head closer to me. "You want immortality. I want my throne."

At last, he was finally answering my question. "Is that who you are, then? A king?"

The faeries floated around him, their light glowing against us as they kissed his skin. *A princeling, a princeling,* they whispered, filling the air with the word. *Princeling of the forest.*

"My name is Hyacinth," he said.

Now I remembered the faeries calling his name the night before. The blue of his eyes certainly matched the flower. *Hyacinths,* my mother had once pointed them out to me at the market, and I'd brushed my hands against their clustered blooms. *Hyacinths are the harbinger of spring and life.*

"What happened to your throne?" I asked him.

The boy named Hyacinth ignored my question. His expression had suddenly shifted from mischief and mystery to something tragic, a flash of sadness that cut through his trickery. It disappeared as quickly as it had come, but the ghost of it lingered at the corners of his face, pulling me closer to him.

I looked at my notebook. "And why did you take this?" I asked.

He started to play again. I breathed deeply at the music. "You made a wish, Nannerl, and so I have come to you. You'll discover that your notebook will now serve you in more ways than simple lessons at the clavier. Use it as your path to me. You can always find your way to me, Nannerl, if you speak to me through your music."

If you speak to me through your music. I imagined this boy listening to the secrets in my heart, his eyes peering through the web in the woods. His hand taking mine and leading me down an enchanted forest path.

"What way is that?" I asked him.

"Why, to my kingdom, of course," he answered.

Hyacinth's words reminded me of my brother's question from last night. "You say you seek your throne. Are you the guardian of the kingdom, then?" I whispered.

He turned to me with his secret smile. His eyes glowed against his skin. "I am *your* guardian, Nannerl. Tell me what you want. I will find a way to give it to you."

Tell me what you want.

No one had ever said those words to me before. A slow, creeping cold began snaking its way down my fingers, until my arms grew heavy with numbness. The boy's eyes hypnotized me.

"But be wary of what you wish for," he went on. "Wishes have a habit of surprising their makers."

I closed my eyes and swallowed hard. The cold crept farther up my arms and to my shoulders.

When I opened my eyes again, he was gone.

I looked around in bewilderment at his sudden absence. I was alone in this strange grotto, my notebook still sitting on the clavier's stand. With a burst of panic, I grabbed the notebook before it could disappear again, and then I sprang from the bench and turned back toward the tunnel. I called out for Woferl, but only silence greeted me. My stomach turned. He must still be in the main shop—I had to go back to him. Sebastian must have come for us by now.

"Woferl!" I shouted, running faster as I went. "Woferl, answer me! Where are you?"

And then, just as abruptly as I'd entered the grotto, I stepped through the door and stumbled right back into the shop.

Everything looked unchanged from when I'd left it, the hazy air golden under the sun, the shop's shelves stacked heavy with trinkets. But the tremor of whispers and music no longer lingered in the air. It was replaced instead with the smell of aged wood, the bustle of everyday life outside the shop's walls. I stood still for a moment, trying to regain my sense of place.

Woferl looked over from where he was loitering near the windows. "There you are," he said.

I rubbed my eyes and glanced behind me. The tunnel had vanished, leaving behind nothing more than a tiny closet overflowing with empty crates.

Perhaps the dust in the shop had made me sleepy, and my mind had woven for me a web of illusion. The ice-cold burn of the clavier's keys, Hyacinth's glowing blue eyes . . .

"Are you all right?" Woferl asked, his eyes turned up at me in concern. "You look pale."

I shook my head. "I'm fine," I answered.

His eyes darted next to the notebook I clutched in my hand. "Oh! You've found it!" he exclaimed.

I blinked again, still surprised to be holding it. Had it been in my hand seconds ago? *Was* it all truly a dream?

"Was it the boy?" he asked rapidly. "Did you see him again?"

Sebastian came to my rescue before I had to answer. He ducked his head out from below the baker's signpost, caught sight of us, and nodded. "Fräulein. Young Master. We have prolonged this trip enough."

Woferl let the question drop as his attention turned momentarily to coaxing a sweet from Sebastian's pockets, and I gratefully let him go. My mind lingered on his questions, though, so that the rest of the trip home passed in a fog.

Everything about the grotto seemed so distant once I was back in the familiarity of the Getreidegasse. But even if it had been a dream, it was a dream that persisted, the same world that kept returning to me day after day, year after year.

As Woferl pranced around Sebastian trying to make him laugh and give him another candy, I looked back down at my notebook. My fingers closed tightly against the pages. I had left our apartment without it and would return with it right here in my hand.

The music in the princeling Hyacinth's voice still played in my mind. It was possible that the grotto was a part of this continuous dream . . . or, perhaps, it was also possible that everything was real.

By morning, Papa had already spoken to Herr Schachtner about Woferl's newly discovered talent. Not a few months afterward, as if the princeling had sent them himself, letters began to arrive from Vienna. The royal court wanted to hear us perform.

THE ROAD TO VIENNA

WE WAITED UNTIL THE WORST OF WINTER had passed before Papa began preparations for our first trip. The cold days dragged by one after the other. Outside, the Christmas snow fell. The Bear and the Witch and the Giant roamed the Getreidegasse and children ran squealing from them in delight. Sometimes as I watched from the window, I thought I caught a glimpse of Hyacinth walking with the wild bunch, his blue eyes flashing up toward me. Then he would disappear, leaving me to think I must have imagined it all.

During those short winter days, Papa sat at the clavier with Woferl for hours, praising his swift memory and his accuracy, clapping whenever my brother finished memorizing another piece or added his own flourishes to a measure. Woferl hardly needed his instruction. One day, I came into the music room to see my brother holding Papa's violin, his shoulder barely big enough for the instrument. He was not only teaching himself the strings, plucking each one and figuring out the correct notes as he went—but *inventing* a tune. He was already composing.

I'd heard my father call other musicians prodigies before, but they were men in their teens and twenties. My brother was just a child. I stood frozen in place as I watched him. His eyes stayed closed, and his fingers fluttered as if in a trance.

With me, too, our father turned more serious, extending my lessons, noting my every mistake and nodding in approval each time I played flawlessly. I savored every moment of his attention. Even when I wasn't at the clavier, I sat with my notebook in my lap, poring over the pages in search of whatever magic Hyacinth had cast on it.

I could see no visible change in the pages, but something *had* changed. I could feel the tingle of it in my fingertips whenever I brushed the paper.

On a day when the spring thaw dripped from the trees, Woferl and I stood outside the arched entrance to our building and watched Sebastian and our coachman drag trunks of clothing across the cobblestones, throwing them unceremoniously into our carriage's boot. Mama chatted with Papa as they worked. I could see her unfolding and refolding her arms in barely disguised anxiety. She did not want to leave home.

Their Majesties Emperor Francis I and Empress Maria Theresa. I kept my hands folded in my skirts and repeated their names silently. *Vienna's royal court.* Mama had said that kings and queens are remembered. Perhaps being remembered by royalty was the same.

Beside me, Woferl shifted his weight from one foot to the other, trying in vain to keep his excitement subdued. His eyes were bright with anticipation this morning, and his brown ringlets brushed past flushed cheeks.

"What are we going to do for two weeks in a carriage?" he asked me.

I leaned down toward him and raised my eyebrow. "Having no clavier on the road does not mean you can get into mischief. Papa and Mama will not have it, do you hear?"

Woferl pouted, and I patted his head. "We will find something to pass the time. The countryside will look beautiful, and soon you will get your instruments back."

"Will they have an ear for music, do you think?" he asked me curiously. "The emperor and empress?"

I smiled at his boldness. "Best not ask that question at court, Woferl."

He tucked his hand into mine and leaned against me. I noted how much thinner and stretched out his fingers already felt. The softness of his youth was rapidly disappearing from his tiny hands. "Herr Schachtner said the emperor likes a spectacle," he said.

And a spectacle they would get. Woferl's improvement on the clavier only quickened with each passing week. Papa had to commission a tiny violin for him. Before Woferl, it was unheard of for a child his age to play the violin at his level. That quality of instrument simply did not exist in the shops. Our names now regularly circulated the Getreidegasse, whispers on the tongues of the curious and skeptical that Herr Leopold Mozart's two children were in fact both musical prodigies. They would say my name first, because I had played for longer. But they saved Woferl's name for last, because he was so young.

I tried to keep my unease at bay. I woke up early every morning of the winter to practice, staying at the clavier long after Papa had left for the day. I'd play and play until Woferl would tug on my sleeve, begging for his turn. When I was not at the clavier, I tapped my fingers against the pages of my notebook and hummed under my breath. I spent my days wrapped in the music, lost in its secrets. When I dreamed, I dreamed in new measures and keys, compositions I would never dare write down.

I was, after all, not my brother.

Sometimes, over the long winter, I'd also dream of Hyacinth whispering in my ear. *Desire is your lifeblood, and talent is the flower it feeds.* I'd wake and play his menuett on the clavier, the tune I'd heard in the grotto, wondering whether it would call him back again. Perhaps he was watching us right now as we stood outside our home, his pale body washed warm by the light. Out of instinct, I tilted my head up toward our windows, certain I would see his face there behind the glass.

"Nannerl."

I looked down to see Papa approaching, and straightened to smooth my skirts. He placed one hand on Woferl's messy head of curls. "Time to head into the carriage," he said gently.

Woferl released me, then ran off to hug our mother's waist, babbling affections all the while.

Papa touched my shoulder and led me over to the corner of our building, so that we stood partly in the shadow of the wall's edge. I looked directly at him. I did not do this often; my father's eyes were very dark and frequently shaded by furrowed brows. It was a stare that dried my throat until I could not speak.

"Nannerl," he said, "this will be a long trip. I'll need you to keep your wits about you and conduct yourself like a young lady. Do you understand?"

I nodded quietly.

Papa's gaze flickered over my shoulder toward the carriage. "Woferl's health has been delicate lately. All this winter air." I nodded again. My father did not need to tell me. I had always known this about my brother. "Two weeks in the carriage may wear him down. Take care that he does not catch a chill. The emperor specifically requested his presence, and if we are to perform again in Europe, we will need Woferl's reputation to precede us." He put a hand on my shoulder. "Be mindful of your brother."

I waited for him to tell me to take care, too, but he did not. My hands brushed at the edges of my petticoat. Specks of dirt had spoiled the fabric's light color. "I will, Papa."

He removed his hand. His expression changed, and he started to move away. "The coach is almost ready. Come along, Nannerl."

As I headed after my father, I watched Woferl pull Mama's arm down toward him. I could not hear him, but his words coaxed a tear from her eyes, and she gathered him into her arms.

Years later, I learned that Woferl had asked, *Mama, will you be sad when I grow up?*

We traveled along the upper rim of the Alps, where the terrain changed to gently rolling hills and patches of forest. Papa and Mama chatted together on one side of the carriage, while I sat with Woferl on the other. The ride was so bumpy that I had to press him against the carriage wall to keep him from sliding around.

While our parents dozed, I spent my time looking at the

ever-changing landscape. The houses had grown sparse, and the sun shifted in the sky so that it peeked in just below the carriage window, bathing us in light. I smiled at the warmth, leaned closer, and narrowed my eyes. The passing hillsides transformed into a stream of colors, gold and peach and orange, hazy layers of billowing silk. Tree trunks blurred by.

Beside me, Woferl's eyes were half closed, and his lashes glowed white in the sunlight. His slender little fingers danced, composing in the same way I'd seen him that morning with Papa's violin.

"What are you thinking, Woferl?" I asked in a soft voice.

He opened his eyes. "I am writing a concerto," he whispered.

I nudged him affectionately with one elbow. "How are you writing a concerto, silly, with no paper?"

"I can write it down in my head and remember it." He rolled his eyes upward, thinking, then looked at me again. "I am imagining the kingdom."

His mention of the fantasy otherworld sent a familiar thrill through me. Woferl had, for months after the incident at the trinket shop, asked me exactly what I'd seen that day. I'd told him about the clavier and the notebook, the grotto and the princeling. All I'd left out was the conversation between Hyacinth and me. It had seemed like a secret meant for no one else.

"Are you, now?" I said. "What does a concerto about the kingdom sound like?"

He turned his large eyes on me. "You want to hear it?" he asked eagerly.

I hesitated for the space of a breath. "Of course," I replied.

He cleared his throat and hummed a few bars. It sounded light and airy, not like the perfect music from my dream or the grotto, but instead like the scenery we passed. Somehow, I felt relieved that it was so different. Perhaps the music from the otherworld was something only I truly understood. My mind returned for a moment to the moss-paved tunnel, Hyacinth's bright eyes and polished fingernails. Now and then, I thought I could see an upside-down tree flash by our window, although I could never quite focus on it.

"I like it," I said to him when he fell silent again.

"We should give the kingdom a name," Woferl announced. I shoved him, glancing pointedly over to our sleeping father. "A *name*," he repeated in a whisper.

"All right. A name. What do you want to call it?"

Woferl closed his eyes. I watched his sun-soaked lashes resting against his cheeks and wondered for a moment if he had fallen asleep. Then he opened his eyes and flashed a grin at me. "Let's call it the Kingdom of Back," he declared.

"What a curious name," I whispered. "Why?"

Woferl looked pleased with himself. "Because it's all backward, isn't it?" he replied. "The trees turned on their heads, the moons where there should be sun."

Now he was turning playful from restlessness. "And does that mean the people are backward there too?" I teased him. Here, in our sunlit carriage, the kingdom seemed just a figment of our dreams, Hyacinth a fleeting memory. "Were we backward?"

He giggled. "Everyone is backward." At that, he offered me a mock frown, an imitation of a backward smile, and tried to flip the syllables in his name. It sounded so garbled that I covered my mouth, trying to stifle my laughter.

He shifted in the carriage seat toward the window. "Backward," he repeated to himself. "I'm going to put that into my concerto."

"Are you going to write it down when we reach Vienna?"

"Yes."

"The whole concerto?"

"I am almost finished with the first movement."

I shook my head gently at him, disbelieving, then patted his knee. "Surely you can't remember all of that. I would not be able to hold such a long piece in my head."

Woferl simply shrugged. "I can." Then he uttered a contented sigh and rested his head against my shoulder. As he did, he hummed under his breath, so softly that I could barely hear him. But I did. And this time, the sound struck deep within me. I recognized the kingdom in his melody—at once sweet and beautiful, newly formed, in a minor key that made it sound like a place that could never quite settle.

At first, I heard echoes of Papa's rigid teaching. But I could also recognize the parts that my brother drew from my own playing, the pauses and crescendos in his measures. I could make out the way he was turning my inspiration, the sound of my yearning, into his own. How, in a way, he was taking what I could do and improving upon it.

How silly I was for thinking, even for a moment, that Woferl could not create something beautiful enough for the kingdom. I closed my eyes, dizzy, envious, wanting more. This was not the composition of a child. Within it was the wisdom of an old soul, not the innocence of a young boy. No child could create a piece like this and keep it all in his mind.

And as I thought this, as I listened in awe to my brother's raw concerto, the memory of Hyacinth came to me in such a strong wave that I opened my eyes, shivering, certain he would be sitting in the carriage with us.

But he was not there. Papa still sat across from us, dozing with his chin resting against his hand, while Mama leaned against his shoulder, swaying in her sleep. Still, I felt the ripple of something strange in the air, the heady sensation of a new presence. Through the window, something pale flashed by among the trees. A glimpse of glowing eyes.

In an instant, the kingdom and the princeling no longer felt like a faraway dream. They were very real, and they were here.

Suddenly, the carriage lurched to one side. Mama gasped. Papa startled awake with a curse on his tongue. I cried out—my hands flew to the carriage wall to keep myself from falling forward. Our trunks clattered free of their ties and careened out of the boot, landing with a crash in the dirt path outside. We settled to a halt in a cloud of dust.

Papa was first up on his feet. He scrambled against the slanted floor until he could pull himself out of the suspended door, then reached over to hoist my mother out.

I grabbed Woferl's hand. He was crying. He rubbed at his head, and between his fingers I could see a pink mark growing from where he'd hit his forehead against the edge of the carriage window. When I called for him, he reached obediently for me, and I lifted him around to Papa's outstretched hands. Together, we helped him make his way up. I went last.

We found ourselves at the edge of a forest dense with oak and spruce, its canopy so thick that hardly any light reached its floor. I peered around for a moment, blinking dust out of my eyes. The road we'd traveled wound in a slender arc, its bordering trees trailing off into mist, and ahead of us stretched more of the same. In the middle of the road lay one of our carriage's wheels, which had somehow come completely off its axle. Faint lines carved in the dirt curved up the path from where the carriage had passed.

I trembled. My eyes searched for a lithe figure crouching in the trees.

While Papa retrieved the wheel and helped the coachman lift the carriage enough to replace it, I studied the tangle of trees surrounding us, looking up on an impulse into the thickness of their leaves. Beside me, Woferl bent down to pick something out of the dirt.

"Nannerl, look," he said, holding an object out to me. When I peered closer, I saw a blue rock in his hand, glowing faintly as if lit from within, its grooves striped and strange like a seashell's. I turned my attention to the other rocks strewn across the path, and when I did, I noticed that hidden underneath their coats of dust were glimpses of bright blue, an entire spread of shining, sharp fragments that had jolted our travels to a full stop. At first glance, they looked strewn in a random pattern, careless stones clumped with twigs and fallen leaves.

But looking closer, I realized that the tracks our carriage had left behind looked like two long sets of lines in the dirt, unmistakably reminiscent of the kind that make up bars of sheet music. The shining blue rocks lay glittering across these lines, round notes winking at us in the afternoon light. My lips parted. I found myself humming the tune on the ground. The notes littering the path changed to music on my tongue, then drifted into the air and faded away. Woferl listened in wonder, his eyes fixed on the sight.

I looked over my shoulder into the forest. An uneasy feeling seeped into my chest, coating my insides and pushing against my ribs until I could hardly breathe. Was it simple coincidence that the wheel had come off from the side where Woferl sat? Coincidence that this happened right as I felt Hyacinth's presence in our midst, after Woferl hummed his composition?

My brother still rubbed at his bruised forehead, although he made no more mention of it. Deep in the forest, the wind blew a tuneless song through the leaves, and in it I thought I could hear snatches of a voice I recognized all too well. Something quivered at the edges of my sight, a blurred figure. I knew that it would vanish if I tried to turn toward it.

I took Woferl's hand and made him drop the blue rock. He uttered a sound of protest and looked up at me with wide eyes. "Let's stay near the carriage," I said, guiding us back toward where Mama stood. "They are nearly finished."

And so they were. We began to board the carriage again, and when I glanced back at the rocks, they seemed quite brown and ordinary. The lines carved in the dirt looked more like carriage tracks this time. Only the whisper in the leaves lingered. Even as I shrank from the sound, it beckoned to me, tugging at my dress with an irresistibly coaxing song, the words almost comprehensible and yet completely foreign. As if spoken in a backward language.

You have not forgotten me. And I have not forgotten you.

The Secret Page

Two weeks later, we finally arrived in Vienna on a stormy Wednesday evening, with the rain pouring in sheets down the sides of our carriage. I held Woferl's hand and waited in our seats with Mama as Papa helped the coachman bring our trunks inside the inn. My breath clouded in the air and drifted out into the wet world.

"Are you warm, Woferl?" I asked, pressing a hand to his forehead. His cheeks were pale, but at least his skin did not seem feverish.

My brother only stared at the inn. "The emperor lives in an awfully small house," he declared.

The four of us stayed together in one room, Papa and Mama in the larger bed and Woferl and me in the smaller one. Papa had requested a clavier to be brought up to our room, so that we could practice for several days before seeing the emperor. We listened to our parents talk about how we were to deliver notice to the palace that we had arrived in the city, how Woferl and I should be presented. This was our first performance outside of Salzburg, and our reputations—as well as that of Herr Schachtner, who had arranged it all—depended on how this concert would go. As our reputations went, so would our fortunes. Everything needed to be just so.

"Woferl needs new shoes," Papa said.

"Nannerl needs a new petticoat," Mama added.

The more I heard them talk, the more my thoughts churned. I slept poorly that night. My nightmares fed on one another, visions of a clavier with no keys, in a room with no audience. Of my hands, cracked and scarred, unable to dance to the music in my mind. Applause in another chamber, far away from where I was playing.

The clavier came the next morning. It was a frightfully worn little thing, but Woferl clapped his hands in delight and immediately asked Papa for sheet music paper so that he could write down the concerto he was keeping in his head. He spent the rest of the day bent over pages scattered across the instrument, alternately scribbling and playing. When Mama finally had to pull him away so that he could eat something, tears sprang to Woferl's eyes.

"Look at this child, Anna," Papa said to my mother. "It's as if you are tearing his heart from his breast."

The sight of the paper tempted me too. Perhaps I could compose my own variation of the melodies that haunted my dreams and days—but, of course, I could not ask for such a thing. So instead I pretended not to notice as Woferl scribbled his notes down, smearing the ink across the page with the ball of his hand. When he went to play them on the clavier, I recognized them as the harmonics that he'd hummed to me in the carriage. They were intact, the very same measures.

He had told the truth after all. He remembered every bit of it.

Papa watched him with a bright light in his eyes. He seemed unable to speak, lest he interrupt his young son's brilliance. If I had left the room and wandered away into the streets, I did not think he would have noticed.

Finally, after several hours, Papa left for the palace to make sure all our arrangements were in place. Mama had taken Woferl to look at the market that sprawled in the streets outside our inn. I was alone with the clavier, which sat unused and quiet.

I took a seat at it. The bench let out a loud creak. The keys looked yellow and scratched, the black paint nearly gone, and as I skimmed a hand across them, I noticed several notes sounded horribly out of tune. The highest E did not work at all. But it was still a clavier, and better than drumming my fingers on a wooden carriage wall. I opened my notebook, set it on the stand, and began to play.

I started with my scales and arpeggios, then transitioned to a menuett. My fingers warmed. I closed my eyes and finally let myself sink into this safe place. In the darkness appeared a rolling line, sloping upward and then back down like a painter's brush, smooth and bright and liquid, a climbing trail of notes that floated into open skies. My senses grew heavy with their sweetness.

It took me a moment to realize that I was no longer playing my menuett, but the tune I'd seen strewn across our carriage's path in the forest.

I stopped abruptly. My eyes fluttered open. A sudden longing seized me then, and I looked over to the quill that Woferl had set on the edge of the clavier's stand, ready for him to compose.

My heart tightened with fear and confusion.

Composition was a man's realm. Everyone knew this. It was the world of Herr Handel and Herr Bach, of my father and the kapellmeisters of Europe. It was the world Woferl was already discovering. I had never questioned this rule before I'd heard the kingdom in my dreams or the princeling's perfect voice. Composition was not my place, and my father had never hinted otherwise.

So why could I not look away from the quill atop the clavier?

I heard the music in my mind, the song from the kingdom that shifted the longer I dwelled on it. My throat turned dry, and my hands trembled. When Hyacinth had told me that he would help me, was this his intent? To give me this desire?

Papa would not approve, if he saw me. What would he do? Take away my notebook, perhaps. He might even ban me from future performances and let Woferl go alone. But most likely of all, he would destroy my composition as punishment for my disobedience. A daughter who went around her father's lessons, who stepped into a realm that he never gave her permission to enter? He would be embarrassed at my brashness and angry at my rebellion. I imagined him tossing the music into the stove, both of us watching the delicate paper curl into ash.

To create something, only to see it destroyed. The thought of that risk stabbed the sharpest at me. I tore my gaze from the quill, almost ready to abandon it.

But the melody from the forest lingered in my ears, beautiful and alluring, coaxing me forward. I felt the ache of it with the same intensity as the night of my first dream, when I'd woken with my hand outstretched, wanting to be a part of that world. The rain tapped a muffled rhythm against the roof, the pulse right before a song.

What would Hyacinth say? The glimmer in his eyes told me he would urge me on. And Woferl? He would clap his hands in delight and ask to hear the melody. Slowly, slowly, the threat of my father's punishment began to fade against the steady desire to write it down.

Finally, with one bold gesture, I took up the quill and dipped it into the inkwell. My hands reached up as if of their own accord toward my notebook. I turned the pages until I'd nearly reached the end, and then I stopped on a blank page that no one would think to look at.

For a moment, I hesitated. *I am done with it,* Hyacinth had told me when he'd returned the notebook. *Use it as your path back to me.*

What had he done to it? The blank page seemed unremarkable. Yet the longer I stared, the more I felt it staring back at me, as if the princeling had touched his fingers to the paper and soaked his otherworldly being into the fibers. He was watching me, waiting.

Tell me what you want, he had said.

So I began to write.

The room was silent, save for the scratching of quill against parchment and the roar of the music in my mind. The strokes of my ink shook slightly against the page, but I forced myself to steady. The palms of my hands turned clammy with sweat. It wouldn't be long before I would hear someone coming back up the steps to our room.

But I couldn't stop. A wild joy rushed underneath my blanket of fear. This moment was fleeting—and mine.

The melody from the carriage path looked back at me from the paper, suddenly made into reality. I continued, writing as quickly and quietly as I could, nurturing the little tune, knowing that at any moment I would hear someone coming back up the steps to our room. When I finally put the quill down and ran my finger across the page to see if the ink had dried, I noticed how warm the paper felt. My breaths came shallow and rapid.

I hurriedly replaced the quill, then closed my notebook carefully so that it would not flip open to my page. My heart beat wildly at the thrill of this secret. The air, the light, *something* shifted in the room. It was as if the princeling were watching me through the paper that connected us.

I had never disobeyed my father before. From now on, there would forever be my life before this moment, and my life after.

By the time Mama came back upstairs with Woferl, I had returned to my usual lessons. They listened to me play for a while. From the corner of my eye, I could see Woferl's large grin, as if he could not contain his joy, and my mother's face, a calm canvas, even as she smiled and nodded along to my playing. On the floor beside her, Woferl fidgeted as he hunched over his papers, scribbling.

I felt as if he already knew about my secret, that he would jump up at any moment, flip through my notebook's blank pages, and expose my little tune to the light. I could hear his familiar laugh in my head. But he remained beside my mother. I continued to play. Long moments passed.

Finally, when I finished and Woferl took his turn at the clavier, Papa returned home with a rumpled powdered wig and a whirlwind of words, hardly able to contain his excitement. "Anna," he said breathlessly, gesturing to my mother. I looked up from where I sat on my bed. Only my brother continued to play, as if lost to his surroundings.

Mama laughed. "You look flustered, Leopold."

"Word has spread throughout the city," Papa replied. I could not remember his eyes ever looking so reflective. "On the streets, in the palace square. Everyone wants to hear more about our arrival. They call the children miracles. We are being talked about everywhere."

Woferl and I exchanged a quick look. Mama clapped her hands together in pleasure.

"The sentries tell me that the empress has taken ill," he continued, then quickly added, "Just a cold, nothing to fret over! Woferl and Nannerl are to play for them in three days, at the Schönbrunn Palace, at noon."

And so our first performance was decided. Woferl looked up from his writing and announced that he would name his concerto after the empress. I thought of my secret page and waited for my father to look at me and see it imprinted in my gaze.

What have you been up to, Nannerl? he would ask.

But he didn't. Instead, he went on about the court's excitement over our concert, the reactions of those in the streets. His eyes crinkled with pleasure. I stayed where I was and watched the way he took Mama's hands in his. He did not know. My secret hummed in the back of my throat.

Somewhere in the air, unseen, Hyacinth watched me and smiled in approval.

I knew that he had heard it. And I wondered what he would do next.

Mama woke both of us early the next morning. I startled out of a dream of wandering down a dark path through the upside-down trees. Woferl sat up in bed and rubbed sleep from his eyes. Through a crack in our window, I could already hear the bustling sounds of Vienna's streets waking to greet the day.

"Hurry now, children," she said, patting both our cheeks and giving us a wink. "You need to look the part you will play."

We ate a quick breakfast of cold meats and poppy seed bread, and then I put on my white cap and left the inn with Mama, Papa, and Woferl. Compared with Salzburg, the streets looked wider here and paved with newer cobblestones. It was still early, and the wet air bit my cheeks with its chill. I could smell the honey and wheat from the bakeries. Ottoman merchants in layered coats and shining sashes gathered near the Fleischmarkt's coffeehouse, conversing with one another in Turkish. Men hawked walnuts and colorful ribbons at intersections.

I held Woferl's hand. Papa walked on my brother's other side, distracted by the sights, and when he walked too fast, I picked up my skirts and hurried along behind him. To the Viennese, it must have been obvious that we did not live here. I looked nervously away from several curious passersby. It seemed like a long time before we finally arrived at the tailor and dressmaker shop, adorned with a sign that said DAS FEINE BENEHMEN.

"Welcome, welcome," said the man that opened the door for us. He blinked blue eyes at my father. "May I ask your name, Herr?"

I glanced behind him at the shop. It was very tidy, lined with elaborate caps and leather shoes on models, stays and stomachers trimmed with braided silk, rolls and rolls of fabrics in all colors and patterns, petticoats and gowns with beautiful embroidery. In one corner stood a full mantua gown cut in the latest fashion, made out of an elaborately patterned yellow silk that sloped elegantly at the hips. I found myself admiring its repeating floral images, the way it bunched in and then straight at the back. It was the kind of dress one wore before royalty. What *I* would wear soon enough.

Papa shook the man's hand. "Herr Leopold Mozart." He bowed his head slightly. I caught a glimpse of my father's eyes—and saw a hint of pleasure coiled within them. He was waiting for the tailor to recognize his name.

To Papa's satisfaction, the man's grin widened. "Ah, Herr Mozart of Salzburg!" he exclaimed, taking my father's hand with both of his. "I've heard a great deal about your arrival to our city, friend. Your children are to perform at the palace soon, yes?"

When Papa nodded, the tailor's eyes turned round. I watched him carefully. More and more, he reminded me of Hyacinth—a hint of blue in his skin whenever he turned; a slight slant of his eyes, a trick of the light; a flash of his bright teeth. I wondered if he had leapt into the body of a man in order to prepare us for our debut. His eyes darted between Woferl and me. "Rest assured, children, that you will leave here looking like royalty." The man's smile had grown so large by now that I thought it might fall from his face.

"Thank you, sir," Papa replied. "Spare no expense. I want them at their best."

I glanced quickly at our father. Mama would scold him for this later, if he ever revealed how much he was willing to spend on our clothes. Already, I could imagine her arms crossed and her lips tight.

The man bowed. "Let me fetch several others. We will begin

straightaway." With that, he hurried off. Woferl made a move to dart after him, but I grabbed my brother's arm and spoke sharply. I did not want Papa to think I could not mind him.

The man soon returned with his help. Two dressmakers, one clothier, and one assistant. They approached me with a new garment and I held my arms up so that they could pull the stay tightly around my waist, the inner boning pressing against my ribs. The gown itself was made from deep blue satin, smooth and soft and cold to the touch, open in front to reveal the creamy layers of petticoats buried underneath. The collar was high, concealing the skin of my clavicle and throat. One of my hands stayed pressed against my blue pendant, deep in my petticoat's pocket.

I stared at my reflection as the dressmakers and clothier worked, my eyes locked on the mirror standing before us. My cheeks looked flushed from the cold streets.

Nearby, I could see my brother wriggling throughout his own fitting. At one point, he hopped down from his dais and ran to our father to hold up his shining cuff links, forcing his clothier to hurry after him. I looked on in silence, unable to mind him. The rigid structure of my dress dug into me, holding me back. Even if I wanted, I could not move my arms as freely as Woferl could.

Would this be how I performed before royalty? Barely able to move?

When they finished, Papa guided me to the nearby wig shop and parlor. There, the wigmakers pulled my hair back and away from my face, fitting me with a curled wig that piled high on my head and then tumbled down my shoulders in a cascade. They patted the hair with white powder until the fine dust floated in the air around us. I wrinkled my nose at its stale scent. The weight of it made me keep my head and neck at a strict, straight angle.

I tried to puzzle out how to lean into my music as I played while wearing such a thing.

Noon approached, and finally we finished our fittings. As we thanked the clothier and made our way out, I cast one final glance over my shoulder. The tailor smiled back, his tall figure cutting a long shadow on the floor. His teeth were very white, his eyes so blue they seemed to glow.

"I'll see you soon, Nannerl," he said. I looked at my father for his reaction, but he did not seem to notice. Only Woferl tightened his hand in mine. I tried to remind myself that this must be part of Hyacinth's plan. In order for me to perform, I must first look the part.

The rest of the day passed in a blur of practice, as did the next.

I could not sleep the night before our performance. Instead, I stared at the ceiling in silence, drowning in thoughts. My glass pendant lay tucked underneath my pillow, so that I could feel its slight bump against the back of my head. I let myself take comfort in its presence. A reminder of my wish.

"Nannerl?"

I turned to look at my brother. His eyes blinked back at me in the darkness. I propped myself up on one elbow and smiled at him. "You should rest," I whispered.

"So should you," he protested, "but you're not." He glanced over to where our father slept, afraid that he would stir.

It had not occurred to me that Woferl might be nervous too. I reached over and took his hand in mine. I was small for my age, but his fingers were tiny even in my palm. "You have nothing to be afraid of," I said gently. "All of Austria is excited to hear you. The emperor requested you personally. You will not disappoint."

Woferl closed his little fingers around one of mine. "I'm not afraid," he said.

I smiled again. "Then why are you awake?"

Woferl scooted closer to me, buried his head in my pillow, and pointed toward the clavier. I followed his hand until my eyes rested on my notebook. It was closed.

"What is it?" I asked.

"The notebook is singing," he whispered. "I can't sleep."

I turned my head quickly back to the clavier. We fell silent. I heard the sound of a late-night coach from the streets below, the whisper of wind, Papa's gentle snore, a trickle of water from some mysterious place. I did not hear the music.

"Are you sure?" I whispered to Woferl. "What do you mean?"

He wrinkled his nose at me. "Nannerl!" he exclaimed in a quiet hiss. "It is singing right now—you can't hear it? It is very loud."

It must be Hyacinth. He has done something to my notebook. He is here.

I waited for a minute, forcing my breathing to stay even, until Woferl began to squirm. Then I swung my legs over the side of the bed, rested my feet on the floor, and slowly made my way to the clavier. Still I heard nothing. The floor numbed my feet. I took care not to tremble.

I should be in bed, I thought. *Our performance.*

When I had moved close enough, I picked my notebook off the clavier's stand and clutched it to my chest. Gingerly, I made my way back to bed.

Woferl sat up straighter, eager to see. "It keeps repeating the same lines," he insisted. "Over and over and over."

My skin tingled. We both froze for an instant as Papa stirred.

I kept my eyes on him until he turned away from us, and then I relaxed my shoulders. I opened the notebook quietly. "What does it sound like?" I whispered.

Woferl hesitated for a moment. "Like this." He hummed a few notes as softly as he could.

I swallowed hard. My initial excitement, my sudden thoughts of the princeling, all vanished. Woferl must have discovered my secret composition, I thought, the little wisp of music I'd written down several days ago. I felt an abrupt rush of anger. "You're making it up," I whispered harshly. "The notebook is not singing at all. You are."

Woferl burst into a fit of giggles. He threw himself facedown into his pillow. I closed my notebook in disappointment and hid it inside our blankets. His quiet laughter stung. "This is my notebook, Woferl. You shouldn't take what's not yours. You aren't going to tell Papa, are you?"

His giggles died down. He looked at me solemnly. "Well, why are you hiding it? It's beautiful."

His words were so serious, said so truthfully, that any anger I might have had flitted away. "Young ladies do not compose," I told him.

He shook his head. "Why?"

I took his hands in mine and squeezed them once. How much and how little he understood of my life. "Please, Woferl, let it be our secret. Promise me you won't tell anyone else."

It was Woferl's turn to look upset. "But who will hear it, then?" he whispered, horrified. "You're not going to let it stay there forever, are you?"

"Yes, I am." I gave him a firm look. "If you love me, then promise me."

Woferl stared at me for a long time. When he knew that he could not sway me with his defiance, he flopped back down in bed. “I do love you,” he declared grudgingly. “So I promise I will never tell.”

I settled down into bed. We drifted into silence, but Woferl’s teasing brought back the undercurrent of fear, my muted excitement from when I wrote the music down. What was Hyacinth up to, coaxing me into this? God will punish me for hiding such a thing from Papa. It would mean that I was the kind of girl who disobeyed her father, who would go on to disobey other men—her husband—in her life. So many stories already circulated about us. What would this story become, if it began to spread?

They would say that she was the kind of girl who did not listen. She was the kind of girl who had her own ideas.

I pulled my blankets higher until they reached my chin, and then imagined the princeling turning his head this way and that, his bright eyes watching me from the other side of the room. I hoped he was.

I clutched my notebook closer to my chest and stared, searching the darkness, until I drifted off to sleep with the image still branded in my thoughts.

The Princeling in the Palace

THE NEXT MORNING DAWNED WITH A FLURRY OF activity.

I forced myself to nibble on some bread while Woferl played with the cut meats on his plate. After our quick breakfast, we hurried to the tailor shop to collect our new clothes. I sucked in my breath as my mother helped me pull the boned corset of my new gown tight until I could barely breathe. When she finished, my waist tapered thin and straight in the mirror.

Beside me, Woferl shrugged on his new coat and shoes. We looked less like the brother and sister who arrived to the shop huddled and whispering together, and more like the rumor of us that had been circulating the city. We looked like the Mozart children, musical prodigies. Fit to play for a king.

By the time we arrived at the Schönbrunn Palace, I was trembling slightly from my nerves, and a cold sweat had dampened my hands. The palace stretched for what looked like miles in each direction, white and gold, with countless rows of framed windows and stone pillars. A guard greeted us at the front of the courtyard. I walked carefully, so as not to ruin my new gown, but Woferl flitted in front of us like a restless bird, chatting with the guard, asking him his name and how long he'd worked at the palace, until Papa finally gave me a stern look and I hurried over to pull Woferl back to my side.

We walked through halls of towering pillars and carved banisters, walls covered with sheets of gold. The ceilings were painted in every room, and in every room, I felt as if God were looking down at me, laying bare my secret page I'd written. I kept my head down and hurried forward. My leather shoes echoed on the marble, and I felt oddly embarrassed. My steps did not sound graceful. I reached for a moment into the pocket of my petticoat, where I'd stashed my pendant. My fingers found its smooth surface. I tried to let it reassure me.

Finally, at the last doorway, we paused to let the guard walk ahead of us. He bowed to someone I could not see.

"Your Majesty," he said. "I present Herr Leopold Mozart and Frau Anna Maria Mozart, and their children, Herr Wolfgang Amadeus Mozart and Fräulein Maria Anna Mozart."

The first thing I saw when I entered the chamber was the clavier.

Larger than ours in Salzburg and certainly larger than the one from the inn, it had white keys instead of dark and was covered in baroque art. It looked like the clavier I'd seen in the trinket shop, surrounded by a cavern of moss. The sight struck me so dumb that I nearly jumped when they announced my name.

My eyes swept the room. The marble floor was decorated with thick rugs, and the half a dozen men who comprised the emperor's council sat facing the clavier, with the emperor and empress themselves centered between them.

The emperor sprang to his feet at the sight of us. "Ah! Herr Mozart! Frau Mozart!" he called out. Beside him, Empress Maria Theresa gave us a warm smile.

"Your Majesties," Papa said, bowing low to the ground. Of course he would do so, but the sight startled me—I had always seen my father with his head held high, the master of our household. It never occurred to me that he would behave differently in front of those with greater power.

Mama joined in with her curtsies, and Woferl and I followed their lead as we were each introduced to the emperor and the empress, then to their children.

It was only at the very bottom of my curtsy that I noticed the hooved shape of the emperor's shoes. I blinked. When I looked again, they looked normal. My eyes darted to the rest of the audience. For an instant, I thought that one of the seated men appeared thin and willowy and blue.

I tried to get a better look, but then the emperor knelt to look at my brother. My attention shifted back to them. "And this! Ah!" the emperor said. "This boy must be the one we've heard so much about! How old are you, Herr Mozart?"

"Five years old, Your Majesty," he said loudly. Though his skill was impressive enough, Papa had still instructed him to take two years off his age.

"Five!" the emperor said, rubbing his chin in mock wisdom. "I suppose you must be too small a boy to play the clavier, *ja*?"

At that, Woferl puffed up in defiance. "I am not, Your Majesty," he declared. “I can reach all the keys, and if I stand I can reach the pedal, and I can play the violin as well.”

Behind him, the others in the audience chuckled. I caught the eye of the emperor’s smallest daughter, Princess Maria Antonia, who smiled at me. When her eyes darted to Woferl, she blushed and looked away.

The emperor let out a hearty laugh. “You all heard the boy. He can play the clavier! And the violin, as well!” Then he turned to me. “You must be Fräulein Mozart,” he said.

“Your Majesty,” I replied, sweeping into another deep curtsy.

“Lift your head, Fräulein, so that we may have a better look at you.”

I did so—and found myself staring straight into Hyacinth’s eyes.

I did not know what kept me from leaping away in shock, but there I stood, transfixed by the boy’s beautiful face. Did the others not see him? Here, he was no longer dressed in his assortment of leaves and shining twigs. Instead, his hair was pulled back into a low tail at the nape of his neck, tied with a neat length of white ribbon, and layers of trimmed lace adorned the breast of his shirt. The white-and-gold royal coat he wore made the tint of his skin appear even paler.

“Hello, Fräulein,” Hyacinth said, and the music of his voice was unmistakable. I turned my eyes down, terrified and elated to be the only one who might be seeing him. Above me came his chuckle.

“Herr Mozart,” he said to my father. “Are you sure your daughter is not an angel stolen from Heaven?”

Papa and Mama bowed their thanks. All they saw was the emperor. I said nothing as I curtsied again. My eyes flitted to my brother, but he had his back turned to me, his eyes already settled eagerly on the clavier. If he could see Hyacinth, he did not appear to notice him.

"We will hear young Herr Mozart first," Hyacinth declared, then rose to his feet. I found myself turning to join my parents in the first row of audience seats, where my mother now sat beside the empress herself. I took my seat with them. My father remained standing, perhaps too anxious to sit, his eyes locked on Woferl as if there were no one else in the chamber. I pressed my hands into my lap.

Hyacinth had told me that the notebook would connect my world to his, and so I had written my music into it, on my secret page. I must have summoned him here.

Woferl struggled a bit to climb up onto the clavier bench, but once he was there, he needed nothing else. *What did he see?* I wondered. I kept my eyes on his bright red sleeves as he played. He added flourishes into his performance, and his fingers danced across the keys without effort, nimble and quick. In the corners of the chamber, a light mist had begun to build, gathering along the floor until it seemed as if we were shrouded in a dream.

My hands fidgeted endlessly in my lap, yearning to reach into my pocket for my pendant. I tried to predict how my own performance would go, whether I could bring the same wonder to my music as Woferl did to his. Would anyone want to hear me when my brother finished? Or would it be like my first song for Herr Schachtner, who had barely heard me at all? Would it please Hyacinth, to watch me perform? Had he come here to witness what I could do, so that he could begin his work on my wish? Was this a test of my talent for him to see if I was truly worthy?

Woferl played one piece, then another. Papa stepped up and

covered the keys with a cloth, and Woferl continued to play as easily as before. Our audience gasped in delight.

Then, too soon, Woferl was finished, and suddenly all I heard around me were exclamations and applause, the clack of shoes as their owners rose to their feet. I joined in the clapping. Woferl hopped down from the bench and bowed low to his audience, then looked at me with a toothy grin. He hurried back to his chair. His eyes never went to the emperor—to Hyacinth—and I began to wonder if this time the princeling did not mean for my brother to see him. Perhaps he was only here for me.

My heart began to beat rapidly as the room's attention shifted to me. The emperor nodded for me to step forward. "Fräulein Mozart," he said.

I rose from my chair in a daze and walked to the clavier. There, I took my position on the bench and rested my hands against the keys. I'd memorized my music in those long morning practices and late-night lessons. It was easier this way to see the notes how I wanted to see them, without the papers in front of me. But here, with the phantom mist curling against my feet, I became unsure that I could see anything. The fog seemed to touch the edges of my mind, softening my nerves until I felt like I was not a part of my body at all. Until the world around me looked like a dream.

In the corner of the room, the princeling sat in the emperor's chair and watched me with a curious tilt to his head.

My breath seemed to hang in the air. I reached for the stillness in my mind. My fingers curled against the keys and I began to play.

The notes that emerged from the clavier did not seem like music of the real world, what my menuetts were when I practiced them at home. Instead, they sounded like what I'd heard in the hidden grotto, like the voice of Hyacinth on the wind, suspended on the breeze. They sounded like they did in my head, perfect and fully formed, the same way they seemed right before they usually emerged into the world flawed and distorted. I floated above myself as I played, immersed in the dream of it, intoxicated, unwilling to leave.

I was not playing for an audience. I was playing for myself. And slowly, gradually, the chamber around me began to change.

The marble pillars turned wooden. Leaves budded from the stems of the chandeliers and candle sconces. Ivy crawled along the spectators' chairs, their tendrils spiraling in long, tight patterns around the legs, winding their way across the floor until they reached my clavier. The light filtering through the windows dimmed, turned cool and blue. The mist thickened, mixing with the sound of notes in the air. My hands numbed, moving seemingly of their own accord. My body leaned into the music, swayed by the sound.

Hyacinth left his seat and began to walk toward me. The others did not seem to react to this, and when I chanced a look at my father and Woferl, I realized that they sat motionless in their chairs, as if time had chosen to suspend them entirely.

I finished the menuett and promptly began another. From deep within the notes emerged the rise and fall of a breeze, ripples of rain in a pond, the high trill of a bird's call. Sparrows flitted overhead, settling into the branches of a thick forest's canopy that had grown over the ceiling's paintings and chandeliers. The wind curled around me, stirring the edges of my wig. I felt the cooling mist of rain against my cheeks.

Hyacinth finally reached me. He sat down on the clavier bench beside me and studied my dancing hands. The glow of his eyes cast blue light against his face. "Very good, Nannerl," he said, his voice bright with approval. "Play another."

I did. *Why are you here?* I asked Hyacinth. I must have spoken the words in my mind, for they sounded muted, and my lips stayed closed.

He heard me all the same. Hyacinth smiled and stayed beside me, inching closer as I played. His boyish face glowed with joy. Then he stood and wrapped his arms around me, his slender young hands covering my own. He moved his fingers with mine and the music changed to something new, the notes as crisp and clean as rain against glass.

"I am here because you wished it," he replied. "The music you wrote into your notebook called for me. Did it not?"

And I nodded, because it was true. Our fingers danced along the keys in a perfect scale.

"I have brought you to this performance. And look at you now, dazzling with your true talent. The court will remember you." He leaned closer to me. "Now that I have helped you, it is your turn to help me. Are you ready?"

I nodded again. I was not playing to an empty room with no audience, like I'd seen in my nightmares. Here I was playing before the royal court, their attention rapt as they listened to me, their tongues silent and suspended. The boldness I'd felt as I composed my secret page, as I made my secret wish . . . it filled my head so fully now that I swayed from the rush. What more could come in the future? How much farther could we go, if I helped Hyacinth in return?

I'm ready, I told him.

The princeling's fingers skipped across the keys. "Return to Salzburg. Wait for me there. I will come for you. Bring your brother." He smiled at me, his teeth perfect and sharp. "The time has come for the next step in our plans, Nannerl."

Bring your brother. I instantly wondered what Hyacinth needed with Woferl. But I did not ask him what he meant by it. It made sense, I suppose—for I could not be here at this performance without my brother in attendance. So must it be in the kingdom. I leaned back into the music, grateful for Hyacinth's presence, afraid of what he might say if I refused.

And then, finally, I finished.

Just like that, the world vanished with the end of my performance. The ivy and moss, the birds and blue light, the mist on the ground. Hyacinth. My dreamlike trance. I blinked and they all went away, leaving in their wake the palace chamber, the marble pillars, the paintings adorning the ceiling, the emperor and empress, my father and mother, my brother.

For a moment, there was utter silence. Then the emperor jumped to his feet with the empress and clapped loudly. The princesses and princes stood and joined in, followed by the royal council. In the midst of the ovation came praise, words like *splendid* and *prodigious* and *ethereal.* I rose in a daze, bent low, and curtsied. Mama clapped in delight, and beside her, Woferl applauded so loudly that he seemed hardly able to contain himself. My father smiled at me in approval, pride and surprise clear on his face. He had never heard me play like this before. A surge of joy seared my chest.

As I rose from my curtsy, I chanced one more glance in the emperor's direction. No princeling to be seen. The emperor stood firmly in his place instead, not a young, lithe boy but the merry gentleman who had first greeted us. Hyacinth had disappeared so thoroughly that I couldn't be sure he was ever here. But the magic he'd left behind still hung thick in the air, in the sound of the applause, in the echoes of the music he had inspired me to play.

I shivered at the touch of it, the memory of his fingers against mine, dancing across the clavier. He had given this to me, and in return, I had promised to offer something to him. My glass pendant felt very heavy in the pocket of my petticoat.

"How long will you stay?" the empress asked my father.

"Another week or two, your Majesty," Papa replied. "We will play for some of the nobles before we return to Salzburg."

She glanced toward Woferl and me. Her eyes were gentle, her lashes very pale. "Herr Mozart," she said to my brother. "Your hands produce such clean, crisp notes, and with such grace. Thank you, my child, for honoring my court with your presence." She looked at me. "As with you, Fräulein Mozart. You and your brother show talent far beyond your years. I will send you both a present tomorrow, as a token of my thanks."

Woferl, unable to resist her kind words, leapt up from beside me and wrapped his arms around the empress's neck. He kissed her soundly on her cheek. Before the empress could react to his embrace, he turned to face Princess Maria Antonia, the little girl who had blushed for him earlier, and asked her promptly if she would marry him.

The others laughed heartily at his display of affection.

I often think back on this moment. Sometimes I wonder whether this was when the princeling first began to work his magic on Woferl, and whether Hyacinth ended up haunting the little princess too. Years later, Maria Antonia would become the feather to break the backs of an already-broken people—France's young queen, Marie Antoinette.

The Night Flower

THE EMPEROR AND EMPRESS, PLEASED WITH US, showered us with gifts. Empress Maria Theresa gave Woferl a dark lilac outfit once worn by a young archduke, lined in gold braids and buttons and cuffs, a beautiful justaucorps over a matching waistcoat. I received a violet taffeta dress embroidered with ripe flowers and soft silver blossoms, adorned with snow-colored lace. Then came gifts of snuffboxes, four of them, along with three hundred ducats, almost a full year of Papa's salary in the archbishop's orchestra.

It was more than my father expected. I knew this because he hummed as he packed the gifts away in our luggage, the edges of his eyes crinkled with the possibility of what our future prospects might hold. Mama squeezed my shoulders and gave Woferl and me proud smiles. A year's salary earned in the span of a day. Papa could buy his new coat. Mama could have a new porcelain set.

We could keep respectable company, occupy our space at dinner parties with our heads held high.

When Papa told Woferl and me that we had both done well, Woferl only snorted. "The emperor has no ear for music," he said. "Papa, he didn't even hear the string on my violin that was out of tune. He said instead that I play very accurately."

Papa threw his head back and laughed. "Well, he certainly knows talent when he hears it," he said, winking. His eyes settled on me. Everything in me brightened at the approval in them. "Isn't that right, Nannerl? The emperor said you were truly blessed by God."

Even better than our royal gifts were the frequent mentions of our names on the streets, scattered in conversations, all murmuring about *Herr Mozart and his remarkable children.* We heard it from every corner as we prepared for our trip home. The gossip kept Papa in high spirits all the way back, despite the snow we encountered.

By the time we arrived home, invitations had poured in. News of our performance in Vienna had already spread throughout Salzburg and beyond. All the nobility of Austria and Germany wanted to keep up with the latest trend, and we were it.

Papa made the arrangements with a resolute face.

"You will exhaust the children," Mama said to him one evening as she looked over the schedule he had laid out for us.

"The children will not be worth watching forever," he said. "This is only the beginning, Anna. There is no time to waste. We need to play for the royal courts, for all Europe." He turned then to look at me and Woferl. "You love the performances, at any rate. Isn't that true?"

I nodded, because it *was* true. Beside me, my brother clapped his hands at the thought of traveling beyond Austria's borders.

My father smiled at Woferl's reaction. "Miracles from God," he said. "And so long as they can, it is their duty to perform God's work."

Woferl brightened like a star, drinking in Papa's words hungrily. *Miracles from God.* I took the words and folded them away carefully, letting the weight of them sit in my chest, savoring the memory of my father's pleased expression turned in my direction.

This was my worth. Without it, I was simply a child. With it, I would be what Hyacinth promised.

Immortal.

Several weeks passed. I started to think that perhaps Hyacinth wouldn't return after all—that there was no task in store to repay him for his help. A part of me was relieved. Another part of me missed him, yearned for the wild song of his voice.

Finally, my wait ended on a cold night in late April. A flurry of snow blanketed our windowsills. Woferl slept soundly beside me, but the cold kept me awake. I rubbed at my toes to warm them, letting my eyes wander around our room before settling on the slit beneath our door. Every night, I half expected to see the light of fireflies again, like we'd seen that night when Hyacinth had first appeared.

"I wonder if it feels like winter in the Kingdom of Back too," Woferl murmured. I looked down at him, surprised to see him awake. He blinked sleep from his eyes and snuggled closer to me.

"I'm sure it does," I whispered. "Now, hush." The late hour and the chilly air had turned me grumpy.

Woferl didn't seem to care. He sighed and folded his arms over our blanket, then turned his eyes up to the ceiling. "I bet winter in the Kingdom of Back is different from winter in our world, isn't it?" he said. "I bet it doesn't feel as cold, and the snow looks prettier."

"Woferl, please." I frowned at him. "Do you want Papa to scold you tomorrow for practicing poorly? Go back to sleep."

"*You're* not sleeping." Woferl grinned. "You always get to make up the stories about the kingdom. This time I'll make it up, and you can listen."

I sighed. There would be no quieting him. "Very well, indulge me. Tell me something we don't know yet about the Kingdom of Back."

Woferl cleared his throat, then furrowed his brow in concentration. I watched him without saying a word. It was the same expression he made when he performed for an audience. I wondered suddenly if he saw the kingdom as I did, as sheets of music in his mind, an entire world laid out in neat measures and round notes on paper. I wondered if he heard what I heard whenever he played, if we had access to the same secret world.

"In the Kingdom of Back," he began, "the snow layers the forest in white, like frosting on the cakes at the bakery. And the ocean never freezes over. Its water feels warm even in the winter."

"Yes," I whispered, listening half-heartedly. His childlike voice had started to lure me into sleep. "Naturally."

"And the ocean has a guardian too, just like the rest of the kingdom has the princeling." Woferl paused to think. "A faery queen of the night, trapped in an underwater cave."

As he went on, I drifted away. The room around me blurred, I sank without protest into the early fog of sleep, and in my dreams Woferl continued with his faery tale. I thought I could see light at the bottom slit of our door, and hear something that sounded like music from a clavier. *My* music. "Woferl," I whispered, shaking my brother.

He halted in his story. "What is it?" he asked.

Before I could reply, I saw him sit up and turn his attention toward our door. He heard it too. "It's coming from the music room," he whispered. His hand automatically found mine.

"It's my music," I said, suddenly afraid. "From my notebook. You recognize it, don't you?"

"Of course I do." Woferl swung his legs over the bed and tilted his head so as to hear it more clearly.

We stayed like that for a long moment, silent, as the music continued. I shivered.

"It's coming closer," Woferl whispered.

My hands went to the candlestick on our dresser. I lit it, then held the light out before us.

The door squeaked, then opened into a tiny sliver. Both of us froze in our places and my face grew hot with fear. I knew it was not our parents, and not Sebastian.

It was Hyacinth.

The princeling came accompanied, as always, by the dim blue glow of faeries flitting about him in tiny pins of light. He peered into our bedroom and looked idly around before settling his gaze on us. On me.

"Nannerl," he said. His voice wrapped itself around me in an embrace. "I am so pleased to see you again." And before I could wonder if he'd made himself visible to Woferl too, he turned to my brother and offered him a smile. "The little one is still awake, waiting for an adventure."

Woferl grinned back, delighted. "It's you!" he exclaimed.

"Yes, it seems so," Hyacinth said.

"Have you come to steal something again?" Woferl asked.

I swallowed at his words, afraid of angering the princeling, and elbowed him in the ribs.

Hyacinth only laughed. The sound pierced my ears. I thought that it would certainly wake Sebastian or our parents. When he stopped, he fixed his eyes on Woferl. "I've come to ask a favor from both of you," he said. "But first, you must follow me. Quickly now." He frowned at the candle in my hands. I noticed the way he shrank from its warmth, as if, even from a distance, it could scald him. "Leave that. You will not need its light in the kingdom." And without another word, he vanished from the doorframe.

Woferl leapt up first. "Let's go," he whispered eagerly. Before I could refuse, he had risen and hurried to the door.

"Woferl, wait—" I started to say, but it was too late. He had already rushed out. I slid my feet into my pair of slippers and followed his path, into our living room and out our front entrance, down the flights of stairs that would lead us to the main street.

My feet crunched on snow. It surprised me so much that I cried out and stopped in my tracks.

I stood in the middle of the Getreidegasse with a sky full of stars above my head, illuminated by the brilliant light of two moons hovering over opposite ends of the street. The city was deserted, the dim streetlights fading into the darkness around us. The snow did not look like how I remembered it from earlier that evening, dirty with mud and ice, shoved onto the sidewalks in heaps. This snow was clean and white, untouched. I looked up. All the windowsills were covered in this pure snow, so soft that I thought it would feel like a warm blanket to the touch. I reached down and put my hand against its surface. It fell apart against my fingers.

The snow layers the forest in white, like frosting on the cakes at the bakery.

Woferl's voice echoed somewhere ahead of me. When I looked in its direction, I realized that the little crooked path I'd once seen from our window had now reappeared at the end of the Getreidegasse. It led away from the buildings and toward the dark forest of upside-down trees, and at the forest's entrance stood the same sign that had been there when I'd last seen it. Now, though, I could read the words.

"To the Kingdom of Back," I whispered.

Next to the sign was Woferl. He waved at me. Somewhere in the forest behind him, I glimpsed Hyacinth's lean figure heading deeper in. I gathered up the bottom of my nightgown, shook snow from my slippers, and hurried to my brother.

We walked in silence. The path started with cobblestones, but as we continued, the cobblestones began to fade away, growing sparser, until we walked on dirt lined with blankets of snow. Woferl pressed against me as we passed the trees. Their roots reached up toward the stars and cut the sky into slivers. Their leaves curled at the bottom of each tree, and in them were pools of still, black water, with no bottom that I could see.

I remembered my own warning about the pools and pulled us away from their edges, lest we fell in. Our surroundings had grown so dark that I could barely make out the path ahead. I tried not to look behind us. Shadows crept into every crevice when there was no light to push back their edges, breathing life into things that shouldn't exist.

"Are you afraid?" I asked Woferl.

His small shoulders trembled. "No," he lied. "Where do you think the path will lead?"

"Well, I can't be sure," I said, trying to keep him calm. "It is your turn to tell me a story, remember? Tell me, where would you want this path to lead?"

Woferl smiled. "To the shore!" he exclaimed in a hushed voice. "To the white sand and warm ocean."

As he spoke, pinpoints of light caught my attention. They flitted from tree to tree, clusters that glowed blue, the same tiny faeries that had appeared in our music room on the first night. With them came the curious sensation that Hyacinth must be near. Sure enough, one of the lights came to rest in my hand. It felt like a feather.

This way, it cried. *This way.*

And with their light and that of the two moons, our path was illuminated just enough for us to see it winding deeper into the woods.

We walked for a very long time, until the forest grew darker and darker, and the trees grew closer and closer together. I wondered if perhaps we had missed a trail that could have branched away, that Hyacinth may have gone a different direction. The tiny faeries had faded away too, leaving us wandering alone through a colorless world.

Finally, when I was ready to turn back, the darkness of the forest began to fade and I saw what seemed like a strange blue light appear on the tree trunks. "Do you see that, Woferl?" I said to him. "Maybe we're almost there." He didn't answer. It was just as well—I did not want him to ask me again whether or not I knew where this path could lead.

The end to the forest was so abrupt that I stumbled on my slippers. The last of the upside-down trees now stood beside us, and in front of us stretched a shore of white sand that hugged the edge of a deep sapphire ocean, its color interrupted by two perfect silver reflections of the moons.

I caught my breath at the sight of it. This was the ocean from my very first dream.

Dozens of blue seashells lay winking against the white sand. Woferl noticed me admiring their color and, on impulse, picked one up and shoved it into his pocket.

Hyacinth was waiting at the edge of the water. Tonight, he looked more like a boy than ever, his tall, slender frame covered with skeleton leaves, his hair rumpled. His eyes reflected the ocean. "Are you cold?" he asked me.

I shook my head. The winter chill that had clung to us on the Getreidegasse and the dark forest path did not exist here, and the ocean's water lay as still and flat as a mirror's surface.

"Good." The princeling nodded at us. "I have a task for you both."

"What is it?" I asked.

Hyacinth gave me a sidelong smile and gestured toward the water. "I need a night flower," he replied. "You can find them at the bottom of this ocean, inside a hidden cave. I'm unable to get there myself. You see, I cannot swim well."

"These flowers grow inside an underwater cave?" I said.

"Yes," he replied. "This cave is a lovely grotto, and inside lives an old witch, with wrinkled hands and long white hair. I sealed her in the cave long ago with the rising waters, and she has remained there ever since. She has stayed there for so long, in fact, that her feet have become part of the cavern floor. She cannot move from her spot, and her powers, although terrible, weaken when the twin moons are not aligned. Still, you must be careful. She can call great golden fire with her hands and engulf you with its flames. She feeds on the night flowers that grow along the cave walls, and anything else she manages to reach."

Woferl's vision of a guardian for the ocean, I thought. A sudden sadness filled my heart. "She must be very lonely," I said.

Hyacinth turned his eyes to me. "Do not take pity on her. She will try to lure you to her with a sweet song, the most beautiful music you've ever heard in your life, so potent that sometimes sailors can hear it across many oceans. They call her the Queen of the Night." He stepped closer to us. "Do not approach her. Do not look into her eyes. Do not talk to her. She is not what she seems."

I swallowed, distracted by his nearness, and promised that we would not.

He turned away from us and pointed out toward the still waters. Not far from the shore lay a series of rock formations, carved from limestone, and when I looked from a different angle, the moonlight washed them into silver.

"The grotto lies under the water, in those rocks," he said. "The waters are low right now, so you and your brother will have a bit of time to get the flower. Do not be fooled by this peaceful ocean. It will rise so steadily that you will not realize it until it's too late."

Woferl listened with a determined look on his face. "We are very brave," he said, looking up at me. "I'm not afraid."

The princeling smiled at him. "You are indeed very brave," he replied, and looked back toward the rocks. "It's why I've chosen you both for this task. Now, you must hurry."

I did not feel brave, but Hyacinth looked so calm, and Woferl so eager, that I nodded and started toward the water. The echo of applause from my performance in Vienna, the look of pride on my father's face . . . they came back to me now, filling me with the memory of joy. I had told Hyacinth I was ready, and so I was. My mind lingered on the night flower that we needed to retrieve.

I removed my slippers. Then I waded carefully in, holding my breath in anticipation of cold ocean water. But the instant we dipped our feet into it, I realized that it was as warm as a bath, just as Woferl had said. I smiled in surprise. Woferl let out a giggle at the warmth and splashed right into it, getting water all over my nightgown. I looked back to the shore once we'd gone in waist-deep. Hyacinth watched us from where he had sat down in the sand, his stiltlike legs crossed over each other.

We swam until the rocks in the distance became very close, so that I could make out their jagged edges and the carpets of moss that grew in clumps on their backs. When we were near enough to touch the walls of the rocks, I wiped water away from my face and looked down into the ocean. The water looked lit from below, a brilliant blue. I took a deep breath and submerged myself to get a better view.

Not far from the surface appeared a crevice in the rock.

I came back up. "Woferl," I said breathlessly. "I've found the entrance to the grotto."

Hyacinth was right. The ocean had not risen yet, and the waters were low. We did not have to dive far to reach the grotto's entrance. The crevice looked dark when we approached it, but as we swam farther inside it began to grow lighter, the same strange blue light that we had seen when we first saw the beach of white sand. I took a giant gulp of air as we surfaced inside the cave. The water tasted sweet, like diluted honey.

I could hear nothing in the grotto except the sounds we made—the splash of water, our breathing. The light came from hundreds of flowers that grew along the sides of the limestone walls, black and violet in color, each with a glowing spot of brilliant blue light in its center. The limestone itself looked wet and crystal-like, almost clear. I saw Woferl stare in wonder at the scene. Garlands of heavily scented flowers hung so low from the grotto's ceiling that I could touch them.

Then we saw her. She stood in one corner of the grotto where the lights from the flowers shone the brightest, leaning her head against a silver harp, and there she wept silently. A long, tattered gown of white and gold draped against her slender figure. Her hair was snow-pale, like Hyacinth said, and dotted with tiny black flowers. Her skin seemed delicate, the knot of her eyebrows thin and dark. I had been afraid to see her, picturing her as a withered, bony, witchlike faery, but now I felt drawn to this poor creature. Wrinkles on her skin, soft in the blue light, gave her a fragile appearance. Her feet melted into the cavern floor, so that I could not be sure where her legs ended and the rock began.

Woferl pointed quietly at her shoulders. There, I saw that her wings were faded and torn, hanging limply against her back. She must have fought hard to escape this grotto.

Woferl guessed what was going through my mind. "She is not what she seems," he whispered, repeating Hyacinth's warning.

Do not look at her. Do not talk to her.

I turned my face away, my heart pounding. As I did, I noticed the black ivy trailing along the cavern wall behind her. The night flowers that blossomed there were much larger than the rest, their vines coated with angry thorns. These were the flowers Hyacinth had sent us for.

The witch heard Woferl's whisper too. She lifted her head and looked around, bewildered, before settling her gaze on us.

I froze. She must have been very beautiful when she was young, and even now her eyes were large and liquid like a doe's, framed by long dark lashes and mournful shadows. She stopped crying.

"Hello," she said. Her voice sounded very weak.

Do not speak to her.

Hyacinth's warning seemed to echo in the cavern. But the witch's eyes were so mournful, so intent on me, that I heard myself reply. "Hello," I echoed. Beside me, Woferl gasped at my disobedience. We hoisted ourselves out of the water and onto the rocky floor. "I am sorry to disturb you."

The witch smiled at us. "Not at all, dear child," she said. "Come closer, please. My children were stolen from my side and I was sealed in this cave. I have been so lonely here, trapped for centuries without a soul to keep me company. Oh! Tell me, little one. What do the twin moons look like outside?"

I swallowed hard. The words fell from my lips as if compelled by some enchanted force. "They are bright as coins," I said, "and sit at the opposite ends of the sky."

She shook her head back and forth. "Ah, it's no wonder I am so weak. When they align, my magic shall come back to me, and I will find a way to return home. Have you come to free me?"

Her voice sounded so hopeful that I immediately felt ashamed. "I'm sorry," I said. Woferl squeezed my hand tightly. "We haven't come to free you."

The witch's smile grew wider. "No matter. I have missed the

sound of another voice. Come here, children." She held out her arms to us. "Come here, so that I may see you better."

Woferl looked at me with a frightened face. "She is a witch," he whispered. "Remember what the princeling said?"

I shot him a warning glare. The faery blinked at us, then giggled. Her voice was strangely lovely, as if she was younger than she appeared. "Don't be afraid, little boy," she said to my brother. "I will not hurt you. I know you will not stay long, but I only want to see your faces closely, to touch another's hand before I return to my prison."

My thoughts fluttered, frenzied, through my mind. I did not know how we could pick one of the flowers from behind the witch. One of us would need to distract her, and the other would have to take it. I felt a pang in my chest at the idea of stealing from this lonely creature. Hyacinth's warnings still lingered in my mind, but they were starting to turn numb.

Woferl and I exchanged a pointed stare. Then I released his hand and started to walk closer to the witch. She smiled.

"What is your name, child?" she asked me. Her words had begun to sound like musical notes, as if she sang each sentence she spoke.

"Maria Anna Mozart," I said. "I'm called Nannerl."

"Nannerl," the witch repeated. "What a beautiful little girl you are. You remind me so much of my daughter."

From the corner of my eye, I could see Woferl walking alongside me, each time a tiny step farther away. He was going to steal the night flower. I kept my gaze locked on the witch. "Thank you." I wanted her to stay focused on me. "What is *your* name?"

"I no longer have one," she said. Her voice caressed me in its folds, full of sweet melodies and muted violins. "I'm afraid I have been here so long that I cannot remember anymore." The notes in her voice turned tragic, so that they tore at my heart with their sadness. I steadied myself.

"You look young and strong, child," she went on. She did not notice Woferl's widening distance from me—she was too interested in keeping my attention. "You could help me escape."

"How would I do that?" I asked. "You are bound to this grotto's floor."

"All you would need to do is take some of the water from the pool," she said, gesturing toward where we had come in, "and pour it on my feet. It will loosen them from the stone." Her eyes flickered toward Woferl. He stopped in his tracks, feigning innocence. The witch smiled at him, and I let out a breath.

It would be difficult for us to sneak around her.

"We have nothing with which to hold the water," Woferl said. "I am only wearing my nightclothes. We have no shoes, or thick aprons to use as a vessel."

The witch frowned for a moment at this problem.

"Perhaps we can use one of the night flowers behind you," I suggested. I pointed toward them. "They are very large. They may be able to hold enough water."

Her eyes lit up. "Yes," she said. "You're right, clever girl." She twisted herself around in place, bent toward the wall behind her, and picked one of the night flowers from the wall. It glowed more brightly in her hand, perhaps in fear, and I saw its thorny stem move slowly in her grasp. Woferl watched it with wide eyes.

I started to move toward her. I could see her wrinkles more plainly now, the circles under her eyes, the creases and folds, the frailness of her skin. She continued to smile at me.

"Nannerl," she whispered as I drew closer. The night flower

glittered in her hand. “Help me escape from this grotto, and I shall repay you in ways you cannot imagine. I can answer your wish. I can keep you from being forgotten, like I have been.”

I swallowed hard. “How do you know about my wish?” I said. My eyes darted toward Woferl.

“Your brother can’t hear me, Nannerl,” the witch said. “Only you can. I know who you are, and I know what you want. If you free me, I can help you.”

I was so close now that the cold blue light of the night flower reflected against my skin. The witch’s eyes bore into mine. “Did Hyacinth tell you?” I asked her. “He must have mentioned it.”

Her lips turned down in a menacing frown. “You sound like you’re very fond of him.”

I hesitated, unsure whether I should answer her.

“And what is it that makes you so fond?” she said. The sweet violins in her voice now turned bitter, the nostalgia shifting into a dark memory.

“Should I not be?” I asked her. A seedling of doubt against Hyacinth was planted in my mind. *Careful what she says to you,* I reminded myself, alarmed.

“That isn’t something I can tell you. Do you trust him?”

“I don’t know.”

She held out the flower to me. “You would do well to trust me instead, Nannerl.”

I took a deep breath. Behind me, Woferl stood unmoving beside the cavern pool. I turned back to the faery, reached out, and took the night flower from her hand.

Her touch, colder than the wind of a winter night, froze me in place. I wanted to cry out. Instead I found myself staring at her, overwhelmed by the sound of music that came from deep in her throat. The melody flowed through her body and into my hand, wrapping itself around my skin, refusing to let go. I closed my eyes, unable to tear myself away from her.

I wanted to be a part of the music so badly, ached for it to swallow me. It sounded like it came from everywhere—from her throat, from the air, from deep inside me. But the ice of her touch turned to fire. It threatened to scorch me from the inside out, until I turned to ash against the walls.

Woferl's voice came from somewhere far away, another time and place. "Nannerl!" I thought he said. I could not move. The music roared in my ears.

The white of the witch's eyes had completely filled with blackness. She did not smile any longer. The music that flowed in me turned deafening, shaking my limbs. Pain shot through my chest. It was too much.

Then I felt a warm hand grab mine and pull me away. I gasped for air. My other hand stayed locked around the night flower. I glanced around in a daze and realized that Woferl had broken her hold on me and was running with me toward the pool.

Behind us, the witch shrieked.

"Help me!" she screamed. "Fill the night flower with water, and pour it on my feet. Free me!"

Such anguish pierced her voice that a part of me yearned to go back to the music that flowed in her. My heart tugged against her magic. *No.* I struggled against it, then forced myself away with all my strength. The witch tried to lunge toward us, but we were too far now, and diving back into the water where she could not reach, where her grounded feet prevented her from following. The warmth of the water washed away the last of her icy magic's pull.

The instant we plunged into the pool, I lost my grip on the night flower. Woferl saw it happen and grabbed it. We swam away from the grotto and out through the crevice, where we could no longer hear the witch's cries. We swam away from the rocks, back toward where the deep blue ocean became shallow and hugged the edges of the beach's white sands, away from the cave of night flowers.

Hyacinth was standing exactly where we had last seen him. I trembled as he approached us. Water dripped from my eyes and down my face. Woferl shivered beside me, clutching the flower to his chest. When the princeling saw the night flower in my brother's hands, his eyes lit up and he let out a laugh of joy.

"Splendid!" he exclaimed. He gingerly took the night flower from Woferl's hands, then patted my brother's cheeks twice in affection. "You've done very well."

Woferl beamed, pleased at his praise, and wrapped his arms around himself with pride.

"Why do you need the flower?" I asked Hyacinth.

He glanced at me, then leaned down very close and kissed my cheek. His smile was sweet and grateful. "It is part of what I need to reclaim my throne in this kingdom," he replied. "Soon you will see."

I frowned. "I don't understand."

But Hyacinth had already turned away and motioned for us to follow him back into the woods. I looked down at Woferl again. He seemed tired, and his long lashes dripped water. A bright spot of blood pooled on his thumb. He had pricked his finger on the flower's thorns.

I woke with a start. Gone were the ocean and the faery queen and the night flower. Gone was the princeling. I was lying in my own bed with Woferl breathing gently next to me, his eyes quivering ever so slightly underneath his lids. He had one of his thumbs in his mouth, a habit that returned whenever he dreamed. I stared at the ceiling.

A dream. The words echoed in my mind. But it had all felt so vibrant. The shore had been so white, the shells so blue, the water so warm. The witch's eyes so dark. Still, it must have been a dream. *Had* to have been. What had Hyacinth said he needed the flower for? I couldn't remember. Nor could I think of what we had done after we surfaced again. Hyacinth had praised us, hadn't he?

I looked at Woferl again, then gently pulled his thumb out of his mouth. There, along the top of the tender flesh, was a small cut and a drop of blood. He must have bitten his thumb a little too hard during the night, I thought.

But the vision of the night flower's thorns and Hyacinth's smile had not yet faded away. I continued to stare at my brother until the room began to blur again, until this time I sank into a dreamless sleep.

The Castle on the Hill

WOFERL SEEMED QUIETER THAN USUAL THE next morning. He lay in bed beside me, round cheeks flushed, sleep still glazing his eyes, and listened without a word as I told him about my dream.

"We went together to the white sand beach," I said. "We saw the Queen of the Night, a witch trapped in an underwater grotto there. She was very frightening."

Woferl murmured his astonishment as I told him about how we escaped from the grotto and ended up giving the night flower to Hyacinth. But his wonder felt muted, his attention scattered. The glaze in his eyes gave them a feverish shine.

"Are you all right, Woferl?" I asked when I finished my story.

He shrugged and curled up tighter in bed. I glimpsed a tiny scar on his thumb from where he'd bitten it in his sleep. "My throat is just a little dry," he said, and dozed off again.

Papa always left early and did not come back until later in the day, when Woferl switched from practicing on his violin to the clavier. So the morning hours were mine, a time when I could play without being disturbed, when no one came in to check on my progress or how many times I had run flawlessly through a menuett.

Mama was out, so the only person with me this morning was Woferl. Now that he knew about my composition, and had managed to keep it from Papa, I felt safer with him nearby, someone with whom I could share the burden of a secret. He sat with me on the clavier's bench, his elbows propped up on the keys, watching intently as I played my scale.

After I paused to turn to a different piece in my notebook, he said to me, quite abruptly, "I wish you would write more music."

I stopped to look at him.

Woferl flipped to the second to last page of my notebook, and pointed out the few measures I'd written. "You never finished this one," he said.

All I could hear was my father's voice in my head, and the words he'd spoken to Mama over dinner yesterday. *Nannerl makes an excellent companion for Woferl. Together their fame is twice what it could be. Can you imagine the spectacle we could create if one day Nannerl performed one of Woferl's compositions?*

Mama had listened and nodded. Of course, it would be preposterous to suggest that I could compose my own pieces.

In truth, I was an excellent companion. But I would be nothing more than a performer for my brother's compositions. If I wanted immortality, it would not come from my writing. The words hung weighted around my neck. Composition was for men. It was an obvious rule. What would others think of my father if they knew I composed behind his back? That he could not even control his own daughter? What kind of girl shamed her father by secretly doing a man's work?

The image of my compositions burning in the fire flashed before me, the thought of my father's stern eyes . . . I had seen Papa toss letters in a rage into the stove, remembered watching the embers light the edges of those papers. The memory made me wince. Even seeing my little tune exposed here on the page was making my heart quicken. I glanced nervously toward the door, half expecting Papa to step in at this very moment, and then turned to a different menuett.

"I can't," I replied to Woferl.

He frowned. "Why not?" he asked.

"I'm afraid to."

"But don't you want to?"

"Of course, but it is different with me."

"Music is music. The source of it does not matter so much."

I sighed. "Woferl," I chided him, and he had the grace to give me a guilty look. "I cannot do what you do. It is something you will never ever understand."

He pouted at me in frustration. His tongue had sharpened when it came to composition. *Everyone fancies himself a musician,* he'd complained to me. *No one respects the soul of it.* I'd seen him turn Papa's face red with embarrassment when he once scoffed at the skills of a local noble who had given composition, in his words, *a whirl.*

Charlatan, Woferl had called him to his face. I would have been reprimanded harshly for saying such a thing to a nobleman, but Papa just chuckled about it later.

My brother did not reply again. Instead, he hurried off, and I returned to my lessons.

Minutes later, he returned with a quill and inkwell.

"Woferl!" I exclaimed, pausing in my playing. But he did not apologize. Instead, he adjusted the writing instruments and pointed the quill's feather in my direction.

I began to tremble at the sight of it. This was not Woferl at work. This was God taunting me, tempting me to write again. Or, perhaps, it was Hyacinth, his will bubbling up from my brother's sweet eyes. Was I hearing the words of the princeling on Woferl's lips?

"Will you do it?" he whispered eagerly to me.

"Woferl, this is Papa's," I said. "How will we explain that it is not in its place?"

Woferl simply closed the notebook and gestured to a loose sheet propped against the clavier stand. "I have started to write," he said. "Papa will know the ink is here because I use it daily. How would he know about you?"

I felt my cheeks grow warm at the thought. "But, Woferl," I protested. "Where will I write mine? I cannot continue to compose in my notebook. Sooner or later Papa will see, and that will be the end of it."

"Write on loose sheets," he said. "Then you can fold them up and hide them in our bedroom."

My music, my measures, each one painstakingly written, curling into ash in the fire. The fear lingered, holding me back. But my brother's eyes were still on me, and with them, I felt the ache again to write, his encouragement pushing me forward.

If Papa discovered I was composing, he might burn my work as he burned the letters that upset him.

But he couldn't do that if he never found it.

Woferl finally shrugged, impatient with my long hesitation,

and wandered off to continue his own compositions. I stayed at the bench and stared at the quill in silence, thinking. Ink dripping down the side of the well had touched the clavier stand, staining the fibers of the blank parchment.

In the mornings I would find the quill and inkwell on the stand, along with the smudged pages of music that Woferl had composed the afternoon before. Papa saw them in the evenings and would show them to me in a merry temper, as if I did not know what they sounded like.

Woferl was right. Our father had no reason to think that I would also compose. Every morning, when he was not at home and Mama had left for errands, when only Woferl and I were in the music room, I would take the folded sheet of music from the bottom of my bedroom drawers and add measures to it.

My pages were not as clean as Woferl's. Under Papa's watchful guidance, Woferl wrote more than I did. He composed so quickly, in fact, that he produced stacks of paper at the end of every day. Mama was constantly sending Sebastian out to buy more sheets. When I looked at my brother's pages, I would marvel at how sparse his changes were, how much of it already seemed fully formed in his head.

My thoughts shifted more. I would write entire lines and then cross them out. I would take a measure of music and then flip the harmony to see how it worked. I would go over and over a page before I finally produced a finished copy of it. At the end of the day, my work would bleed with ink, a mess of moving thoughts that, to me, told a story of how the music came to be. I would run a hand over the dried notes and hear the early drafts in my head. My heart would keep time with the rise and fall of the melody. In those moments, the room around me faded away. My surroundings changed into a secret world of sound and peace. I would stir out of that dream of creation with tears in my eyes.

Woferl often watched me write. Sometimes he asked me questions, but mostly he sat beside me in silence, his chin propped in his hands as I worked. When he wrote his pieces, I could hear traces of my own style filtering into his, like milk curling into tea.

Woferl's handwriting, childish though it was, looked very much like our father's with its coiled tails. Mine, untrained and unrefined, did not. From what my brother shared with me about his lessons with Papa, I learned the proper format for recording my work. Eventually, my writing began to look as polished as my brother's, nearly identical to his hand.

In case Papa discovered the sheets of music in my drawer, he could assume that they belonged to Woferl. And I would be spared.

Then, several weeks after my dream of the night flower, Woferl fell ill.

At first, he coughed a little at night, nothing much, only enough to wake me for a moment. Shortly after, his skin began to look paler than it usually did. One morning I went to the clavier and found Woferl sound asleep by the windowsill, his breath forming a circle of fog against the glass. The tiny wound on his thumb still had not disappeared, and the skin around it was flushed and pink, hot to the touch. He did not even wake when I shook him. It was only when I began to play that he finally sat up and looked around, dazed.

"Oh, Nannerl, it's you," he said when he saw me. Then he turned, his eyes far away, and touched the window's glass with his small fingers. "Hyacinth was outside."

I looked down at the street, half expecting to see the princeling's familiar smile. But nothing was there. The hairs on my arms rose. Perhaps Hyacinth was appearing to Woferl alone, as he sometimes did to me. What did he want with him?

At first, my parents did not dwell on Woferl's bout of sickness. Children tended to catch illnesses, especially in these cooler months, and Woferl had always been a frail child.

But the illness lingered. It worsened. His dark eyes turned bright with fever, and his delicate skin gleamed with sweat and angry, red bumps.

Papa did not sleep on the first few nights that Woferl's rash broke out. He sat in our bedchamber and looked on with a grim face as Mama patted my brother's forehead with a damp cloth. As the illness hung on, he began to pace.

"I have had to cancel a performance already," he said in a low voice to my mother. "Soon we will cancel a second, and for the Postmaster-General Count Wenzel Paar, no less."

"The count can wait." There was an edge to Mama's voice tonight. "Woferl will recover before you know it, and you can resume your schedule."

Papa frowned. "The archbishop has already cut my salary. Four hundred gulden! How is a family to live on four hundred gulden a year? We can't afford to cancel a third."

"Well, you will just have to wait, won't you?" Mama answered.

Papa turned away from her in a huff. As he passed me, he paused. "If Woferl is not well by next week," he said, "you will have to perform alone."

"Yes, Papa," I replied. If he was upset enough to raise his voice at my mother, then I did not dare add to it. *Alone.* It was a frightening thought, performing without my brother. But somewhere deep inside me, a voice also stirred to life.

The attention will be yours alone.

I joined Mama in Woferl's bedchamber, for I could not sleep either, and watched my brother toss and turn with fever. When she would at last fall asleep in exhaustion and Woferl would wake up, I would hold his hand and tell him more stories to keep him from crying.

Finally, one night, my mother and father left to seek out a doctor for help. I alone remained beside Woferl, turning my pendant in my hand as if to give my brother good fortune.

When he woke to see me at his side, he squinted and began to cry. "My skin burns, Nannerl," he murmured. His hands reached up to scratch, but I forced them back down. He protested weakly. "My knees and elbows hurt."

His joints were swollen. I could see the rounded look of them. It was such a pitiful sight that I squeezed his hand and tried to give him a smile. "It will all pass soon," I reassured him as I wiped his tears away. "And you'll be back in front of your clavier. I promise."

He looked away and toward the window. "Do you think Hyacinth is watching us right now?" Woferl asked.

Hyacinth again. I felt a chill. Why was he in Woferl's thoughts so much these days? I tried to think back to the dream I'd once had. The night flower. The witch. Hadn't Hyacinth given Woferl something? The scar on his thumb had finally faded, but I found myself touching the spot where it had been, trying to remember. The back of my neck prickled, as if someone else might be in the room with us.

Did Hyacinth know that Woferl would fall ill? The thought so unsettled me that I immediately dismissed it.

It was possible, I suppose. But perhaps, more likely, Woferl's fever had been brought on by wind and rain.

"I don't know," I finally said to my brother. He turned toward me, eager to be distracted, and I obliged. "Maybe he is sitting somewhere in the kingdom's forests right now, perched high on the roots of a tree, watching us through a round mirror."

"Do you think he is sad, like me?" Woferl said.

"Very sad," I replied, and reached out to stroke his damp hair. "When Hyacinth weeps, his tears form puddles at the bottom of the trees. This is how the drowning pools form."

"Maybe he isn't in the forest anymore," Woferl said, "but going somewhere else that we haven't seen. A castle in the hills." He burst into a fit of coughs that brought tears to his eyes.

"Yes, a castle in the hills," I said. "Perhaps this was his old home, the palace where the princeling once lived."

Woferl nodded, grumpy. "What happened to him? It must have been very tragic."

Tragic. Woferl's words reminded me of the look I'd seen on Hyacinth's face in the trinket shop's grotto, a moment of sadness that was there and then gone. What *had* happened to the princeling in his past?

"There is a river that surrounds all sides of the hill," I went on as I pondered, "and the grass at the very bottom near the water is lush and green, but the grass higher up is dry and dying, for it hasn't rained in months. Hyacinth has to swim across the river to reach the castle, but because he cannot swim, he can only sit on the banks and yearn for his lost home."

"Why do you think he left?" Woferl whispered.

I thought of little Hansel and Gretel abandoned in the woods, stumbling across the witch's house of gingerbread. I thought of the smith of Oberarl, offering his daughter to the devil in exchange for the healing waters of the Gastein Valley. The faery tales swirled in my mind as I tried to think of what Hyacinth's history might be.

Finally, something came to me. I looked at Woferl. My story came out hushed, dark as the shadows flickering in the corners of the room.

"The castle is crumbling now, for no one has been there in a long time. Years ago, a young king named Giovanni ruled the land with his beautiful queen, whom he loved dearly. In fact, everyone loved the queen, even the great Sun, who bestowed his golden magic of fire unto her so that the land flourished under her warmth. The light and the heat restricted the land's faery creatures to the woods, where their dangerous magic could not harm the kingdom. All was well for many years. When the queen finally announced that she would have her first child, the people rejoiced and counted themselves very lucky indeed. Everyone waited eagerly for the birth."

As I spoke, Woferl grew quiet, his aches momentarily forgotten, so that the only sound in the room became that of my voice.

"But the queen fell ill when the first snows arrived that winter. Her rosy cheeks drained of color, and her shining hair became dark and limp and damp from her lingering fever. The king's doctors would boil their medicines and feed her teas made from strange, exotic roots. The queen began having terrible nightmares. During the day, she saw visions of death and suffering, and at night, she witnessed strange dark figures glide past her windows and her bed. She grew paler, even as her unborn child swelled in her belly."

The story unfolded and my voice began to change. It became wilder and deeper, as if it belonged to another. The words did not quite match with the movements I made with my lips. A strange fog in my head left me feeling distant. The candle burned lower. From the corner of my eye, I could see a shadow at the edge of the room growing, until it looked as slender and graceful as the dancing shape of a princeling.

"The queen gave birth to twins, a boy and a girl, in the spring of that year. Although ravaged by illness, she did not die, but sank deep into a permanent madness. Frightened for her, the king assigned his champion to her side in an attempt to keep her safe. But it did no good. She rose one morning, took her baby boy in her arms, and walked out into the forest, murmuring about the faeries calling for her in the woods."

"What happened to them?" Woferl whispered.

"I don't know," I replied, but through the fog in my mind, I could hear someone else continue the story, as if the answer had existed all along. "The faeries are always looking for something to devour. They find the sadness in souls particularly enticing, and when they happen upon them, they will do whatever they can to get it."

Woferl looked grave. "Then they must certainly have taken her," he said.

I shivered at the finality in his words as I went on. "When the queen and her son did not return, the entire kingdom went to search for the pair. No one ever found them. The grieving king lost all will to live, leaving his castle unguarded and untended, so that it began to crumble from disrepair. He ordered his daughter locked away in the tallest tower of the castle, so that the faeries could not steal her away too. He lost his memory over time. He forgot his wife and his two children. The Sun, devastated by the young queen's disappearance, abandoned the kingdom and plunged the land into eternal night. The crops withered. When the king finally died, the people fled the land until the castle stood empty and alone."

"What about the princess?"

"No one knows, although some say that she is still locked away in the castle's highest tower, waiting for someone to remember her." My words seemed so true in this moment that I found myself thinking of the faery witch trapped in the cove. She had told me how lonely she was too.

Woferl sighed deeply and sank his head into the bed's pillows. "Hyacinth is the son who disappeared," he added. "He comes back to the edge of the river every morning and every evening to stare at the castle. But he can never reach it. He cannot swim."

He cannot swim. Suddenly, I remembered Hyacinth telling me this in my dream of the night flower. Woferl, instinctively, knew this too.

I hesitated for a moment, surprised at where the story ended. In the corners of the room, the shadow that had been a princeling now faded to a flicker. I could still taste the wind of his voice on my tongue, as if he had told his story through me. For the first time, I thought I understood why Hyacinth had chosen to become my guardian.

He had been abandoned too, left behind like I feared I might someday be, forever yearning for a world to which he cannot return.

"That is what happens to children who are forgotten, Woferl," I ended, leaning over to kiss his forehead. I was exhausted now, and the castle had suddenly become too real in my mind. When I looked down at my brother, I saw that he had fallen mercifully back to sleep. "They stay forever trapped," I murmured to myself, "nothing but a lost memory."

Woferl's doctor, a man named Herr Anton von Bernhard, came late that night in a quiet flurry to stir a cup of medicine for my brother.

He looked at Woferl's eyes, listened to his heart and his lungs, and studied the angry rash on his skin. In the candlelight, I could see Mama's face. She wanted to say something, perhaps ask Herr von Bernhard a few questions, but each time she looked at Woferl she closed her mouth, as if she'd forgotten what she had to say.

"Scarlet fever, I imagine," Herr von Bernhard told my father afterward. "See the bright red color of his tongue. Keep him in bed, and in a room with fewer windows." He rubbed his temples as if it were a nervous habit. "I will come again tomorrow and give him a dose of angelica. He should drink warm water, boiled thoroughly."

We moved Woferl into a smaller chamber the very next morning, one with only a single window and dim light. Each evening Herr von Bernhard came to give him medicine, and each morning Sebastian would open the window briefly to let the sick air out. Sometimes Mama would sit with Woferl while I retreated to play the clavier, Papa at my side, absently correcting my mistakes.

The halt to our performances stretched on. Papa's bright mood from our successful Vienna trip darkened. I performed alone before the Elector of Bavaria. Without my brother, the court's attention stayed only on me, and when they burst into applause at the end of my performance, I felt so stunned that for a moment I forgot to curtsy in return. The tips of my fingers still tingled from the thrill of my music.

My eyes went automatically to the space beside Papa, where Woferl would be. But only a court official stood in his place. Behind him, I thought I saw a pale, lithe figure stride through the crowd, his blue eyes fixed on me in approval. I turned my gaze back down to the floor. The thought of my brother lying stricken in bed clashed with the sound of thunderous applause.

"Nannerl, listen to what the newspaper says," my father said several nights later, as we sat for supper. "The girl played the most difficult sonatas and concertos quite accurately, and in the best of taste." He smiled at me. This was such a rare sight that I did not react quickly enough to smile back. My heart soared. "They do not say this lightly, Nannerl. Well done."

A newspaper article! Papa had praised me for it! I reread it in bed that night, going over the words until I fell asleep with it still clutched in my hands. In my dreams, I curtsied before an applauding opera house covered in night flowers. My mother and father sat in the front row. My brother was nowhere to be seen. Up in the balconies, Hyacinth leaned against the banisters and watched me with a look of pride.

I shook awake, my eyes still turned up in his direction.

But Woferl stayed ill. My enthusiasm dampened every time I passed his room and saw him delirious with fever in his bed. Papa muttered frequently under his breath. Sometimes I heard him speak harshly to my mother in the next room, then lower his voice in apology.

"Your father is simply frightened, Nannerl," Mama would say when I asked her about it. "He loves Woferl dearly, and he worries for us."

"You worry for us," I replied, "and you never speak harshly."

Mama smiled a little. "I am your mother, darling. What good would that do?"

One night, when both of my parents sat by Woferl's bed with me, I saw something stir in the corners of the room. The shadow flickered, turning sharper and sharper, until finally it materialized into the shape of Hyacinth.

He had grown taller and more slender since I last saw him, matching me in height and paleness, as if he too were edging out of his youth, growing from boyhood and into the body of a young man. As I watched him, he came to kneel beside me. His eyes, serious and silent, lingered on my brother.

"I know what you are thinking, Fräulein," he said to me. "You think that perhaps I did this to him when he pricked his finger on the night flower."

"Did you?" I finally asked.

"Your brother pricked his finger only so that a drop of his blood could spill inside the kingdom," he replied. "I wanted to be sure that I had a link to his lifeblood and his talent, just as I have a link to yours through your notebook."

The thought of Woferl's blood being his link to the kingdom sent a shiver tingling through me. "What will happen the next time we step into the kingdom?" I said. "What other dangers will we have to face? Will the kingdom require more of our blood?"

Hyacinth shook his head. "The next time you come, you will come alone. I only need you for your next task."

"Why is that?" I asked.

"Because, Fräulein." Hyacinth looked at me, his eyes pulsing in the dark. "The fulfillment of your wish has always been about *you*."

Me. I remembered my name printed alone in the newspaper article, my father's praise only for me. It had all happened after fetching the night flower for Hyacinth.

When I didn't answer, Hyacinth turned his attention back to Woferl and heaved a wistful sigh. "Poor little boy," he said. "Look at the color of his face, as if he is hovering between two worlds."

Two worlds, the kingdom and ours. In his musical voice was such a sound of pining that I felt truly sorry for him. I thought of his past that I'd told Woferl, the way the kingdom's forests had claimed him along with the queen, and wondered if he missed the home that he had been stolen from. Perhaps this was why he'd chosen to help me. One forgotten child to another.

"What will my next task be, then?" I asked, looking again at him—

—but Hyacinth had disappeared as quickly as he'd come, leaving nothing but emptiness beside me.

I blinked, disoriented to be alone all of a sudden. Beside me, Papa and Mama looked undisturbed, their heads still bowed in prayer.

I waited, half expecting Hyacinth to appear again. Had he ever been in here? The air seemed chillier now, and when I glanced outside the window, I noticed a ghostlike figure shrouded in dark colors drift by.

Hadn't the queen from the Kingdom of Back seen those

figures too, floating shadows in the mist, the pieces of her dreams? Hadn't they surrounded the castle on the hill and reached for her as she lay ill?

I trembled as another glided by, then another. Now their shapes condensed, turned solid, so that I could make out their red eyes and black hoods, their twisted fingers on long, spindled hands. They grew as the candle burned low by Woferl's bedstand.

The queen had not recovered from her nightmares, and instead had fallen into madness. I pictured her in my mind again, afraid and alone, lost in a world that those around her could not see. Perhaps they had frightened Hyacinth away too.

I rose from my place and left the room, returning with two more candles. I replaced the dying one, then set another right beside it. Mama watched me silently.

"It is too dark in here," I said to her. "Now it is better."

The figures outside the window slowly faded away in the brighter light, until they looked like nothing more than the flag that flapped against the glass.

A Family of Dignified Reputation

My brother remained weak with fever for several more weeks, catching a second illness before he finally began to recover.

He had now been absent from the clavier for nearly two months. I had not written any music during this time, either, as I would not have been able to explain the quill and ink sitting beside the clavier stand. Instead, I spent the months practicing, while the unwritten notes ached in my mind, hungering for paper to rest on.

Momentum, once slowed, was difficult to pick up again. Our invitations dwindled. I still played alone before my smaller audiences. I still relished knowing that the applause was for me. But newspaper headlines were not the same as gifts of coin. My performances did not draw the kind of patrons we needed, their pockets lined with gold.

Papa began to speak more and more about money, sometimes

about nothing else. Even I had already learned that three hundred and fifty gulden a year as the vice-kapellmeister of Salzburg's court was no large sum. We could earn more with one night's worth of performances.

"A pittance," Papa complained, "for such a position. There is no respect for the creation of music anymore, and certainly not by the archbishop."

"We could dismiss Sebastian for a time," my mother suggested in a quiet voice, so that Sebastian would not hear it from where he was tidying their bedchamber.

Papa made an irritated sound. "The Mozarts and their famed children, unable to keep a manservant? Imagine inviting a member of the court to our home, only to have my own wife serving him tea and cake. Who will send us invitations, then?" He waved a hand in the air. "No, no, I will arrange for their portraits to be done."

"A portrait each?" My mother's eyes flickered, and I could see her doing the calculation of the cost in her mind. "Leopold—"

"What's the matter? Are we not a family of dignified reputation? Do our children not deserve the best? Let our guests see how fine and young they both are, how well we are doing. Do you want to be the laughingstock of Salzburg, Anna?"

My mother pressed her hands tidily into her lap, in the way she always did when she knew she could not bend my father's ear. I thought of her asking me what good it would do for her to speak harshly. "Of course not," she said in an even tone.

My father went on, talking of taking us to France, to Paris. He had already begun soliciting, asking for the names of the kapellmeisters in each French town we could pass, and whether or not the townspeople cared for musical performances.

"They will not stay young forever," Papa finished. "The older they are, the less magnificent their skills will seem." Then he turned away, mumbling, in the direction of Woferl's closed bedchamber door. He stepped inside, then shut the door behind him.

In the silence, my mother looked at me and noticed my expression. She sighed. "You must forgive him his anxiety in times like this," she told me. "He is only looking out for our well-being. He says such things because he is desperately proud of you and your brother, and wants to ensure you are spoken of in high regard."

"Are these times really so bad?" I said. "Why is Papa so worried?"

Mama gave me a stern look. "These are not questions a young lady should ask. Concentrate on what your father expects of you, and nothing more."

I followed my mother's example and did not speak again. No question was ever one a young lady should ask. It was useless to bring up my performances, that I had been tiding us over all this time. Papa was still waiting to hear my brother again. And I could feel it, my father's mind pulling away from the memory of my talent. I was retreating into the dark spaces of his attention.

I thought of Hyacinth. My fingers ached, longing for the chance to write again. He had been gone so long. I needed him to return, before he forgot me too.

Slowly, to my relief, Woferl began to emerge from his bed. He started to chatter once more. I would find him in the music room in the mornings, seated on the clavier. A pink flush came back to his white cheeks. I took comfort in seeing him return to us, in all the familiar scenes that had been absent for the past month. Perhaps my worries about Hyacinth's involvement were just nonsense, after all.

And then, one morning, the quill and ink were out again, and he was scribbling away.

My heart leapt. It meant that I could start writing again too.

Papa spent longer hours at the clavier with Woferl and with me, as if to make up for the time we'd lost. He made Woferl play so late into the night that my brother could not concentrate anymore, then slapped Woferl's hands when he saw my brother's eyes drooping at the clavier.

I'd seen Papa's temper strain many times before. But Woferl had suffered so long during his latest illness, and I'd been so truly unsettled that Hyacinth had done something to him, that now I felt the urge to defend him rise in me.

"Maybe he should rest now, Papa," I said as Woferl wiped tears away, the dark circles prominent under his eyes. His small shoulders hung low like a wilted plant. "I'm sure he will play better in the morning."

Papa did not look at me. He watched as Woferl started once again to play through the beginning of a sonata. The music did not have its usual joy, and Woferl's hands could not play with their usual crispness. After several measures, Papa stopped him.

"To bed, both of you," he said wearily. Shadows hid his eyes from me, so I could not guess what they looked like. "I've heard enough for today."

That night, Woferl curled up tight beside me and fell into a deep, exhausted sleep. He had recovered, but his strength had not fully returned to him. It was strange, his quietness. I draped my arm around him, touched my lips to his forehead, and let him rest.

I continued to write my music in secret. By now I had accumulated a small stack of papers at the bottom of our bedroom's drawers, little sonatas and whimsical orchestra pieces, and had started to scatter them around in different spots so that they did not all sit together. I would write when Woferl and I had a rare moment alone.

One day Woferl, who sat near the clavier and watched me work, spoke up again, a welcome respite from his silence.

"You should show Papa your music," he whispered. He rose from the windowsill and came to sit beside me on the clavier's bench, pressing his small, warm body to me. When I looked down, I saw that his feet still dangled some distance from the floor.

"Papa will not like it," I said. "I've told you this before."

"You don't know that," he replied. He turned his eyes to the sheet of music before him, his face full of wonder. "How can someone not like this?"

I sighed. "Woferl, it's very kind of you to like my music, but you are not Papa. What can he possibly use it for? He certainly would not let me publish them, or perform them for an audience. He may tell me to stop writing altogether then. He will think it is a waste of my time, when I should instead be practicing for our performances."

"Why?"

I always disliked this question from him. "I am a lady. It is simply not proper for me," I decided to say. "I would need to have your fame, and your ability to draw a crowd, to even risk such a thing."

Woferl frowned. I had never spoken of our performances like

this before, as if we did not play together, as if we were separate. "But you have my fame," he said. "You draw your own crowds."

I looked down at him. He meant it in earnest, I could see that, but I knew it was not true. Still, I put my arm around him and squeezed his shoulder once, my silent thanks, before I turned back to the half-filled sheet before me. "Let me finish," I said. "It is almost your turn at the clavier."

I saw Papa alone that evening. It was very late at night, and Mama had already retired to bed, and I had carried a sleeping Woferl from the music room to our own bedroom. I had returned to the music room to fetch a candle. On my way back, I caught a glimpse of Papa sitting by himself at the dinner table and paused.

Both of his elbows were propped up on the table, and his head sat in his hands. I watched him for a moment, my face partially hidden behind the edge of the wall. Papa had rolled his sleeves up to the middle of his arms, messily, and gotten a stain on one of them. Mama would have to wash his shirt in the morning. His powdered wig lay forgotten on a nearby chair. I saw his dark hair in mild disarray, combed through with his fingers, loose strands everywhere. He seemed not like the stern figure often standing beside the clavier, but a tired soul, vulnerable and small.

In this light, I could see what he might have looked like as a young man, wide-eyed and smooth-faced, how my mother must have seen him before the weight of family and fortune carved lines into his skin. Perhaps he had been carefree in his youth too. A teenage Woferl.

I could not picture my father playing childish pranks on his peers, though, or clapping his hands with laughter at a story. He must have always been serious, even when he was handsome and charming enough to have coaxed my mother into his arms. And something about the intensity of his presence, the gravity of him, made me feel bold.

What if I did as Woferl suggested and told him about my compositions? Would it cheer him? Surely, he could feel some sense of pride, however fleeting, in knowing that his daughter could write music as competently as his son? I remembered Woferl's words, and then my own. Perhaps he would let me perform them, just now and then, a small refrain in the middle of a private concert, some opportunities before my performing years ended.

An urge rose in me then, to tell my father about my secret. To hear the approval in his voice. My pendant felt suddenly heavy in my pocket. I thought about what Hyacinth would say. Did he want me to do this? I took a deep breath, wondering how I could word it to Papa.

I am composing my own music. It is mine, from my hand alone. I wrote it for myself and I wanted to share it with you. Do you see me?

Papa must have heard me take my breath. He lifted his head from his hands and looked around as if in a momentary daze, then settled on me. For a moment, his eyes softened, as if in guilt or sorrow. He looked like he wanted to say something to me. I waited, my heart pounding, my entire body tilted in anticipation toward him.

Then my father's gaze retreated behind a wall. My courage wavered. I held back the words from the tip of my tongue.

"Have you stood there long?" he said.

I shook my head quietly.

He looked away from me and rubbed a hand across his face. "For heaven's sake, child, go to bed. It is not polite to stand in doorways, spying on others."

The moment passed as quickly as it came upon me. I could no longer remember what I wanted to say. What was I thinking? Seconds earlier I had nearly spilled open my secrets to him—now, it seemed absurd to mention such an idea to my father. He would have torn my music in half, tossed the ruined sheets into the fire. The thought made me pale.

I murmured an apology, stepped away from him, then turned toward the bedroom. Silver light sliced the floor into lines. When I moved past the windows, I thought I could glimpse the twin moons of the Kingdom of Back hovering in the sky, moving slowly and steadily closer to each other. Exhaustion suddenly weighed against my chest, and I wanted nothing more than to lie down and sleep. In the morning, I would forget this moment. So would he.

Someone was waiting for me right behind my bedroom door. He was so quiet that I startled in the dark.

It was Hyacinth.

He had grown a little taller since the last time I saw him, and his face was more beautiful than ever. He had heard me calling for him through my music, however unconsciously I had done it. I stared at him, caught between fright and joy.

Whatever expression he saw on my face, it made him shake his head in sympathy. "You've performed well on your own," he said. His fingers brushed against my cheek. "Why are you sad, my Fräulein?"

His words were so quiet and gentle, his look so attentive, that I felt an urge to cry. I swallowed, waiting until my eyes dried, and then whispered, "My father is unhappy." I looked to where my brother lay curled in bed, withered from exhaustion and already asleep. "Woferl has been very ill lately. It has made him withdrawn and quiet."

Hyacinth's glowing eyes roamed around the room before they finally settled on the window that overlooked the Getreidegasse, and the lands beyond.

"Perhaps I can do something to help your family," he said to me, taking my hand in his. "It is time for your next task. Are you ready?"

I thought of my father with his head bowed in his hands, the dark circles under my brother's eyes. I thought of my mother, wringing her hands. I looked down at his smooth, elegant fingers wrapped around mine, his presence here with me when others were not.

"Yes," I replied.

The Ogre and the Sword

THE AIR WAS COOL AND ALIVE TONIGHT IN THE Kingdom of Back, as if all this land had taken in a deep breath, stirring, at the return of their princeling. The west wind caressed us, delighted by Hyacinth's presence, and Hyacinth smiled at its touch, tilting his face up so that the wind could kiss his lips. As I followed him through the trees, I wrapped my arms tighter around myself, shivering beneath my lace and velvet. My dress dragged along the ground, picking up bits of dirt and grass.

Hyacinth! Hyacinth! the faeries called as we went, dancing excited circles of blue light around their princeling. *He's here!* They drew close to him, kissing his cheeks and skin affectionately, but he waved them off, his breath fogging in the midnight air.

"Away tonight, loves," he cried, gripping my hand. "I have my Fräulein with me." I couldn't help but smile, secretly pleased by his singular attention.

The faeries hissed their disapproval at me, scattering as Hyacinth waved his hand at them and then coming back together to tug sharply at my hair. I scowled, batting them away. "You must be firm with them," he said to me, the glow of his eyes reflecting against his shoulder. "It is hard for faeries to understand subtlety."

We paused in the middle of a clearing in the forest. Here, I gasped aloud.

The twin moons of the land hovered at either end of the clearing's sky, where the trees' roots reached up against the night. The moonlight illuminated the stalks of edelweiss that filled the field, painting them all in a silver-white glow. I'd never seen so many flowers in my life. They carpeted the entire clearing, transforming it into a scene of snow. Overhead, the sheet of stars was so brilliant that they seemed to be raining stardust down upon us.

Hyacinth smiled at my awe. "Look closer. Aren't they lovely?"

When I peered more closely at the flowers, I realized that their glow did indeed come from a thin layer of glittering white dust that coated them. When the moonbeams cut through the forest around us at just the right slants, I could see it—the shine of dust in the air, floating gently down by the millions, and when I looked down at myself, my arms and dress were sprinkled with starlight.

I smiled and, on impulse, leaned down to touch the shimmering edelweiss growing around me. Each time my fingers brushed their petals, a note sang out, so that running my hand through them sounded like a soft chime of bells. I closed my eyes for a moment to savor the sound.

Hyacinth turned his face up toward the stars. They seemed to lean down toward him in response. *The princeling,* came the whisper, echoing around us.

"Why are we here?" I breathed, mesmerized.

He took my hand, then pulled me toward him, pressing his palm against the small of my back. I blushed at the warmth of his skin against mine. "Because I'm taking you up to see the stars," he replied.

A peal of bell-like laughter answered from above, and a moment later, a dozen threads dropped from the sky. Upon their ends were silver hooks that winked in the night, each one's curve large enough for a person to sit upon.

Hyacinth grabbed the one nearest to us, tugging twice on it, and up in the sky, a star winked back. He lifted one foot to press down against the hook, as if to test it. Then he pulled me up with him so that we stood together on it.

I started to open my mouth, but my words scattered to the winds as the hook suddenly yanked up, slicing us through the cool air in a shining line. The forest shrank into a mass of tiny limbs below us. I squeezed my eyes shut. Hyacinth laughed at me as my arms tightened in panic around his waist.

When we finally stopped moving, and the pit of my stomach had settled, I hesitantly opened my eyes.

The forest was gone. The meadow had long ago vanished somewhere far below us. We now stood on a hook suspended in a world of clouds, white wisps drifting around us in a mist. Overhead, the stars that I'd always seen as dots were now bright balls of light, blue and gold and scarlet, a sheet stretching to infinity in every direction. They looked so close I thought I could reach out and pluck them from the sky.

The bell-like laughter echoed from somewhere overhead, and I looked up to see the line of our hook disappearing around the top of a glowing star hovering right above us.

“Starfishers,” Hyacinth said. “They like to tempt the gullible, who always take their bait. Then they pull them up into the sky and dangle them there for weeks on end, taunting them until they let them back down again.”

He smiled when I shied away from the sharp hooks. “But they will not harm their princeling, or you,” he added. Still, I noticed him ducking instinctively away from the burning balls of light overhead, as if afraid that the stars might scorch him. “Come.” He stepped off the hook and leapt onto the soft blanket of clouds below us, then held his hand up to me. I took it, stepping down too. The clouds felt like moss made of air between my toes. “I brought you up here so that you can see what I need,” he said.

We walked over to the edge of the clouds. Here, Hyacinth lowered into his usual crouch, while I lay down on my stomach to peer over the side. He swept a hand out toward the world under us.

My gaze followed his gesture to the landscape below, where the expanse of forest spread out beneath us like a darkly woven rug, its edge cut by a ribbon of white sand and the still, silver ocean beyond. I recognized the shore where the witch’s grotto of night flowers had been. Edelweiss covered every field and clearing in sight, their snow-white patches billowing like magic under the moonlight. It was an untouched land, strange in every way—but recognizable too. Here and there, I saw a coastline peeking through the clouds that reminded me of the boundaries of Europe I’d seen on maps. The sight so surprised me that I tugged on Hyacinth’s hand.

“There!” I gasped.

“Everything here is backward.” He smiled. “The kingdom is a mirror of your world.”

A mirror of our world. My eyes roamed across the rolling land. In the northern regions, I could see fires lighting hearths in a village buried in the snow, golden menorahs in their windows. In the east, a massive field of soldiers in blue and white charged across a plain, their movement like the ripple of a flag. I looked on elsewhere as a line of young women, their limbs bound, stood on the gallows as guards held torches to their feet. And far out, where the land shifted to an endless expanse of black ocean, ships as small as dots sailed from the shores in pursuit of the New World.

"How is it a mirror?" I asked him.

"When you hold an image up to a mirror, you see every detail of that image exaggerated, things happening all around you that you might have missed in your everyday life." He nodded down toward the ships crossing the Atlantic Ocean. "Look hard enough, Fräulein, and the kingdom will show you every truth that your world doesn't."

A new wind blew past, numbing my fingers and toes. I turned my gaze back down to the mirrored world. Dots of dark-red blood stained the ships' decks and the snow piled against the village houses' charred windowsills. Fire consumed the gallows, and cries came from the battlefield.

Every truth. I swallowed hard, but my eyes stayed on the scene, determined to remember it.

Hyacinth's hand stopped. There, the forest ended abruptly at the banks of a river that encircled the entirety of a castle and its surrounding cluster of villages.

"The castle on the hill," I whispered.

A massive jungle of black thorns grew on either side of the river, unbreakable and unrelenting, a wall separating man from forest. No lights flickered in the villages. They were abandoned and empty. On the hilltop, the castle loomed, its dark stones decaying, and just as in the story I'd told Woferl on his sickbed, dark phantom figures floated around the castle's tallest tower, their wispy trails visible even from here.

"I wanted you to see the full expanse of our world," Hyacinth said when I looked expectantly at him. "It was not always so. The kingdom was once prosperous, in harmony with the land around it. After the queen's death, though, the king ordered the thorns grown around the river and cut it off from the kingdom within."

I looked down to see the guard towers dotting the length of the river. They looked long abandoned, just as the villages had been, their turrets overgrown with thorns and black ivy. In the river itself swam something enormous and sinister, its fins billowing out of the dark water.

"You are the lost prince," I said, turning to him.

His eyes had a faraway look now as he stared at the castle. "Over time, I have grown up with the faeries, and each year I spend with them, I become more like them."

"Do the thorns keep you away from the castle?"

"The thorns and the river," he replied sadly. "That is what I need your help with. There is an old sword so sharp that it can cut away the thorns around the castle. This sword can only be found in an estate"—here he paused to point out a house in a crescent-shaped clearing in the woods—"where a great ogre lives. He wears nothing but black, so he blends into the shadows, and if he finds you stealing his sword, he will slay you where you stand."

I looked at him in horror. "An ogre?" I asked. "You expect me to get past such a creature?"

"You survived the Queen of the Night, didn't you?" Hyacinth

looked at me with his glowing eyes. "I have tried to reach the sword before, but the ogre has a particular nose for those of us from within the Kingdom of Back. He can smell the wind and night on me. You, though, are from another world, and he cannot recognize the scent of that world on you."

I turned my attention back to the house in the clearing. Ivy draped its walls, while a puff of smoke floated lazily up from the home's chimney. Not a single light glowed from the windows. I tried to imagine what the ogre would look like when he turned his eyes down toward me.

"We have a bargain," Hyacinth said to me as he tilted his head. "Can I trust you?"

I nodded. "I will do it."

He smiled at me warmly. As he did, a breeze picked up, combing its delicate fingers through my hair before it turned to blow in the direction of the ogre's home.

"The west wind will carry you there," he said to me. "The ogre sleeps very soundly, and if you keep your wits about you, you will find the sword without waking him and return on the wind's back before he is any wiser."

I tucked my nightgown tightly under my legs as the wind blew harder, the breeze turning into a gust until it finally lifted me from the clouds. I wanted to cry out, but all I could do was look back once at Hyacinth before the wind bore me away from the sky and swept me downward across the nightscape.

The ogre's house was so quiet that the wind dropping me in the silver clearing sounded deafening to my ears. I crept toward the entrance, my heart pounding. The door itself was slightly ajar, as if the ogre knew that there was no one in the kingdom who would dare steal from him. Still, I stood there, unable for a moment to will myself to do it. What if I went in and never returned? Why would Hyacinth trust me with such a task?

But then I remembered the warmth of his smile, the promise between us. If I completed this task, I would have only one more to finish my end of the bargain. It would bring me that much closer to Hyacinth fulfilling my wish.

The world around me seemed to surge in response to my battling thoughts. I turned sideways, barely slipping through the open door, and disappeared into the shadows inside the house.

The home was a wreck. Broken things littered the lower floor: the seat of a large stool; the shattered porcelain from a former cup; an enormous table missing half of one leg, as if it had been chopped clean off. Cobwebs draped the inside of the fireplace, the wood in it layered with dust. A half-eaten loaf of hard bread sat on the kitchen counter. Even a rat had decided the bread was not worth taking, a few half-hearted nibbles visible on the edge of the crust.

There was no sword to be seen anywhere.

A gentle snore from the upper floor sent a tremor through me. I hurried into the shadows by the stairs before looking up. The steps were each twice as tall as the ones I knew in our home along the Getreidegasse, their middles sagging as if used to bearing a great weight. I waited until the snores were even. Then I climbed up the steps, one at a time.

They led into a bedchamber strewn with old clothes, open drawers, and discarded armor.

I could hardly see anything in the darkness, except for a shapeless mass lying in the enormous canopy bed, obscured behind translucent black drapes. From inside it came the snores, each so loud that it seemed to rattle the floorboards.

The sword. I looked at each discarded piece of armor on the ground. A breastplate, covered in old grime. A dented armguard, a shield emblazoned with a magnificent burning sun on its rusted surface. A forgotten belt, the scabbard at its side empty.

What if Hyacinth was wrong, and there was no sword here at all?

A particularly loud snore made me jump and whirl to face the bed. The figure behind its drapes stirred, rolling onto its side with a sigh that sounded half-labored, half-mournful, a sound full of tears. The creature was enormous, a fearsome shadow in the night that blocked out the moon.

As it moved, something metal glinted in the dim light. It was the sword, its hilt in the ogre's massive clutches, steel still sharp enough to cut a line in the bed's sheets.

The ogre suddenly gasped, and I ducked to the floor, certain he'd awakened and seen me crouching here. He asked a question I couldn't understand, then continued muttering to himself without waiting for an answer.

A dream. As he shifted again, he let out another strangled gasp and sighed.

"I've searched for years," he muttered this time, and in his voice was a song of mourning, the ache for summer when winter has already settled in. "Where are you?"

Hyacinth had been afraid of this ogre, and so was I, but even monsters must dream of fears and wants, and the sadness in his voice drew me closer. Now I could see the faint outline of his face in the night. What I'd imagined as the jutting jaw and ivory fangs of a beast, I now saw was a thick beard, aged and unkempt.

"Where are you?" he repeated.

Something told me I should answer, so I did. "Here," I whispered.

He stilled, then turned his closed eyes toward the window, in my direction. I froze. "I heard you," he said, wonder seeping into his voice. His lips, hidden beneath that hard beard, tilted into a hopeful smile. "There you are! Are you near the trees?"

I crept silently around his bed until I was on his other side. "Yes, near the tree line," I answered.

"Ah!" he exclaimed, and turned to face me again, still asleep, his hands shifting against the sword's hilt. His fingers twitched. "Are you safe?"

Who was he searching for, I wonder? I cleared my throat, then dared to step closer. "I am safe," I replied, "although there is a great beast here, right across the river! Lend me your sword, so that I may fend him off."

The smile wavered, and the ogre's brows knotted. He hesitated, his grip still tight on the sword's hilt. "It is very heavy. Will you be able to wield it?"

"I can," I said, creeping closer. I stood right at the edge of the black drapes now, my hands poised near his. From here, I could see all the sword's details, its red pommel stone and the fine curls of writing etched into its blade. "You will just need to throw it a bit farther."

He murmured again, his voice too low for me to catch. Then, suddenly, his fingers loosened against the hilt.

Now. With a surge of strength borne of panic, I reached past the drapes and toward the hilt, my hands grasping it right as his hands started to close around it again.

"What is this?" he grumbled, his brows knotting deeper above his eyes.

I yanked the sword out of his loose grip before he could move again. The weight of it surprised me, and instead of pulling it to me,

I stumbled and dropped the blade with a loud clatter on the floor.

The ogre startled, stiffened, and grew quiet. His eyes fluttered open.

I did not hesitate. I scrambled forward and hoisted the blade with both hands, then half ran, half stumbled toward the stairs. Behind me, I heard the bed groan as the ogre shot up and let out an angry snarl.

"What are you doing in my house?" he growled.

I didn't look back. I rushed down the stairs, the blade bouncing heavily at my side, my arms already aching from carrying it. Heavy footsteps landed one after the other behind me. The door was wide-open now, blown askew by the wind, and I struggled to move faster.

A hand landed on my shoulder, yanking me backward. I cried out in terror.

"There you are, thief," he said.

I shut my eyes tight and tried desperately to pull away.

A great wind surged underneath me then, and when I opened my eyes, I saw every window in the house blow open, the ogre falling backward as the west wind came to my aid. It carried me up into its embrace again, and then, as I clutched the sword closer to me, it lifted me out through the door and up into the sky.

A strangled cry of fury came from the ogre as he raced out into the clearing. When I looked over my shoulder, I saw him standing there, his figure growing steadily smaller, his face turned toward me in shock and rage, the woods and river and land around him turning back into blankets of darkness and silver ribbons. I trembled all over. The sword in my hands glinted in the night, reflecting starlight as the wind carried me up, up, back up to the clouds where Hyacinth waited.

His eyes brightened in delight when he saw me. "How brave you are, my Fräulein!" he said, taking me into his arms and kissing my hands. He marveled at the sword. "Very well done."

I smiled, but the memory of the ogre's dream lingered like a ghost in my mind, keeping me from feeling pleased. "Do you know anything about the ogre?" I asked Hyacinth as he ran a long finger down the sword's blade.

"Oh?" Hyacinth said idly.

I told him about the ogre's dream, the way he stirred and startled and called out in fear. "Who was he searching for, that he wanted to find so badly?" I said.

Hyacinth's glowing eyes found mine, and for a moment, he straightened, touching my chin once. "The ogre hungers for flesh," he explained in his wild, gentle voice. "He hunted the kingdom's villagers, when they still lived here. All feared him. The ogre was surely dreaming of his hunt, and how to devour it." He shook his head. "It's a terrible thing to talk about. Let's keep it between us, Fräulein."

I thought about the way the ogre had tossed me his sword when I seemed in distress. It was not the response of a hunter to his prey. Still, I nodded and said nothing. Hyacinth was pleased, and I had fulfilled another part of my side of our bargain. Neither the grief of an ogre nor the crown emblazoned on his shield was something I was here to dwell on.

The Boy in Frankfurt

When I slowly stirred out of my sleep the next morning, Woferl was already awake.

I turned against my pillow to see my brother's eyes open, tentatively studying the ceiling. For a moment, I watched him. The sobs of the ogre still seemed to tremble in the air around me. I wondered whether Woferl could hear him, but he said nothing. In fact, he looked dazed, as if he had spent his night tossing and turning.

When he saw me looking at him, he reached out and squeezed my arm with his little hand. "Am I awake?" he said to me in an urgent whisper.

His question made me blink. I pushed his curls away from his forehead. His skin was not hot, but his eyes seemed fever-bright, as if he was still not entirely here. "Yes, Woferl, of course," I reassured him, and put an arm around his shoulders. "Why are you trembling?"

He didn't say. Instead, he scooted closer to me and wrapped his arms around my waist. There he stayed for a quiet moment, slowly coming out of whatever dream must have had him in its throes.

I wanted to ask him if he had dreamed of Hyacinth, and I wondered whether I should tell him about my dream. But he seemed so quiet this morning that I didn't have the heart to frighten him with stories of an ogre. In the air, the sobs from another world still echoed, along with the whispers of a princeling.

Let's keep it between us, Fräulein.

So I let the silence linger until Woferl finally straightened, recovered, and crawled out of bed.

"It is time for me to write," he said as he went. His voice had shifted from one of frightened urgency to determined focus. His fingers were already in motion, as if resting against clavier keys. "I've thought of the perfect introduction for my sonata."

I watched him go. Underneath my pillow, my pendant felt cold and unused. Something stirred in the base of my chest, a strange, ominous rhythm. I could not shake the feeling that there was something in all of this that I didn't quite understand. That there was something Hyacinth was not telling me.

When summer arrived and Salzburg finally shook the cold from her fingertips, Woferl had recovered enough so that Papa could have us resume our tours. This time he had no plans for us to hurry back home after only a couple of months. We would head to Germany, then to France and England and perhaps more, if we were successful. It was a trip that could stretch for years.

When I asked Mama how long we would be gone, she only smiled reassuringly at me and patted my cheek. "You will have an excellent time on these adventures, Nannerl," she said. "Aren't you looking forward to it?"

"I am," I replied. And I was. My bones had grown restless, and my music ached to be heard again.

But the same nagging feeling I'd had the morning after my dream of the ogre still lingered with me. Hyacinth had not appeared to me again since then, weeks ago. I wrote my music and waited for him. Beside me, Woferl composed reams and reams of new work. He would hand them to Papa, would beam as my father beamed. I'd look on, and then hide my music in my drawer.

Woferl did not ask me about Hyacinth. So I began to wonder if the princeling was appearing separately to Woferl. Would he do such a thing? Was Hyacinth only my guardian, or did he have others to whom he made secret promises?

There was also another reason for my uneasiness. My monthly courses had arrived.

The first time it happened was at a Wasserburg inn, and the blood had startled me so much that I wept. Mama tried to console me, helping me change out of my stained petticoat and undergarments, sending a maid out into Wasserburg to buy new clothes. She fussed over me and brushed my hair, helped me bathe, did not comment on the fish I let sit on my dinner plate, and sang to me in bed.

"The pain will pass in several days," she told me. "Don't be afraid. I am delighted for you."

I liked to see my mother happy, so I smiled for her. "I'm not afraid, Mama," I said. Neither of us talked about how I could no longer pretend I was anything but a girl slowly becoming a woman, that it was a reminder of my dwindling years performing before the public.

By the time we left for the small town of Biberich, I'd begun to notice small changes in my body. When Mama helped me dress in the mornings, the lacing of my clothes cut my breath shorter than usual. The inner bone of the bodice pressed harder against my breasts. My cheekbones looked more pronounced, and something about my face made my eyes look larger than I remembered them, dark ponds set in snow. I had also grown taller. Mama had to fix my dresses twice in the course of six months.

My father's past words stayed with me. The older we were, the less magnificent we seemed. The approach of my eighteenth birthday, the end of my years as a child prodigy, suddenly seemed very close.

We traveled through the summer, stopping throughout Germany at Biberich, and then Wiesbaden, and then Kostheim. Our days became a blur of inns. The Three Moors. The Golden Wheel. The Giant. The Red House. Spectators would crowd into the inns' main rooms, jostling one another in order to see us. We performed at palaces whenever we secured an invitation. Newspaper headlines followed us as we went. *The Mozart children will perform tonight,* they'd always say. *Look how young they are. Look at their skill.*

Our travels continued and at some point, I could no longer remember which town we had come from or even which we were currently staying in. At night, I lay awake in bed and tried to imagine what our trip looked like from the clouds in the Kingdom of Back, whether we resembled the tiny villages in the snow or the troops rippling across the battlefield. I wondered what flaws the kingdom's mirrored world showed about us.

I didn't dare compose music during this time. Papa was watching us very closely, staying beside us during our clavier lessons late into the night. So I had to indulge myself by watching Woferl write instead.

He had gotten it into his mind recently to compose a symphony; and while he loved our father, he stubbornly told Papa one evening that he preferred to write his music alone, free from his watchful gaze. Papa had raised an eyebrow at him. But he did not linger near the clavier the next night when Woferl began to write, and instead went downstairs with Mama and Sebastian.

Only I was allowed to watch Woferl as he composed.

"Do you hear the violins in your head, separately from the others, and then the cellos and basses?" I asked him.

He glanced at me, but his attention stayed focused on his music. "Sometimes," he said. "But I also hear them at the same time, as if in four different lines. Each sounds very different." He shook his head. "Remind me to give the horns something worthwhile to do."

I watched him write down another measure. "This is not meant for horns," I said. The piece was light, full of playful footsteps and dancing scales. I had to giggle. "You are cruel. The violins will have a hard time keeping up with you."

Woferl shook his head. He was serious, wholly absorbed in his music. "That's because Hyacinth is running away from them, and they cannot catch him." He reached over and pointed to a measure. "You see? He is sprinting through the forest, up a hill, higher and higher, and then when he reaches the top, he slides all the way down. He likes to lead them deep into the forest, so that they cannot find their way out, and then to reward himself, he naps in one of the trees." His finger guided me across the lines of music, so that I could hear the scenes he explained.

I smiled, but the mention of Hyacinth unsettled me. Woferl hadn't forgotten about him. Again, I wondered whether the princeling had been appearing to my brother in his dreams too. Why else would Woferl be thinking of Hyacinth so much that he was writing him into his music? The envy that came with the thought was like a poison in my mind.

"You are a tease," I said.

Woferl dipped his quill into the inkwell again and scribbled faster, so that large droplets of ink splattered on the page and he had to wipe it away with the ball of his fist. The ink smeared across the page, like a child's painting. "*You* are a tease, Nannerl. You write music, and then you hide it away."

My brother's words hovered in the air, hung there as if the starfishers from the Kingdom of Back had caught them in their hooks. Suddenly, I felt as if we were not truly alone in the room. A slight movement by the window caught my attention, but vanished when I turned to look directly at it. It had seemed like a ghost of a familiar face, a sharp smile and a pair of bright eyes.

"I tease only you," I said to Woferl, nudging him once. "Because only you know it exists."

When Woferl laughed, it was someone else, the sound of wind through reeds.

It was not until we arrived in Frankfurt that I began to understand what my monthly courses and longer dresses truly meant.

Our first performance in the city happened on the Liebfrauenberg.

Woferl and I did not play the entire time at this performance. The local orchestra performed first, for some time, and then a young woman sung an aria. Woferl played, expectedly, more than I did. I accompanied his violin concerto on the clavier, and performed two other pieces with the orchestra. But for the most part I remained quiet on the sidelines next to my father, looking out into the crowd, and this is when I caught sight of a boy.

There were many young children there, restless and tugging at the coattails of their parents, and adults, but there were few in between—and he was in between. My eyes skimmed right past him the first time and returned to Woferl, who played on the clavier with a cloth tied over his eyes.

I looked at the boy the second time because Papa had announced to the crowd that they could test Woferl's talents for themselves. He challenged the people to sound out a note, any note at all, and see Woferl name it correctly on the clavier. The shouts came fast and furious. I watched my brother take them with a smile, sometimes even with a roll of his eyes, which always seemed to get a laugh from the audience.

The boy joined in this game too, and that was why I looked at him. He would call out a note, and my brother would name it correctly without a moment's hesitation. But the boy glanced at me whenever he spoke. I found this curious, even humorous—and did the same, looking in his direction each time I recognized his voice. He wore a faded blue justaucorps, with bright brass buttons that winked in the light, and a simple white wig that came down into a tail behind his neck. He was very pale, like my brother. His brows appeared raised each time I looked at him, as if he were perpetually surprised.

I found myself unable to linger on his face. Every time I did, the flush would rise on my cheeks, and I would glance away.

I lost sight of him after the performance had ended and the audience had started to disperse, some of them gathering near the orchestra to speak to us. Papa greeted each person with a smile and a handshake. They would take my hand and bow or curtsy to me. The largest crowds clustered around Woferl, of course, and he continued to perform for them in his own way, climbing up onto the clavier's bench and singing a tune for them, and then laughing when they cheered and clapped. Each bit of attention he coaxed from the audience made him desperate for more, and his antics grew as his audience demanded them, until he had everyone roaring with applause.

Somehow, his new tricks prickled my nerves. I felt the tightness of my new dress, the ache in my belly. I felt keenly my height against his, his impressively small stature beside mine. While he soaked up the crowd's attentions, my hands stayed folded obediently in my lap. My smile remained demure. *The older, the less magnificent.* My father's words echoed in my mind.

I felt a tap on my shoulder. I looked away from my brother and saw the boy from the crowd standing there, his eyebrows still slightly arched, a smile on his mouth.

"Fräulein Mozart?" he said, as if he could not be sure who I was.

Now that I saw him closely, I noticed that his eyes were a light brown, almost like honey. I curtsied to him. "Nannerl," I said. "I hope you enjoyed our performance."

He bowed to me. "My name is Johann. My family lives here in Frankfurt. I've heard about you and your brother for a long time, and when I knew you would be coming to Frankfurt, I had to see both of you." His smile grew wider. "You were spectacular. I held my breath the whole time."

The air was warm in August, but I had not felt it hot against my face until now. I curtsied again in an attempt to hide my blush. My heart fluttered in my chest like a trapped creature, and I worried for a moment that he could hear it.

"Thank you," I said in a soft voice. "I'm flattered."

"Are you staying long in Frankfurt?"

I shook my head. My eyes darted nervously to where Papa was still busy greeting others, and then back to Johann. "I think we might be here until the end of the month."

"Then I shall try to attend another of your performances." He smoothed down the edges of his jacket with hesitant fingers.

I smiled, embarrassed by my silence. My own hands hung awkwardly at my sides. I finally decided to hold them together against my petticoat, even though they felt exposed there. Everything of me, my face and neck and arms, felt exposed. "Thank you," I said at last. "I would like that."

He grinned. "It was a pleasure, truly, to hear you play."

Before I could respond, Papa saw us. He looked at Johann first, then back at me, and his eyes flashed in the light like fleeting fire. I swallowed. My father said nothing, but his eyes continued to linger as he approached us, and the line of his jaw had tightened.

Johann bowed to my father first. "I have never heard a performance like this, Herr Mozart," he said. "I wish that my parents had come with me. I think they would have enjoyed it."

Papa's expression did not change. "Thank you," he said, his voice clipped and cold.

When he said nothing more, Johann bid us a hurried farewell and returned to what remained of the crowd. His eyes darted at me before he left. I did not dare return his look. Papa's attention was fixed entirely on me now, the others around us forgotten.

"Who was that, Nannerl?" he said to me.

I kept my head low, and my eyes downcast. "I don't know, Papa," I murmured. "He said his name was Johann. He said he and his family live here in Frankfurt."

"I will not have you carrying on a casual conversation with boys like that. Surely you must know better. Do that often, and rumors will spread about you, especially in places like Frankfurt, and especially about a girl as well-known as you. Do you understand?"

"Yes, Papa," I said.

His gaze wandered away into the crowd. I knew he was searching for Johann, to see if he lingered nearby. "Young ladies with no manners," I heard him mutter. "Perhaps I should not take you on these trips, if you are going to learn such poor behavior from the locals."

On the stage, Woferl was still entertaining the crowd, winking at a group of women to earn laughter and coos from them. The audience responded with delight. My father was unbothered. His frown stayed on me.

"Papa," I started to protest. "He only wanted to tell us that he enjoyed our performance. He said nothing else."

My father shot me an angry stare. I shrank away at it. "Do not be naïve, Nannerl," he said. "All men are villains. They want only to benefit. Remember that, and do not speak again to a stranger unless I have given you permission to do so."

My heart was beating very fast now. "Yes," I replied quickly. "Yes, Papa."

"Good." With that, the argument ended. Papa looked away from me and back toward the dispersing people.

All men are villains.

He was afraid, I realized, and I wonder now if it was because he knew his proclamation made him a villain too.

Who Directs the Orchestra?

Papa was pleased with how we performed in Frankfurt. Our purses were full again, our expenses for the trip covered. My father spent the night counting out the coins, nodding and smiling at Mama, and in the morning, he bought her a necklace hung with a sapphire teardrop at its base that shone like starlight. For Woferl he bought a tidy new notebook of paper, so that my brother could continue his relentless composing.

For me, he bought a new cap to match my dress.

He was so pleased that when a local count invited him to the opera, my father paid for us all to come with him.

Woferl and I had never been to an opera before. Papa had always been too worried that we could not sit through a performance without wriggling in our seats. So I tried to keep my composure and remember the lessons Mama had taught me. I needed to behave like a proper young lady. Still, my eyes wandered up to the opera house's grand, arched entrances and white pillars, and down to its veined marble floors and rich velvet carpets. Gold banisters, curved stairways, and ceilings covered in rich paintings. The nobles attended these every week. I wondered if they still gaped in awe each time, and if the sights and sounds could still take their breaths away.

Woferl held my hand in earnest and stared so hard at the gentlemen and ladies we passed that I feared I would need to catch him if he fell. We settled with Papa and Mama in our own balcony. In a private box near us, a group of spectators had already taken out their playing cards and started a game, while down below, young men filed through the aisles to flirt with the ladies. They were all beautiful, I thought, women in wide, sweeping skirts and ruffled half-sleeves, their headdresses adorned with feathers, and gentlemen on their arms with shining jackets and bright, blinking canes.

As they flitted about below, I began to imagine that we were in the kingdom, and that I sat alone on a giant root of an upside-down tree, quietly looking on as the kingdom's creatures—these colorful birds—gathered below me. I imagined them turned in my direction, staring back up at me, and smiling. I glanced at Woferl, who in turn watched the opera stage in anticipation. When I told him about my vision of the plumed birds, we grinned together at the absurdity of it and tried to think of strange names for them.

"Papageno," Woferl declared one of the more ridiculous headdresses, and mouthed the name so comically, *Pa-pa-pa-papageno,* that he dissolved into giggles.

I hushed him even as we laughed conspiratorially. "You will get us kicked out."

"No, I won't," he replied as we rose with the orchestra for the conductor. "I'll be down there one day, and it will be me they clamor for."

"What do you mean?"

"I will be before the orchestra," he said, clapping along. "Someday I'll write an aria, Nannerl—the most difficult aria ever written, and they'll clap for me even louder than this."

I laughed. "You put yourself above Herr Handel. Don't you know the king of England himself once stood in delight for his oratorio?"

A grin of delight crossed his cherubic face. "When I play, the kings of Europe will all stand for the entirety of my opera."

He would be directing the orchestra, I realized, and the premonition in his words appeared before me in all its future splendor, him a young man in a red coat, weaving his music to life. I would be on the ground, staring upward to see my brother at the top of the upside-down tree. I would be a lady with feathers in her wig and no quill in her hand, looking on in silence.

Suddenly, I felt angry at Woferl, though I knew it was not his fault. I thought of his outlandish antics onstage, the way people adored him for it. They would relish him even more when he was grown, fawn over his winking eyes and quicksilver smile.

And I . . . it was impossible for me to do the same. The truth of that burned in my chest, hollowing me out from the inside. No matter how talented I was, no matter how well I performed or how much I charmed—I could never stand where Woferl would.

From a higher balcony, I spied Hyacinth with a hand of cards. He turned to watch us, his blue eyes glowing. I looked up by instinct and met his gaze. At long last, he was here. He stared at me for a moment, reading the weight in my eyes, tapping the cards thoughtfully against his cheek. Then, finally, he smiled.

Woferl waited for me to respond to his declaration about putting on an opera, but I pretended that I could not hear him through the applause.

The Arrow

When Hyacinth visited our bedchamber that night, I was already awake and waiting for him. Somehow, I'd known at the opera that he would come for me tonight. He looked once at my brother, but this time he did not bother to address him. Instead, he let him sleep.

"You've grown taller," Hyacinth said to me.

So had he, I realized, his lithe, boyish shape now transformed into something leaner and stronger, and the forest hue of his skin paled even further, white seeping into his hands and arms like frost curls over dew.

"Why have you come back only now?" I asked him in a hushed voice.

"I needed to know exactly how to help you," he replied, flashing me a quick smile. "I was waiting for a sign from you. I finally got it at the opera."

He had been waiting for my anger to rise? "How will that help me?" I asked. "Or you?"

"It is time for you to complete your third task." He looked over his shoulder, jewels clinking in his hair, toward the moon hanging over the city's rooftops. He beckoned to me. "But we must hurry tonight, Fräulein. You have only a short time to retrieve what I need."

I could feel the threads of his urgency tugging against my heart. My legs swung over the side of my bed, my bare feet crept across the cold floorboards. I followed him out of our inn and into the street, where drunken revelers were still staggering home. None of them seemed to notice me, although one man squinted in confusion as Hyacinth passed him, as if he had seen some kind of shadow rippling against the wall.

As we went, moss began to cover the street's cobblestones in a silver blanket. Ivy trailed out from the cracks between the rocks. In the sky, the twin moons shone bright and round as coins, separated now by only a couple of arm's lengths. Crooked trees arched in between buildings. When I turned to look at them, I noticed that their branches were bare, as if they were roots reaching up to the sky.

They grew thicker and thicker, until soon they crowded out the buildings altogether, leaving us hurrying along a mossy path that wound through a now-familiar forest. Faeries dotted the night air, illuminating our path with their light.

Hyacinth broke into a run. I struggled to keep up with him as he darted along the path, barely visible in the darkness ahead of me.

Finally, we arrived in a meadow blanketed with bright silver flowers, their petals dancing in the breeze. Among them lingered the faeries, and when Hyacinth arrived, it was as if all of them came alive at once, their light surrounding us in excitement and their tiny teeth nipping at my ankles. The blue grass beneath our feet waved and sighed.

"There," Hyacinth told me, pointing at a yawning arc of stone that connected two cliffs. Rock pillars formed a large circle underneath the bridge, a valley heavily overgrown with trees and brush. "Long ago, this was all a cavern. When the oceans lowered, it collapsed, until all that remains is this arc of stone." I followed the line of his finger to the underbrush. "Down there, when the moon is shining directly above the land bridge, you will find a golden crossbow fitted with a single arrow."

I peered at the stone arc, then the thick growth underneath it. Nothing seemed dangerous here. "Why do you not go in yourself?" I asked.

The faeries around me quivered, and a hush spread across the meadow. Hyacinth's playful face looked grave now, even afraid. His eyes lingered on the billowing blue plants that grew within the circle. "It is poisonous ground for me," he replied. "I cannot enter."

I looked at the valley. Then I stepped past the rock pillars and into the circle they formed.

The faeries did not follow me into the circle. It was as if they were fearful of the vegetation here too, as if it were toxic to them as well. I was entirely on my own. The grasses sighed at my approach and whispered at me to turn back. A strange sense of foreboding clung to the ground inside the pillars, and as I went, it dragged against my legs, so that I felt like I was slogging through deep water.

When I looked over my shoulder, Hyacinth was nowhere to be seen.

The grass at the bottom of the valley grew tall enough to come up to my waist, and the blades were rough, chafing against my nightgown. I waded through it, searching for a glint of gold in the shadows. Overhead, the moons shifted slowly closer, half exposed and half hidden behind the arching rock.

I searched in a circle until the waving grasses and rock pillars made me dizzy, then turned my face up to the sky. The twin moons gradually moved into position. As they did, the light in the valley dimmed, and a glowing outline formed as the moons beamed from either side of the land bridge. It formed two arcs of light against the grass, as bright as if the blades were glowing silver, shifting wider and wider until the rock pillars surrounding me were entirely illuminated.

The ground beneath me suddenly shifted. I stumbled, looked down, and there, glowing from within a new crack in the earth, was a golden crossbow with a single arrow notched on it, its tip frighteningly sharp.

I let out a cry of triumph and bent down to pick it up. My hand closed tightly around the crossbow's cool handle. A numbing tingle rushed through my arm. I sucked in my breath at the sensation, but still pulled the crossbow close to me and wrapped both my arms around it.

"I have it, Hyacinth!" I called out, turning to head out of the circle of rock.

As I walked, my arms felt more and more locked around the crossbow, and the weight of the weapon seemed to pull me backward with each forward step I took. A great wind blew through the valley, sending the grass billowing like an open sea. The world swam around me. I shook my head to toss hair out of my eyes. Beyond the pillars, I could see the silhouette of Hyacinth waiting for me, calling my name . . . but the faster I tried to run, the farther away the pillars seemed to get, lost in the waving landscape.

The numbness in my arms began to spread. With it came the whisper of a thousand voices brushing past my ear.

Faeries come, but they cannot leave. They fear the poison of these grasses.

Somewhere through the dullness crowding my mind came the sharp stab of panic. "I am not a faery," I replied, but my tongue felt slow, dragging against the floor of my mouth.

You are the one who poisons the land.

"I am . . ." The words scraped against my lips.

You are not meant to be in the kingdom.

With all my strength, I dragged my thoughts out of me and shouted them into the wind's tide. Words that I suddenly wished I could shout before an audience instead of hiding in my quiet curtsy. "I am a composer named Nannerl!"

All of a sudden, the wind gave way—disappearing as abruptly as it had come. I stumbled forward and fell into the grass. As I pulled myself up, I noticed the grass had gone still again, and before me loomed the circle of rock pillars. The whispers were gone, the air lighter.

I clutched the golden crossbow tightly to my chest, lest it vanish, and ran the final few steps past the pillars. A great gasp burst from my lips as I passed the rocks. I could breathe properly again; my limbs no longer felt crushed under an invisible weight. I turned in the direction of Hyacinth and hurried to him.

He'd grown tall enough that I had to tilt my face up to him. "You have done it, Fräulein!" he said, wonder in his voice. Then he placed his cool hands against my face and kissed me.

I froze, caught like a butterfly in his hands. His lips seemed dusted with sugar, sweet and ice-cold, cleansing away the last of the sacred valley's pull. *This is what it's like to kiss a boy,* I thought through the shiver that washed over me.

Johann flashed unbidden through my mind. His raised brows, his quick smile, the way he'd made my heart dance in my chest. But where heat bloomed on my cheeks for him, Hyacinth's touch brought winter with it, the glitter of fresh snow, the feathers of frost that lined a frozen river's surface.

When he finally pulled away, I swayed in place, unable to speak for a moment. My fingers came up to brush against my lips. They tingled, cold to the touch.

"Why," I whispered at last, "did the valley speak to me?"

His smile wavered. "What did it say?"

I repeated for him what I'd heard. *You are the one who poisons the land. You are not meant to be in the kingdom.*

He shivered at the words, turning his face away from me as if in great pain. The glow of his eyes reflected blue soft against his cheeks. Around him, faeries came to comfort him and caress his face. "This place yearns to keep us out," he murmured, casting a glance toward the arching bridge. "Come, Nannerl, let us leave this behind." And before I could ask him anything more about it, he took my hand and began to lead me back the way we'd come.

The Château

In the morning, Hyacinth was nowhere to be seen.

The light beaming into our room had no quaver of the unusual. But the dream of the kingdom seemed startlingly real today. Perhaps it was the memory of Hyacinth's cool hands against my face, pulling me in toward him. The ice of his kiss lingered, so that when I brought a finger up to run along my lips, my skin still felt cool to the touch.

I lay there for a moment, unmoving, trying to remember all the details. Something in my heart felt strangely light and empty. What would happen now? What would Hyacinth do next?

A sudden impulse gripped me and I looked to where Woferl lay at my side. He slept soundly, his small body curled into a ball underneath the blankets. A soft murmur came from his lips. I watched him, noting the flush of his cheeks. When I reached out to touch his forehead, his skin was burning with heat.

For two weeks, a fever wracked Woferl's body. Every evening, he tossed and turned, his brow beaded with sweat, murmuring deliriously until he'd finally fall into a troubled sleep.

Mama blamed the sickness on the fact that Papa had worked us so relentlessly for the past few weeks. Papa blamed it on the cold and the wet air. I sat at Woferl's bedside and watched him quietly. My thoughts dwelled on how my brother had looked when stricken with scarlet fever, how I'd told him the story of the castle and then imagined the shadows floating around his chamber.

The tasks I'd completed for Hyacinth stayed with me. I thought through each one as I watched my brother grimace in his sleep, dark circles bruising the skin under his eyes. Surely it was all a coincidence, the way Woferl's illnesses seemed to line up with these vivid dreams I had.

But I couldn't shake the feeling that his illnesses were linked to the kingdom and to my tasks there. It felt as if my brother's fate and the princeling's and mine were all tethered together as tightly as a violin string. Woferl's hot hand pressed against mine. I held on to him and stared at his pitiful figure, his eyes dancing under their lids. His lips moved silently. Now and then, they seemed to form Hyacinth's name, as if his essence was hanging somewhere in the air. But I heard nothing.

Was my brother dreaming of the princeling? Was Hyacinth visiting with him secretly?

A spark of envy burned in my heart, followed immediately by guilt.

If I were the one lying sick here, I knew my brother wouldn't hesitate to stay by my side every evening, humming to me little tunes that he'd written, kissing my cheeks, and asking me to grow stronger. He wouldn't sit in silence and allow jealousy to invade his mind. The realization made me tighten my grip on his hand.

Would it change what I did for Hyacinth, if I knew that the link between all our fates were real? I lowered my eyes, ashamed that I didn't know the answer right away. He was so small for his age, his body so vulnerable. I thought of all the times he would curl close to me for protection, and my heart softened in affection. I lowered my face to his and whispered for him to get well.

Night after night, I returned to hold Woferl's hand and watch the shadows dance across his face. I stayed until, slowly, slowly, he began to pull out of the darkness. The fog disappeared from his eyes. He began to look alert again. He would wake up in the morning and ask for parchment and ink.

The arguments between my mother and father stopped. My worries about the kingdom's effect on Woferl's health faded away again. And all of us breathed a collective sigh of relief.

I believe that Papa must have felt some regret for his behavior during this time. He had worked Woferl and me relentlessly for weeks, making us go over and over our pieces, watching us practice late into the nights even when Woferl shivered from the cold. His outburst at me over Johann seemed to guilt him too. While we waited for Woferl to recover, he told me that our audiences felt compelled to speak to me, that I was an alluring talent. Sometimes he fumbled over his words, grew frustrated with himself, and turned his eyes away from me.

I don't know if it had anything to do with me witnessing him at his writing desk weeks ago, or if my task for Hyacinth had pleased the princeling enough to earn me a bit of luck.

Whatever the reason, after Woferl recovered from the fever, Papa decided to give us a day of reprieve shortly after we arrived in France, taking us to visit La Roche-Guyon with no performances planned.

La Roche-Guyon was a small commune in the northern part of the country. The La Rochefoucauld family had invited us to visit their château, and Papa never missed a chance to develop new relations with nobility. He lined me up with my brother on the day we were to meet them and warned us not to mention where we would next visit, that the last leg of our journey would take us to Great Britain.

Woferl found this a great source of mischief. "Do you think Papa will be angry with me, if I do mention it?" he said to me.

I gave him a stern look. "If Papa says not to do it, then don't," I replied. "You'll get nothing out of it."

Woferl tapped his shoes in a rhythm against the carriage floor. "How do you know?"

"I just do." I let the conversation end there, and did not reply when Woferl spoke again. I knew perfectly well why Papa would ask us to do such a thing, and the La Rochefoucauld family would be grateful for it, as the end of the French and Indian War did not leave a sweet taste in their mouths for the British.

We arrived in La Roche-Guyon on a bright, blue morning, up to the top half of a large hill where the road ended at a cobblestone walkway. It was a warm day, not unlike the afternoon when we had performed in Frankfurt, and the sun seared my cheeks as we walked, leaving a slight blush on my skin.

It reminded me of the heat on my face when I'd spoken to the boy named Johann. If he were here, would he comment on the sky, the river, the color of my dress against the sandstone walls? Would he take my hand in his, or push loose strands of my hair behind my ear, the way Hyacinth had done?

I shook my head, embarrassed, and pushed my thoughts away. Lately, I'd caught myself dwelling on the dream of my kiss and wondering what such a sensation might feel like in my world, with Johann. I'd seen my father kiss my mother before, although he didn't put his hands against her face and pull her toward him. She didn't lean toward him with wonder in her eyes.

Would kissing Johann feel like theirs? Polite and distant? Or would it feel like the brush of cold sugar, sweet and wintry and intimate, from Hyacinth? Would it be something different altogether?

Papa glanced back at us once. I immediately lowered my head, afraid that he might have seen my daydreams spelled out plainly on my face. The blush on my cheeks deepened.

Madame Louise-Pauline de Gand de Mérode and her husband were already waiting for us. The young lady greeted Mama with delicate, gloved hands. "It is a pleasure to have your company," she said to my mother. Her face looked pale and sickly, like she had just recovered from several weeks in bed, but I marveled at her voice, calming and full of warmth.

Monsieur Louis-Alexandre, a severe man outfitted with a long face, shook hands with Papa and spoke quietly to him before nodding at both Woferl and me. I curtsied whenever someone took notice. Woferl followed my lead in this, thankfully, but I could see his eyes darting here and there, eager to explore our new surroundings in this foreign country, his mouth twitching with curiosity.

"You will behave yourself, won't you, Woferl?" I whispered to him when our parents began to follow the La Rochefoucaulds up the cobblestone walkway. We walked behind them, far enough to talk amongst ourselves.

"I'll try," he declared. "But I need to tell you something."

"Oh? And what's that?"

Woferl lifted a finger and pointed up toward the château that we now headed toward, the castle that belonged to the La Rochefoucaulds. "We should go to the very top," he said. "I saw someone waiting for us up there."

I followed his finger until my eyes rested on the château too. At first, I didn't think much of it, as I simply did not recognize it. It looked like an old fortress tucked into what was once a cliff, with heavy brick towers and tiny, glassless windows. It sat high up on the hill, so that from where we stood we could see the banks of the Seine River.

I looked back at Woferl. He only stared at me, his expression confused, as if he couldn't understand why I did not see what he saw.

"We should go to the top," he said again when we stepped inside the keep's heavy doors. Ahead of us, Papa and Sebastian were listening intently to the monsieur, while Mama talked to the madame in a low voice. I felt Woferl pull at my hand.

"Do not wander off," I whispered to him. My fingers tightened around his.

But Woferl would not listen. "I want to go to the top of the tower."

I took a deep breath to steady myself, and tried to turn my attention to what our parents were saying. Woferl kept his eyes on the stairs. They spiraled up and disappeared around the edge of the wall, partially illuminated by the child-size windows that opened to the river scene below. I could not guess what made Woferl so restless. He was still a young boy, and perhaps today was simply a day of mischief for him.

Without warning, Woferl slipped his hand out from mine and darted toward the stairs. I sucked in my breath sharply. "Woferl!"

Papa turned to see my brother scampering up the stairs, and before he could utter a sound, Woferl had vanished. He shot me a reproachful look. I curtsied in apology to the monsieur and madame, murmured something I knew they could not hear, and then hurried to the stairs myself. I heard Papa stop Sebastian from following me.

"Let her bring him back," he said. "It is her responsibility. At any rate, she will need to learn how to be a mother soon enough."

The words pricked me like thorns as I gathered my petticoats into my arms and ran. Again, I felt my anger shift in the direction of my brother. If he would only listen and do what he was told, Papa wouldn't feel the need to say such things.

The stairs were high and slanted and old, crumbling in some places, the middle of each step worn down into curves from centuries of travelers. My shoes tapped a rhythm against the stones that began to sound like the beginning of a melody. I called out for Woferl again. Somewhere ahead I could hear his footsteps, but they were very far away now.

"Nannerl!" his small voice called back down to me. He sounded like a muted violin. "Hurry, won't you?"

"Come back down immediately!" I shouted up to him.

"But Hyacinth told me I should come up here! Don't you want to join me?"

I froze. *Hyacinth had told him?* Immediately, I thought back to the mornings when my brother would wake with a dazed look on his face, as if he'd had dreams he couldn't explain. I remembered the way his eyes would dart about under their lids in his feverish sleep.

I looked up at the winding stairs. A faint presence of music hung in the air, reaching out from another world. A tremor shook through me, and suddenly I felt afraid. What had Hyacinth been telling him, that he had not told me?

"Woferl!" I called again, finding new strength in my fear.

The stairs continued on. Now and then, as I passed a window, I would catch a glimpse of the bottom of the hill and the moat and the river, and see patches of sky and sunlight. The scene was very familiar to me, and I began to slow down so that I could better see the view at the next window. My shoes rubbed against something slippery. When I looked down, I noticed that some of the steps were wet now, as if fresh from a rainstorm.

I climbed higher and higher. My breaths began to come in gasps, and yet still I could not hear Woferl answer my calls. My irritation grew. I told myself that I would not sit with him at practice tomorrow, to punish him, and that when he would ask me for help in his compositions, I would refuse. Woferl would not remember what he did to me, though. He would simply pout at me later, and ask why I did not care for him anymore.

I paused by a window to rest, careful not to sit on the wet parts of the stairs. Outside I could hear the wind in the trees and the sounds of the river, but they seemed distant too, as if everything in the world was far away from the stairs that I sat on. I gazed out the window, lost for a moment in my thoughts. A melody floated in the breeze and disappeared before I could fully grasp it.

That was when I first noticed it. The sky had grown a little darker, a scarlet tint to the clouds, and the rush of the river suddenly seemed very loud. The moat looked wider than I remembered. The window grew smaller, and I leaned back, suddenly afraid that it would close around my head.

Through the shrinking opening, I thought I saw a dark figure float around the base of the keep, shrouded in black tatters, and shapeless. My hands started to tremble.

The château no longer looked like a château at all. It had become the castle on the hill.

When I looked out the window again, I could see someone waiting on the other side of the river.

The water appeared dark now, its bottom indistinguishable from the murky depths, and strange shadows glided under its surface, fragments of a massive creature with a long tail.

The figure on the other side of the river was Hyacinth.

Even from here I could tell that he looked different. He had grown even taller, white bleeding into his skin like winter stripping the color from a tree, and his glowing blue eyes fixed on me with such intensity that I drew back from the window to catch my breath. When I shifted, my foot brushed past a cluster of edelweiss growing at the base of one step.

I looked back up the stairs. The flowers had appeared everywhere, surrounded by sprouts of grass. I swallowed hard. "Woferl," I whispered, knowing that he could not hear me.

Something called my name from outside the window. "Fräulein. Fräulein."

It was Hyacinth and his sweet, wild, savage voice. The kiss on my lips turned cold again. I trembled and did not reply, although a part of me yearned hungrily for his presence.

"My darling Fräulein," he said. "It is time. We have done what we needed. Now, you must use the treasures you have fetched for me."

My breaths had turned very rapid, and when I looked back out through the window, I could see his arm extended out in my direction. He was too far away for me to see his features, but I knew he was smiling.

"You never told me you were talking to Woferl," I finally said.

He shook his head. "I do not speak alone to your brother," he replied.

A lie, I thought. I could hear it in the air. "What are you telling him? What do you want with him?"

Hyacinth tilted his head at me. "What's this? Are you questioning me?" He laughed a little and opened his arms. "I am your guardian, as I always have been. Come now. Our next task approaches. I must take the next step in helping you achieve your immortality."

I watched him, wary, unsure of everything. Perhaps Woferl is the one teasing me, lying to me about Hyacinth. "What is the next task?" I decided to ask.

Hyacinth nodded toward the giant creature swimming through the river, its fins black and gleaming. "The river that surrounds my castle has been poisoned by a monster that now patrols its depths. The golden arrow you retrieved for me is the only weapon capable of penetrating its scales."

The crossbow I had taken from within the rock pillars was already in my hands. The events of the night came back to me in a rush.

“How do you know this?” I whispered, clutching the crossbow’s handle until my knuckles turned white.

“Because,” Hyacinth replied, fear in his voice, “it has struck me before.”

And when I looked back at the arrow notched in the weapon I noticed the blood on its tip, black and dried.

Down in the river below, the monster spun and its fins flipped, roiling the water. My brother called for me from somewhere far away.

I hoisted the crossbow, resting it against the window ledge, and pointed it toward the moving creature. My hands wouldn’t stop trembling. I had never so much as crushed a bug in my life, and now my fingers froze, refusing to fulfill Hyacinth’s request.

“If you wait too long, you will miss your chance,” Hyacinth said, his voice traveling on the wind.

The scarlet sky made it difficult for me to see where the monster was swimming. I bit my lip and waited. A strange force was holding me back, the deep part of my thoughts that knew something I did not, and as I waited there, I felt my mind cloud. The sky was too scarlet, or Hyacinth was smiling too broadly. I could not remember what my first wish was that had drawn the princeling to me.

“Wait,” I heard myself say in a small voice. “Give me time. I need to think.”

There was a silence. Then, Hyacinth tilted his head at me. “What do you need to think about? I have our sword, with which we can cut through the thorns on the other side of the river. You have the crossbow, so that the river can become passable.”

What was the night flower’s thorns for? Why did the crossbow and the sword exist? Who had the ogre been? What was Hyacinth telling Woferl? The questions mounted in my mind, one by one, until I could hear nothing but their roar. I thought of the black thorns that wrapped around the kingdom's crumbling castle on the hill, the young queen that never returned.

"Are you still waiting, my Fräulein?" Hyacinth said. His voice was still gentle, still amused, but now I could hear an undercurrent of impatience there. "What is it that has you frozen?"

"I . . ." My throat suddenly felt very dry. I thought of Woferl's fever, every worry I'd had of the link between us and Hyacinth. "Do you know why my brother has been ill?" I asked.

A pause. A sharp, off-pitch note disrupted the music in the air. When Hyacinth spoke again, he sounded offended. "You think I am responsible for harming your brother?"

The accusation in his voice was so piercing that I instantly regretted my question. Of course Hyacinth would never do such a thing. "I'm sorry," I stammered, more confused than ever. "I just don't understand. Woferl has been sick lately, often when I speak of the kingdom or come to you to finish my tasks. Sometimes he looks dazed, or appears to whisper your name in his sleep."

This time, Hyacinth was silent. His face stayed turned up to me. The music that hummed around me turned sharp, unsettling.

"What a shame," came his reply at last. Now his voice sounded cold, even sad. "I am here to help you, to answer your secret wish, and you will not help me? Not after all you've already done? After all I've done for you? And for what—because of these small coincidences? A dream that your brother had? Because you think I am doing something to hurt him?" He let out a laugh. "Surely I have served you well up until now. You do not trust me, Fräulein. And you still, even now, continue to think only of

Woferl. But what of yourself? What of your immortality? What has he ever done for you?"

My hands began to shake violently. "I—" I did not know what to say in return. A great fear had risen in my chest.

Hyacinth seemed to look straight at me. "Are you unhappy with me, Fräulein?" he said.

I stayed silent, the ghost of his kiss freezing my lips closed. *Hyacinth, Hyacinth!* came the whisper from his faeries. I had not noticed their blue light filling the dark corridor behind me, their teeth sharp against my ankles. The memory came back to me of my mother and I in the marketplace, when she'd pointed toward the flowers and I'd brushed my hands against their clustered blooms.

Hyacinths are the harbinger of spring and life, she had said, *but they are also poisonous.*

I trembled at the words.

"I have lost my patience. Come and help me cross this moat," he repeated. The sky looked very dark now, scarlet as the wine I had held in my painting.

I opened my mouth to agree to help him, but nothing came out. The words lodged in my throat, held back by the eternal sense that something seemed very wrong. "Please," I finally whispered. "If you just let me think for a moment. If you will just answer my questions . . ."

My words trailed off. I waited, frightened, for Hyacinth's reply, knowing that I must certainly have angered him. But no reply came. Finally, I inched myself closer to the window and peered out toward the river below. My heart sank.

He was gone.

It took me another moment to realize that the crossbow had

vanished too. My shoulders sagged. My hand came up to rest against my chest, and there, I felt the sharp hollow of his absence.

Why did I hesitate? I had been so faithful up until now, and he had indeed been faithful in return. Hadn't he? Why would he lie about visiting Woferl in his dreams? I bit my lip, regretting what I'd done, loathing myself.

In all this time, Hyacinth had been the one who'd appeared to me in my most troubled moments. Had he just lost his faith in me? The tasks I'd done—all for nothing? He had promised to answer my wish. What would happen now? How could I ever hope to be remembered for anything without him?

A surge of panic hit me. If he had been good to me in his content moments, what could he do now that I had upset him?

I dragged myself back to my feet and continued up the stairs, careful not to fall in the dim light. I had to find my brother. The tower seemed more frightening now that I had trouble seeing my way, and the strange shadows that twisted and molded on the stairs made me quicken my steps. The music in the air sounded wrong. I did not dare look out the window again. I was too afraid to see the dark hooded figures floating near the bottom of the keep, or worse, Hyacinth on the inside bank of the moat, running toward the tower to find me.

"Woferl!" I called out again. At the bottom of the stairs would be our parents, I told myself, and the madame and monsieur. If they were near, then the kingdom would disappear again and leave me in peace. "Woferl!"

What if Hyacinth had taken Woferl? And suddenly the terror of it flooded me, the thought that I might reach the top of the stairs and find no boy at all, no sign that Woferl had ever been there.

Finally, I heard tiny footsteps echoing against the wall, hurrying toward me from farther up the stairs. My heart jerked in relief.

"Nannerl!"

Then I saw him, his hair tussled and his shoes stained with dirt, hurrying down the stairs as fast as his small legs would allow. He kept one hand pressed against the side of the wall. I waited until he reached me, then grabbed his hand tightly and began to head down the stairs.

"How dare you run away in front of Papa and Mama," I said breathlessly. My fear had overwhelmed me, and it emerged as anger. "How dare you pull away from me like that, right before our guests."

"I wanted to see the top of the tower," Woferl repeated, bewildered at my temper. He glanced over his shoulder, back up to where the stairs vanished into darkness. "I heard her, Nannerl. Someone locked away at the top of the castle. Hyacinth said it is the princess! I wanted to show you."

Hyacinth said. The young princess, trapped in the castle's highest tower. I opened my mouth to speak and nothing came out, only the silence that now roared in my hollow chest. So, Woferl did speak alone with the princeling. Hyacinth was whispering to him.

What was Hyacinth doing? Why wasn't he telling me the truth? What was he saying to my brother in his dreams, when I was not there to hear it?

I grabbed Woferl's arm. "Tell me what Hyacinth has been saying to you," I demanded.

He tugged against my grip. "He asked me if I like being in the kingdom."

There was more than that, I could sense it, but my brother either didn't want to tell me or didn't seem aware of it. "Woferl, stop," I insisted. "We're not going up there."

"But Hyacinth wants us to go."

"You shouldn't listen to everything he says." My words were hushed, as if fearful that the princeling would hear them.

At that, Woferl gave me an incredulous look. "But I trust him, Nannerl. Why don't you?"

He was so genuine in his words that it sent a chill down my spine. How frequently had Hyacinth been visiting him? I thought of the spark of envy I'd felt at Woferl's bedside, followed by my deep guilt. Now the two emotions tugged again in my chest, coupled with fear. I swallowed, looking up again at the dark steps rising above us, dreading the sight of a slender silhouette.

"I'm trying to keep you safe," I told my brother.

"Of course it's safe," he argued.

"Let's go," I said firmly.

"But I heard her up there!"

"No, you didn't," I said, when I found my voice again, only the voice I found was a hoarse scrape of my own.

"Yes, I did!" Woferl protested. His hand began to squirm out of mine again, but I grimly held on.

"You didn't see anything, and you didn't hear anything." My voice grew louder, more frightened. Hyacinth's words rang in my mind. "I will not have any more of your mischief today. You insulted me in front of our parents."

Woferl scowled. "I heard her, Nannerl, I promise I did. She was behind a heavy door that I could not open."

I ignored him. The music surrounding us grew louder, more discordant. *I am trying to protect you, Woferl,* I thought frantically, although I still could not be sure what I was protecting him from. All I knew was that I had to get him back down to the safety of the real world. We had to leave this castle on the hill.

"Why won't you believe me?" Woferl started to pull his hand out of mine once more.

This time I yanked on his arm more harshly than I wanted to. He stumbled on the steps and fell, hitting one of his knees hard against the stone. He started to cry.

I stopped and pulled him to his feet, too afraid to console him. "You heard nothing, do you hear me?" I cried out. "You're just a child. The Kingdom of Back isn't real. None of this is! Now stop, before you cause trouble again."

Tears streamed down Woferl's cheeks. "But you said we would always go together into the kingdom!" he said. "You said our stories were for us! Our secrets!"

"They are just stories for children! And perhaps you're still a child who loves his childish secrets, but I am no longer one! Now, you will grow out of this silly phase and forget about all this nonsense—or do you want everyone to think of you as a little boy forever?"

Nonsense. It was my father's word. My brother looked as if I'd slapped him. *You are a child,* I'd told him, *and I am not.* The sky outside had slowly begun to lighten again, losing its red cast, and now I could see my brother's eyes clearly. They were wet, but behind that was anger. I glanced down at his knee. The fall had scraped a patch open on his leggings.

"Keep your own secrets," he said. He rubbed his eyes. "I will never tell you anything, ever again. If you go back to the kingdom, go alone and never return."

With that, he yanked his hand out of mine and hurried down

the stairs without me. I opened my mouth to call out to him, to apologize for my outburst, but it was too late.

Woferl's cruelty descended as swiftly and fiercely as his affection. Later that night, I discovered that the pages in my music notebook had been rummaged through. When I flipped through it to the second to last page, where I'd written my first measures of music, my first secret, I found that the page had been ripped entirely in half.

I ran my finger along the frayed edges, then clutched the notebook to my chest and wept.

A Dream Not Lived

Starting the very next morning, Woferl no longer allowed me to watch him as he composed. He did this by letting Papa become his sole companion by the clavier, and Papa would tell me not to stand idle when I could be helping our mother with something. Woferl did not confide his stories in me at night. When we prepared to sleep, he would simply turn his back to me and pretend not to hear my words. He no longer replied when I mentioned the Kingdom of Back.

Perhaps he had taken my outburst to heart, and did not believe anymore.

I took my compositions and folded them into my heart, writing now in complete solitude. Finding the moments to do so became more difficult without Woferl's help, the way he would quietly leave the ink and quill for me at the clavier. I had to be more careful with the precious few moments when I was alone. I would write a few hurried lines before hiding it all away with my other secret papers, sandwiched between the bottom layers of clothing in my belongings. But when I composed a piece that excited me, I had no one to share it with.

My secrets were mine alone now. And I could blame no one but myself.

I kept expecting to see Hyacinth with each passing day—standing in the corner of our inn, smiling at me from our audience, hiding in the shadows of the streets. Fear crept into the crevices of my sleep. I wondered what he would do now that I had broken my promise. Seek revenge, perhaps. Rob me of my ability to compose, or steal my sight so that I could no longer play the clavier. Perhaps he would take it out on my brother instead. Bleed the pink out of Woferl's cheeks until he faded away one day with the morning light.

Or perhaps Hyacinth had turned his back on me entirely and chosen to fulfill my brother's wishes instead. This thought, that my guardian might have abandoned me in favor of Woferl, haunted me the most.

"You should not be so upset with him, Nannerl," my mother said to me one day. We were on our way to London now, having arrived on British soil just a day earlier.

I froze at her words. "Why?" I asked cautiously, unsure if she was referring to Hyacinth or Woferl.

"He is your brother, my darling, and he loves you very much." Mama took my hand. "Try to be patient with him. He is still very young, and his mischief overwhelms him at times. When you marry and have a son of your own, you will understand."

I thought back to the château, the castle on the hill. After a moment, I said, "I am not upset with him, Mama. He is upset with me."

London did not have much sun or sky when we arrived. An oppressive fog settled over the city, dampening everything, and people on the streets huddled into their coats when they went by, uninterested in us. Only Woferl seemed unbothered by the weather. He would grin his broad grin at those we met, sing for them, and tell them little jokes that would make them laugh. He made sure to time his antics for whenever I was ready to speak. The attention would stay on him, as it always did—except now, even my brother ignored me. I'd sit in silence, feeling like I was slowly disappearing into a world that no one could see.

After a week in England, we settled into a small inn near the edge of Bloomsbury, just shy of central London. It was here that I met the boy Johann again.

I saw him one morning when I was outside the entrance of our inn, waiting to see my father come back after his visit to the king and queen. Woferl did not want to wait with me, of course, so he had disappeared somewhere with Mama and Sebastian. I shivered in the cold air. There was the stale scent of fog, and the aroma of beer and salt and vinegar that wafted out from the taverns.

He passed our inn with the bottom of his face wrapped in a scarf. His shoulders were hunched up from the cold, and his hands were stuffed firmly into the pockets of his coat. I only caught a glimpse of his raised eyebrows, and his warm dark eyes.

"Johann?" I said, before I was even sure of it.

The boy had already passed me by, but he stopped in his tracks and looked around in confusion. I dared not call out his name a second time. Papa would be home soon.

I thought for a moment that Johann would keep going, convinced that my voice had just been a part of his imagination. But before he could turn away, he caught sight of me standing in the doorway of the inn. I felt embarrassed for my silence, and the blush rising on my face. Still, I did not turn away.

Johann hurried over to me. He pulled the bottom of his scarf down a bit so that he could speak, and his breath rose in a cloud. "Is it you, Fräulein Mozart?" he said. His face brightened, and he gave me a quick, awkward bow. "I hadn't expected to see you here."

I could not help but smile at him; it was comforting to hear our familiar language. "Neither had I," I replied. "What are you doing in London?"

Johann blinked to moisten his eyes in the cold, and I noted how frozen his lashes looked, the strands beaded with icy dew. He pointed farther down the street. "My father wants me to attend university next year, to study law. We came to London to see the schools." He raised an eyebrow at me, his smile wry. "I may end up back in Germany, as I can't say any here have stirred him. I liked Oxford, but you should have seen his face. He was shocked by the brashness of the students—loud and unapologetic, always protesting something or other."

I put a hand to my mouth and stifled a surprised smile. "It sounds delightful," I remarked, impressed by the idea of such spirit.

Johann shrugged, still smiling. "What about you?" he said. "Are you here with your family? Come to perform for the London public?"

"My father has gone to see the king," I said. "Woferl and I are to play for him soon, I imagine."

"You will be able to see him, without a doubt." Johann put his hands back into his pockets, too cold to gesture with them. "I've heard the Americans are unhappy with the king's taxes and are giving Parliament an earful. He is desperate for entertainment to lighten his mood."

"Then I suppose we must thank the Americans." It was so easy to laugh with this boy. With Woferl gone from my side, and Hyacinth quiet, I found myself savoring the warmth of this small moment.

He told me about his family, then, and about his father. I learned that we had much in common. He and his sister—who was my age, he told me—were the only surviving children of his parents. His father, passionate about Johann's education, had enlisted an army of private tutors and scholars to teach him literature, art, languages, history. He told me that he loved to paint.

I felt a sudden urge to tell him about the Kingdom of Back—all of it, the beauty that took my breath away, the darkness that haunted my waking dreams. He was a painter, someone who also lived in other lands. Perhaps he would understand.

"How long are you staying in London?" Johann asked me.

"I'm not sure. A month, at the least."

"I will try to see you again," he said. His smile turned shy then, and his gaze was full of warmth. "If I cannot, may I have permission to write to you?"

My father will never let you, I thought. But he had slipped past my defenses, and the crisp London air had made me bold. "Yes," I said. I told him about our flat at Getreidegasse no. 9, and the house outside London where we would stay for the next few weeks.

Johann's face glowed. I wondered what I looked like to him—a foolish girl in front of this older boy, unable to think of more to say. I was not raised as the type of girl to keep secrets from her father, and yet, I had so many of them. But I still found myself smiling back at Johann, thinking only of when I could hear from him again.

Johann tightened his scarf around his face, then uttered a muffled farewell to me before he continued down the street. The wind blew his dark hair into a flurry. I was too afraid to return the goodbye, so the word stayed huddled in my throat instead. Finally, when he disappeared into the crowds, I looked the other way, where Papa would come hurrying back.

There, I saw Woferl standing at the edge of the inn, partially hidden behind the corner.

I froze. He must have seen everything.

Woferl's face was turned to me. I wondered how long he must have stood there, and what he may have heard. He did not smile at me, nor did he look angry. He simply stared.

"Woferl," I called out to him.

He did not answer me. I swallowed hard, suddenly wondering if Hyacinth was beside him and had made himself invisible to me. The thought made me tremble. My brother, when he loved me, would keep any secret of mine close to his heart. But the rift between us still felt heavy in the air, like an off-key note, and there was something wary in his gaze that pulled him away from me, something that made me afraid of what he might do.

Then Papa came bustling down the street, his eyes squinting in the cold wind, and Woferl's stare broke. He turned and ran to Papa, gave him an affectionate smile, and tugged on his pockets to see if he had brought any sweets. I watched them carefully. When Papa nodded at me, I smiled back and asked him about his meeting.

"It went well enough," he told me. "We will perform for the court."

But his face seemed tired, his shoulders hunched. I knew immediately that it meant he did not expect us to be paid much for our private concert, that the king must be tightening his purse strings. My heart dropped at the disappointment in my father's voice. England was costing us more than we could earn.

Despite the tempest of my thoughts, I brought myself to nod in response. "I'm glad, Papa," I replied. My eyes darted down to my brother. I held my breath, waiting for the moment when he would speak.

But the moment did not come. Instead, Woferl sucked on a piece of candy and hummed under his breath a tune from another world.

That night, I dreamed about Johann. He and I sat together under the old ivy wall of an English cottage's garden, right next to the door that led out into the countryside. The moon was unusually bright, perfectly halved, and Johann's face was completely lit by its light. From this close angle, he seemed to be the loveliest boy in all Europe.

"Are you happy, Nannerl?" he asked. "Do you like the path that your life has taken?"

"I don't know," I replied. My eyes darted away from his and came to focus instead on the sapphire silhouettes of trees in the distance. A part of me expected Hyacinth to appear, but he never did. I held my blue pendant in my hands, and my thumbs rubbed idly across its glassy surface. When I lifted my fingers and moved them through the air, everything rippled with light. Music played wherever I tapped. The grasses billowed around us in an undulating sea.

Here, this place, this dream, belonged to me.

I turned to him. "Are *you* happy? Do you dream of traveling to a different place in the world?"

Johann leaned toward me until his lips touched my cheek. "We do the best we can." Then he looked past the ivy wall and pointed toward the stars. "If I see you again, and if you see me," he said, "let's run away and marry on a white shore. Let's go to Greece, to Asia and the Americas, where you can perform for any audience you desire. They will love you so. You never need to hide away your music again. Will you come with me, if you see me again? Will you promise me that?"

And all I could say was yes, my heart aching with desire for this world that was mine. I woke with the word still dancing on my tongue.

My hand was clutching my pendant tightly. For a long moment, I lay awake, letting my fingers run against the glass surface. Then I sighed against my pillow, glanced at where Woferl was breathing evenly in his sleep, and rolled over to hold the pendant up to the moonlight.

Something seemed different about it.

I squinted, frowning now, and held it closer. Then a silent cry escaped from me. I dropped the pendant into my lap.

I wanted to shake Woferl awake, but all I could do was stare down at this charm that I had remembered to be a smooth, transparent blue.

Its surface had cracked into a thousand slivers.

Hyacinth's Revenge

Several days later, Papa became gravely ill. At first, he complained of chills, a weary back, and a sore throat, something he waved away as a passing irritation. The next day, he had doubled over on his bed with his hands clutched over his stomach, and Mama and Sebastian had to send for a doctor. Fever settled over him in a heated cloud.

Woferl and I continued our clavier lessons alone, as quietly as we could. I kept my thoughts to myself and did not dare to share them with my brother. My shattered pendant stayed in the bottom of my dresser.

Woferl never mentioned my moment with Johann. My father never found out.

He blamed his illness on the English weather, the fog, and the rain. Without his making arrangements and setting up meetings, several more of our performances were canceled.

We were forced to dig into the money we'd earned in Germany. This only deepened Papa's frustration, which in turn seemed to worsen his state.

I found myself lingering outside my parents' bedchamber, watching my mother wringing out a towel to place on my father's head. I would stare at his pale, sickly face and silently will him back to health. My brother, still reluctant to talk to me, would quietly ask me how Papa was doing. I never knew what to say. Our practice sessions felt strange without his shadow towering beside us.

After several weeks of little progress and performance cancellations, Mama finally moved us to the English countryside outside of London, to a small Georgian house on Ebury Row, so that Papa could recover in peace. The house was plain but spacious, and when we first arrived there I looked out of the carriage window to admire the pastures and estates.

On our first day, Mama requested our clavier be pushed to a corner and covered with a sheet of cloth. We were not to play while our father stayed ill.

This did not stop Woferl from composing music. I saw him working at night, jotting down measures into the music notebook that Papa had given him after our Frankfurt tour.

One afternoon, I found Woferl hunched over his writing desk overlooking the garden and approached him. He did not speak, but his eyes darted up at me, and I noticed the shift of his little body as he turned himself unconsciously toward me.

"May I see what you've written?" I offered.

Woferl did not look up. His hand continued to scribble a fluid line of notes on the page. "After I've finished," he said at last. "I am nearly done with my symphony."

It was a response. My heart lifted slightly at that. He had not spoken to me like this since the incident at the château. Perhaps Papa's illness has finally softened the grudge between us.

I waited. When Woferl finished his page and turned to a new sheet, I tried again. "Tomorrow I am going to explore around the house, and walk in the garden. Will you come with me?"

Woferl said nothing. I looked over his shoulder this time, so that I could see the measures he wrote out. The symphony was light and fluid, with the same liveliness I remembered from its first pages, which I had seen some time ago. I read my way silently down the page, picturing the harmony in my mind. My eyes settled on the last measure Woferl had written down.

It was a chord, three notes played together with no separations. "That does not belong," I said automatically, without thinking.

Woferl frowned. I saw his eyes jump to the same chord, even though I had not pointed anything out.

"You're right," he replied. "It doesn't quite fit."

His agreement surprised me. I reached over, put my finger down on the paper, and drew three invisible notes. It was the same chord, separated out so that each note came after the other. "This would be better," I said quietly.

Woferl looked at the paper for a long moment. He dipped his quill back into its inkwell, and then crossed out the old chord and replaced it with mine. I watched him carefully as he wrote, expecting to hear an edge in his voice should he choose to speak to me again.

But when he looked at me again, there was a small smile lingering on his lips, his satisfaction at a good measure of music.

"It is better," he echoed.

Gradually, Woferl began to ask for my advice again. When I wrote my own music in secret, he would look on, murmuring in appreciation when he enjoyed a measure. He did not come with me to explore the house, but when I wandered the garden, he would watch me through the window. And sometimes, if he were in a particularly good mood, he would slip his small hand into mine, holding us together until some distraction drew him away again.

Papa recovered slowly in his bedroom, with his windows open to the country air and his bedside drawer constantly adorned with fresh flowers from the garden. His mood was better too, now that we were far away from the chilly London streets. I would hear him laughing with Mama sometimes, or them speaking together in hushed voices on warm afternoons. The sound was as sweet as the summer rain.

Woferl had been in good health too. His cheeks were round and rosy, and his childish giggles rang through the house. As we were still forbidden from touching the clavier, we spent most of our days playing together. I invented musical games to humor him and hid trinkets all over the house that he would then have to find.

One day, Woferl dragged Sebastian into our room and begged him to draw us a map of the kingdom. I listened in surprise. The rift between Woferl and me had been because of the kingdom—and yet, now he was asking for it to be drawn as a map. Sebastian did, and my brother laughed and clapped his hands in delight at the funny little boxes he would draw for us, his crooked castle on the hill and squiggly trees.

I looked on, amused but uneasy at my brother's enthusiasm. The kingdom did not look so powerful or frightening on paper. My brother was well. My father's health was slowly returning. And as I watched Sebastian amuse Woferl, I began to wonder whether, perhaps, the kingdom had truly been nothing more than a faery tale. It was easy to think so here, in this rose-scented house soaked in sunlight. I had not seen Hyacinth since the château. Woferl did not have any more nightmares.

Maybe he had left us entirely. I lay awake at night, trying to make sense of it. It had been so long, I began to hope that perhaps Hyacinth had forgotten about my betrayal and wouldn't seek revenge for the way I'd turned away from him.

Perhaps he was never real at all.

Still, now and then, I'd find myself looking into the shadows of my room and wondering whether I saw a slender figure hiding there. I had completed three tasks for the princeling. He had promised, if I helped him, to grant my wish in return.

Was my relationship with Hyacinth really to end this quietly? Was I destined to fade into the air as my brother moved on without me and my father followed him? Would Woferl turn to me one day and point to some empty corner, whispering to me that Hyacinth had returned to him alone?

By the time Papa recovered enough from his illness to bring us back to the city, winter had set into London and the days were darker and even colder. It was a bitter contrast to our sun-soaked days in the countryside. Our concerts were adequately attended, but a far cry from our earlier stops. After several more months of disappointing performances, Papa decided that he had had enough of England and arranged for us to leave.

"There is no love for God's music here," he complained to my mother on our carriage ride to the pier at Dover.

"Perhaps there is too much, Leopold," my mother replied. "Herr Johann Christian Bach himself is the queen's music master."

At that, Papa nodded in bitter agreement. Herr Bach had helped us win an audience before the royal English court in the first place. But how were we to compete with the London master of music? "Ah, Anna," he said with a sigh. "Too many musicians make their living here. We'll go elsewhere. The envoy from The Hague has approached me again. I have already made arrangements with the Duchess of Montmorency."

Mama's expression did not waver, but I could plainly see the disappointment on her face. "I thought that we would not see the Dutch," she said. "We have been away from Salzburg for so long."

"The Princess-Regent Carolina and her brother are anxious to see us," Papa replied. "They wish the children to perform and have requested a bound volume of Woferl's compositions ready for the prince's eighteenth birthday."

"A volume?" my mother asked. "How many?"

"I thought six sonatas could be ready for publication as soon as we arrive."

Six sonatas. I could tell that this was no idle guess, but the number the Dutch had asked for, and that Papa had already promised.

At Mama's frown, Papa lowered his voice into his affectionate tone. "Anna," he said, "it will go better than London, I assure you."

"Do you not remember what happened in Prussia?"

"Prussia." Papa grimaced and waved a dismissive hand. "This is different. The Dutch will pay us in guilders, not kisses. Think of it." He took my mother's hands. "There will be concerts every night, crowded with patrons, and opera houses and gardens overflowing with people who cannot get their fill of good music. Every nobleperson will be eager to receive us. Princess Carolina is a great admirer of ours and insisted on our presence."

I looked down at my brother to see him listening quietly and biting his lip, his face intent. He knew as well as I did that it was no use arguing once Papa had made up his mind. The Dutch envoy knew that our London tour had soured in the end, and it was this weakness he sought to exploit by tempting my father to make up for those performances. Besides—I could see the light in Woferl's eyes, his brightening at the challenge before him despite his exhaustion.

Still. Six sonatas. Woferl had written two during our stay in the country. He would happily write four more. But in such a short amount of time? We must have dipped farther into our savings than I thought, for Papa to agree to such an impossible deadline. Had our landlord, Herr Hagenauer, sent Papa a letter again, asking for our rent?

"Very well," my mother said, and that was that.

So we prepared and packed. Woferl began writing in earnest. I'd wake to see him asleep with a quill still in his hand, an unfinished page of music crumpled under his arm.

On the day we were to leave, Papa helped the coachman drag our things into the boot and paid the last of our fees to the innkeeper. He was in a good mood this morning, humming a strange tune under his breath that I didn't recognize. I kept my face turned down and concentrated on checking my trunks and tidying my dress, tying my new hat securely with a veil.

I watched my father as we rode. He talked in a low voice to my mother, trying to convince her that the payment the Dutch offered was well worth what they asked.

"That is because what others cannot do, Woferl can," he said, turning to my brother with a rare smile. "It is the miracle they seek, and you are it."

I waited for Papa's glance to fall on me too, to include me in his good mood and the miracle that was our family. But he ignored me and went back to his conversation with Mama. I swallowed and looked out the window.

We rested, spent the night at an inn, and crossed the Channel the following day. When our carriage finally clattered over a bridge overlooking one of The Hague's canals and we looked out to see a towering opera house crowded with people, Papa exclaimed how right we were to have come here, how glad he was for all of us.

On our first night in The Hague, Woferl snuggled close to me in bed.

"What's wrong?" I asked him.

He shook his head and refused to lift his head. "I'm afraid of my nightmares," he whispered. As he said it, something shifted in the dark corners of the room.

When I stirred the next morning, hazy with the fog of unremembered dreams, Papa was already bustling about, tugging on his coat while Mama adjusted his collar. "It is the perfect gift," he was saying to her.

I sat up in bed and watched as my father set a book on the room's desk and then hurry out the door. Mama followed behind him.

My eyes went back to the book. Vaguely, I remembered that Papa was planning to bind Woferl's music for the prince and princess into a volume. I blinked, surprised to see the book already finished. Woferl had been writing nonstop, but I thought I knew how much he had finished and how much more he had yet to go. Had he really already composed enough for the book? The volume seemed a good thickness. Papa must have included some of my brother's older works, in an attempt to fill it.

Out of curiosity, I rose from the bed and went over to the writing desk to peek at the volume before Papa and Mama returned. Behind me, Woferl continued to sleep. With delicate fingers, I ran a hand across the front of the book and then opened its cover.

At first, I didn't understand what I was seeing. It was like a mirror, except in a sheet of black notes. I knew these notes. Every single one.

I flipped the first page, then the next, then the next, faster and faster.

I closed my eyes, dizzy, expecting to wake up out of this dream and be back in my bed. But when I opened my eyes, the volume was still here in my hand. My music was still staring back up at me.

My music. Not Woferl's. *Mine.*

My hands were shaking so hard now that I feared I would tear the fine paper. I let out a gasped sob and took a step back—

stumbling so that my legs gave way—and sat on the floor with my dress spilled in a circle around me. In the corner, Woferl stirred slightly in bed and rubbed his face sleepily. "Nannerl?" he murmured. "What is it?"

I didn't answer. I didn't understand.

How could this possibly have happened? I looked in a daze around the room, then pushed myself up and rushed to my trunk. I rummaged through it frantically. My clothes, shoes, hair ties all went flying, until finally I stared down at an empty bottom.

I steadied myself against the trunk.

The neat little stack of my folded parchments, all the compositions I'd created and carefully stored away over the past months. They were gone.

In bed, Woferl sat up now, more awake and alarmed at the expression on my face. "Are you all right?" he said. "You've turned so pale."

The world spun around me. "Did you tell Papa, Woferl?" I whispered, the words springing unbidden out of me.

"What?" Woferl replied. And when I looked him directly in the eye, he did not blink. He was a picture of confusion, pale from the hurt in my words. His gaze flitted to the mess of my belongings strewn around my trunk.

"Did you tell Papa about my compositions?" I said. My voice trembled.

Understanding suddenly blossomed on my brother's face, followed by horror. "I would never," he said.

I leaned against my empty drawer. My thoughts spun over and over until I swayed. It couldn't be. It *couldn't* be. But I forced myself back onto my feet and stumbled over to look at the volume still open on the table. The pages were there. The notes were there.

And my compositions were gone from my trunk, stolen away by my father.

Or by a princeling.

Hyacinth, Hyacinth, Hyacinth. The name tolled like a bell in my thoughts.

I'd been so foolish to think that he had somehow stepped quietly out of our lives. Here he was again, flitting his fingers through the air. He had always known where to hit me the hardest, had been waiting to use this against me should I ever turn my back on him. I had given up my end of our bargain. In return, he had taken my wish and given it to my brother instead.

This was Hyacinth's revenge. The cruelty he had planned for my punishment.

Woferl called to me again from bed, but I could barely hear him. I paged through each piece in the volume until I reached the end.

Six of *my* sonatas, with minor changes. They had been published in a bound volume, like I'd always dreamed of, but they did not have my name anywhere on them. Instead, they were signed by Woferl.

Wolfgang Amadeus Mozart had stolen my music.

The Agreement

I did not scream or cry. I did not answer Woferl when he continued to ask me if I was all right. I did not change my demeanor around Sebastian or breathe a word of it to my father or mother.

What was the use?

Instead, I turned my fury inward and let it consume me.

Later the same afternoon, I retired to bed early, dizzy and sore. By the next day, I'd developed a fever that made my skin hot to the touch, and started to vomit. My muscles ached so much that I had to bite back my tears. Sebastian carried me to my bed that day. My skin turned white and slick with sweat, my eyes grew swollen and tired. Rose spots appeared on my chest. My hair, drenched with moisture, stuck to my neck and forehead and shoulders in strings. I struggled to breathe, my lungs rasping from the effort.

Mama, in a panic, sent for a doctor that the Dutch envoy

recommended and brought him to our hotel the same evening. He hovered over me in a haze of color, so that I could barely make out his grave face. He told my mother that my heartbeat had slowed, that I might be in serious danger. He bled me, then fed me a bitter tonic and left.

I drifted in and out of sleep. Days melted into one another. I had difficulty understanding what happened around me, except that the date to perform for the princess and prince—to deliver the volume of music to them—came and went. Papa and Woferl attended without me.

Sometimes I thought I saw Papa standing near my bed, talking in hushed tones with my mother. Other times Woferl's face appeared, tragic and fearful, and tried to speak to me. I recalled his soft hands in mine. I thought I heard him say, over and over again, that he was sorry, that he didn't know what to do or say. That he had no idea.

I would turn my face away whenever he was near. I couldn't bear to look at him.

I don't know whether Woferl protested to Papa about what he had done with my music. It was difficult for me to recognize when I was awake and when I was dreaming. But no one in our family spoke about it, at least not to me. I did not even question it. I knew the reason why. To my father, it must have seemed like a simple and obvious decision.

We needed the money, Woferl would be unable to finish the volume in time, and here were a dozen finished pieces of music written by me that could never be published under my own name. Of course my father wouldn't hesitate to sacrifice my work this way.

As the weeks dragged on, my sickness grew worse. I began

to have nightmares several times a day, thrashing in my sleep, and Mama and Sebastian would come in and murmur comforting words to me. My father prayed at the foot of my bed. I saw my mother had a pair of faded wings on her back, and her feet appeared molded to the floor as if she were the faery trapped in the kingdom's underwater grotto. She would linger there and cry. My brother squeezed my hand and asked me questions that I couldn't understand. The floor of my bedroom swayed with a blanket of edelweiss, and strange mosses and mushrooms covered my bedposts. Two moons, not one, would illuminate the floor from my window, their positions growing steadily closer together in the night sky.

Sometimes, I saw Johann sitting at my bedside, his face grave. *Are you happy?* he'd ask me. I would open my mouth and say nothing at all.

My thoughts grew muddied and confused. At times I couldn't remember why I was so angry, exactly what had cut into my chest and pried my ribs open, letting my soul leak away.

One night I saw the dark, shapeless figures float past my window, the hooded ghosts from the castle on the hill with their twisted hands and tattered cloaks. I wanted to make them disappear, and bring more candles into my room like I'd once done for Woferl when he had fallen ill. But no one was in the room with me. So I simply stayed there and watched the shapes with growing fear, helpless until the dawn finally chased them away.

On a particularly bad night, I stirred awake with Hyacinth's name on my lips. I had been calling for him in my sleep. The shadows of my room sighed and breathed. I waited in my delirium, dreading, anticipating his return.

✡

As I continued to deteriorate, news came to my father that my six sonatas had been well received by the Princess Carolina, and that everyone marveled at the miracle of Woferl's ingenuity. The Dutch envoy that had pursued Papa from London to France dined with my family, Mama later told me, and during the lunch thanked my father for making his decision to come on such short notice.

Papa returned, his pockets heavy with coin.

Mama did not speak to me about my misfortune, not directly, but she came the closest, telling me the story of the Dutch envoy with pauses and hesitations. She would not have wanted to add to my pain, if I had not specifically demanded to know.

Later that same night, Papa came to see me in my bedroom. I thought he should have looked happier, for the princess had paid him well for my music. Instead his eyes appeared hollow, his brow furrowed. He came in with a hunched back and settled himself down in the chair next to my bed, and took one of my hands in his. I could barely feel it through my haze of fever, but I remembered how cold his skin was.

"You must be brave, Nannerl," he said. "I know your fever must give you much suffering."

I tried to focus on Papa's face, but my vision blurred and worsened my headache. "Am I dying?" I said. A part of me even hoped, bitterly, that it was true, if only to see whether my father would wince.

Papa continued to hold my hand. "The princess sends her sympathy and well wishes. She told me she will pray for you. Woferl tells me repeatedly that you will get well soon. He tells me he has seen to it." He smiled at the thought, and then shifted, somehow uncomfortable. I wondered idly if the chair hurt his back. After a moment, he spoke again. "I do not like to see you in such a state," he said, more softly this time. "I've grown used to Woferl's bouts with sickness, but I am not used to you . . ."

The part of me that was my father's daughter wanted, in spite of everything, to tell him I would be all right, to not worry. But I only lay there and looked at him, unwilling to give him this relief, wishing I could cause him even more pain.

He looked at me for a long time, studying my face. I wondered if he would say anything to me about what had happened, if he would finally acknowledge it. I waited, watching the room grow hazy and sharp and then hazy again, struggling to focus on my father's expressions.

He prepared to say something, then spoke as if he had changed his mind. "Woferl has said to me many times that he wants to stay by you. Why have you not asked for him?"

I did not speak. What was there to say? My father had taken my music and handed it to my brother, yet *I* was the cruel one who did not ask for him.

"Do not be angry with him, Nannerl," Papa said. His eyes were solemn, but not stern. I thought that he even pitied me a little—or perhaps he meant the pity for Woferl. "He loves you and worries very much about you."

When I still did not speak, my father had the grace to look down, embarrassed. After a while, he finally rose and left, shaking his head and muttering something under his breath that I could not hear.

I started to weep. I wept in earnest, silently and bitterly, unable to hold back my grief any longer. I could not stop. My tears formed rivulets down the sides of my face, wetting my cheeks and my ears, soaking my already damp hair. They spilled onto the pillows, forming dark circles.

He tells you to play, so you play. He tells you to curtsy, so you curtsy. He tells you what you are meant to do and what you are meant not to do, so you do and you do not do. He tells you not to be angry, so you smile, you turn your eyes down, you are quiet and do exactly as he says in the hopes that this is what he wants, and then one night you realize that you have given him so much of yourself that you are nothing but the curtsy and the smile and the quiet. That you are nothing.

Days passed, then weeks. We left The Hague for Lille, even though it took all my strength just to sit up. I could feel myself slipping away. My breathing became raspier, my coughs more frequent, as if I could not lift a terrible stone from inside my chest. I could see the knuckles and bones of my fingers very easily now. Woferl would stand and wait by my bedroom door, looking on with large, tragic eyes. Mama wept several times when she came to sit with me. She held my hand, speaking so much to me that sometimes I did not have the energy to understand it all.

"Be brave, Nannerl," she would say, just like Papa. I did not know until later that she meant for me to be brave in the face of death. My parents had already arranged a date for the priest to read me my last rites.

Finally, two weeks later, when I had truly started to believe that I would die without seeing Hyacinth again, he came to me.

I did not recognize him at first. My bedroom had grown very dim, for the candle had burned low and the darkness had crept up to it. I'd become used to seeing the hooded figures floating outside my window. I saw them now, their shapes creating moving shadows on the wall. In the corner of the room grew patches of mushrooms and vines, red and poisonous.

I blinked sweat out of my eyes. Tonight the shadows had real weight to them, like living things. It took me a long time to realize that one of these shadows was Hyacinth.

He did not look like how I remembered him. His once-pale skin and spikes had bled as white as the color of dead birch in winter, and his blue eyes had turned gold. He was even taller than when I saw him at the château. His figure loomed over me, and when he smiled, his mouth grew so large and frightening that I wanted to close my eyes. He had sharper teeth now too, thousands of needles lined up in a row. I could barely see his pupils anymore—the gold color was so pale that it blended in with the whites of his eyes.

Even though he frightened me, his face remained as smooth and beautiful as it had always looked.

"What a state you are in, Fräulein," he said. His voice sounded different, filled with rasps, although still wild and haunting. "Did you call out to me because you missed me?"

I felt too weak to lift my head. My lungs heaved and I burst into a fit of coughing. When he sauntered over to the edge of my bed, I simply stared at him and concentrated on breathing.

Hyacinth's eyes burned into me. "Tell me, my Fräulein, how have you fared since the last time I saw you?"

"You told me that you were my guardian." My voice came out

hoarse and soft. "And then you lied to me. You have been visiting Woferl in secret. You gave my wish to my brother."

He shook his head sympathetically at me. "My poor darling," he said in a voice laced with honey. One of his hands came up toward me and pressed against my cheek. I jumped at the coldness of it. "Your brother was the one who betrayed you. Can't you see that? He has taken from you what history would have praised you for. He will be remembered, while you will be forgotten. That is why you called out for me, is it not? Look at you, Maria Anna Mozart, here on your deathbed and struggling for your next breath. I have already seen it, you know. Your time has come. If you die tonight, history will know you only as your brother's sister, a girl with a beautiful face and modest achievements. A commoner."

I closed my eyes. I'd thought I was ready to see him, but his words stung me.

"Do you still love your brother, Nannerl?"

"Yes."

Hyacinth gave me a reproachful look. "Do you truly still love him, Fräulein?" he asked again.

"I don't know." I frowned, confused by my answer.

The princeling drew close enough so that I could smell the staleness of his breath, the scent of an underwater cave, and he smiled. His breath was cold as snow against my skin. "You and I are one, Nannerl. I am your friend. Friends help each other, and dislike seeing each other in distress. I can help you become what you want to be, help you heal, or I can let you die tonight, mourned only by your father and mother and brother. But I can only be your guardian if you let me help you. Now, what is it you want?"

I thought again of my younger self on the night I'd first

dreamed of the kingdom. I thought of the wish I had sent out into the world, with all the innocent hope of a girl afraid of being left behind by her father.

I had ached so badly to be remembered.

When I spoke now, it came out as a whisper, as harsh and cold as the winter wind. My wish had not changed. It had only grown thorns.

"I want what is mine," I said. My talent. My work. The right to be remembered. The memory of me to exist.

Hyacinth smiled. "I have the flower, arrow, and sword. I can still hear the echo of your first wish. Your immortality." He narrowed his lovely yellow eyes. "Do you want to finish your end of our bargain?"

I nodded. "Yes," I said, and let the word hang on a hook between us. It was time to finish what I started.

Hyacinth tilted his head at me in approval. "Then do not tell your brother," he answered. "Meet me at midnight in two weeks, here in this room, and we shall help each other, as friends do."

The Princess in the Tower

I BEGAN TO GROW STRONGER THE FOLLOWING WEEK. My fever broke, my vision stopped fading in and out, and the rose spots on my chest lightened until they hardly looked different from my skin. A pink flush returned to my cheeks, and my hair no longer hung about my neck in limp strings. Mama wept for joy the first time she saw me pull myself up against my pillows and drink a light soup.

By the time a whole week had passed, I could sit up comfortably and even take the short walk to my window and look down at the streets of Lille. The doctor praised my good fortune. He told me that God had chosen to show me mercy, that He would not take away a girl so lovely as myself.

I smiled graciously at his words. I knew perfectly well who had healed me, and he did not deal in God's pity.

Only Woferl saw the difference. I still practiced at the clavier with my old discipline, obediently following Papa's instructions and criticisms, and I still chatted with Sebastian and told stories to Woferl in our spare time. But my eyes had changed, as surely and sharply as the love between us, as Hyacinth himself had shifted. When I hugged my brother good night, I did not do it with ease and warmth. When he would touch his fingers to mine, I wouldn't squeeze his hand like before. When I watched him write his music, when I knew that a measure would be better with a set of arpeggios instead of a trill, I said nothing.

Sometimes I wondered if Woferl made mistakes on purpose, simply to test me. It didn't matter. My focus was no longer on him.

Two weeks passed. Finally, it was midnight on the day I'd promised to meet Hyacinth, and I lay wide-awake in my bed. After my recovery, there was no need for me to stay in a room alone, and Woferl had returned to sleeping with me while Papa and Mama reclaimed their bedchamber. Our physical closeness didn't change my demeanor. I remained distant, edging as far to one side of the bed as I could. Woferl followed my cue and stayed on his side.

That night, I listened to my brother's shallow breathing in the darkness. He had grown more than I realized, but he was still a petite child who slept curled in a ball. I remembered him telling me once that he did it to protect his feet from ghouls under the bed, that somehow our blankets acted as a magic barrier against the supernatural. At the time, it made me smile in amusement. Now I pulled my feet closer to me and huddled tighter.

Just when I thought that Hyacinth might not visit me after

all, that he had forgotten our midnight rendezvous, something scraped quietly against our door. A sudden compulsion came over me. I needed to slide out of bed and walk across the room toward the sound.

I rose and swung my legs over the side of the bed, careful not to disturb my brother. The floor felt like ice beneath my bare feet. I trembled, hugging my elbows in a pathetic embrace. A strange, silver light—too eerie to be moonlight—spilled in from the crack beneath the door and washed the floor white.

I reached the door and turned the knob. Earlier in the night, when Papa had stepped through it and pulled the latch closed, the door had moaned and groaned like a living thing. Now it swung open without a sound. I made my way along the hall and then down the steps, counting the slices of light and darkness that I passed through.

Hyacinth was waiting for me there, at the bottom of the stairs.

His mouth split his face open with rows of knifelike white teeth. Muscles bulged on his neck and chest. He beckoned me closer. Suddenly, I wanted to run from him, back upstairs and into my bed, and tell Woferl what had happened. But my brother was no longer my friend.

Hyacinth, sensing my fear, touched the tip of my chin with his hand. "Tell me, Fräulein," he whispered. "When did you last see the Kingdom of Back?"

"At the tower," I whispered. "With Woferl."

He regarded me with a careful look. "Yes," he said. "We had a little falling-out then, if I recall correctly."

I swallowed hard, wondering whether I had angered him again.

But he simply smiled at me. "You may notice that several

things have changed in the kingdom since your last visit. After all, a great deal has happened to you in your world, hasn't it?" He gestured around at the streets of Lille, as if to emphasize his point.

"What shall I do?" I asked him.

"Close your eyes, Fräulein," he replied.

I hesitated, then obeyed.

"Now open them," he continued. "And follow me."

When I did, Lille had vanished. In its place stood a forest I did not recognize, under the light of twin moons that now nearly touched each other. The trees were completely black, as if painted with buckets of ink. Their branches reached down toward the ground in the shape of gnarled hands, and their roots tore up from the earth in agonized arches. They grew in torturous rows, each fighting with the next for the bit of space they had.

Above us, the sky hovered low and scarlet and furious.

"Why have the trees changed?" I said, stammering.

"*You* have changed," Hyacinth replied. He leaned close to me to study my face. "Ah, so you've grown fond of this place. You feared it once, and now you ache for it to return. You always want what you cannot have, Fräulein."

He led me down the crooked forest path, the dirt now black like the trees, the lopsided signpost now unreadable from decay. The cobblestones were cracked and covered with ash. The snow piled along the edges black as soot. The trees closed in. I felt their branches claw at the edges of my nightgown, their roots threatening to snatch me from the ground. I looked behind us. Our hotel was no longer in sight. The trees had completely sealed away where I'd come from.

Finally, Hyacinth halted to gesture toward the horizon.

There, not far in the distance, stood the castle—but not as I remembered it. I'd thought the castle looked old before, crumbling from the absence of its king and its people, with its mysterious windows and wide moat. Now the bricks had turned black, like fire had scorched them, and thorny ivy ate at its walls. Even the moat's water had turned to sable, so that I could no longer see to the bottom. Now and then, an enormous shadow glided by, the river monster's fins cutting viciously through the surface.

Hyacinth turned to face me. He was suddenly holding the sword I'd taken from the ogre and the crossbow I'd retrieved from under the land bridge. "Take this sword," he said, "and strap it to your back. Hold the crossbow in your arms."

I knew what he wanted from me. Down by the dark riverbanks, the water churned as the monster passed.

Hyacinth pointed downstream. "The water is shallower there," he said. "You will be able to see better. Take care not to drift far off course as you swim. If you are pulled into darker waters, the river guardian will sense you struggling in the current and tug you under." He brushed the crossbow with one hand. "Aim true, Fräulein, for you have only one chance."

I nodded silently. The weapon felt heavy in my hands. "And when I reach the other side?"

"Take your sword and cut through the thorns," he said. "They will give way to you, but you must keep moving, lest they close in too quickly behind you and catch your legs. Once caught, you will not be able to escape their grasp."

At last, Hyacinth held out the night flower to me. I stared down at its thorny stem, the plant's center still glowing a midnight blue. "Keep this close to you. Do not give it to anyone."

Give it to anyone? "I hadn't thought there were people left in the castle," I said.

"No. Not people." Hyacinth gave me a grave look. "I will be behind you, but as the castle was my home, they can sense my presence more easily. You must go first. If you see someone on the stairs, do not look at them. If they ask you a question, do not answer. They are not human."

I trembled. "What are they, then?" I asked.

Hyacinth did not answer me. Instead, he looked up at the tallest spire of the castle. A desperate longing crossed his gaze. "Make your way to the highest tower. When you reach the locked door at the top, take the night flower and crush it into powder in your hands. Sprinkle it across the door's lock, and it will melt."

The princess trapped in the tower. Tonight, finally, I could free her and reunite her with her brother. But my hands shook as I looked back toward the dark river. "I cannot do this," I gasped. "I am too afraid."

Hyacinth shifted his golden gaze to me. Perhaps the dying kingdom was killing him too, bleaching his flesh the pale color of death. "I do not have much time, Fräulein," he said quietly to me. There was a growl in his voice now. "And neither do you. Do you remember your secret wish? Do you recall our promises to each other?"

I tightened my grip on the crossbow, felt the night flower's thorns sharp in my pocket. I turned away from Hyacinth, then began to walk toward the shallow part of the moat. There, I dipped one foot into the water. Instantly, I hissed and jerked back. The water was cold as ice. I hesitated, then lowered my legs into it, my waist and chest and arms. The icy water pressed in from all sides, seeking a way into my throat. I fought against the current rushing around my legs and started to kick my way across.

From the corner of my eye, I could see the black fin of the

river monster angle in my direction, drawn to my kicking. I struggled to keep my head above the water. As the cold began to numb my legs, I tried to kick faster against the current so that it would not tug me into the deeper water. Hyacinth paced on the shore, watching me.

Suddenly the current swept my feet from under me. My head dipped underwater. For a moment I hung there, no longer in control, a rag doll in the tide. I panicked. My breath escaped from me in a cloud of bubbles, and I kicked frantically. Whispers swirled around me—voices that sounded like Woferl, like my father and mother. In my struggle, I opened my eyes in an attempt to see. Out of the murky distance came a shadow, and when I jerked my head toward it, I realized it was the river monster gliding its way toward me, its eyes white and its jaws splitting its head open.

I screamed and screamed. Bubbles rushed up before me, obscuring my view. Through my terror came Hyacinth's words, clear and cutting as a blade.

Aim true, Fräulein, for you have only one chance.

The river monster sped up. I hoisted the crossbow in front of my chest. In this frozen instant of time, I could suddenly see myself suspended in the water, the gleaming tip of my weapon pointed straight at the creature's gaping mouth.

I pressed the trigger.

The arrow sliced through the water, straight into the monster's jaws, and disappeared into the blackness of its throat.

The creature roared. It jerked away from its path toward me and thrashed, kicking up dirt from the riverbed. Everything around me turned into a haze of darkness. I struggled up, aiming blindly for the surface. My chest threatened to burst.

Miraculously, my feet found their way again, and I came up with a terrible gasp. Behind me, the water frothed with the creature's dying throes. Its shrieks were gurgled now, the sound filling with blood. The smell of metal choked the air. I reached the opposite bank and scrambled up the side. Mud and grime sank deep into my fingernails.

I reached the top of the banks and threw myself to the ground in a heap. When I looked back at the river, I could see the trail of dark blood leaking from where I had been. The river monster was nowhere to be seen. I sat for a moment, gulping air, wiping tears from my cheeks. The crossbow lay beside me, useless now.

On the other side, Hyacinth took a step toward the moat. To my shock, the water now parted where he stepped, as if God had touched the water and split it like the Red Sea. The dry riverbed revealed the corpse of the river monster, which Hyacinth now stepped over without a second look.

I turned toward the forest of thorns, pulled the sword from my back, and dragged myself to my feet.

As Hyacinth said, the thorns parted with a hiss at the first brush of the blade against their brambles. I cut my way steadily through until I could barely see anything around me except their sharp points. They caught against my dress, ripping lines through the fabric as I went. Behind me, Hyacinth had crossed the river and was walking through the beginning of the path I'd carved through the thorns.

A rogue branch lunged for my foot, its thorns cutting a bloody line across my ankle. I cried out, swinging the sword down blindly at it. The blade made contact, and the bramble shrank away as if from fire. Other branches reached for me, hungering for skin. I hacked at them even as they closed in.

I missed one branch. It twisted around my ankle, tightening, its thorns digging hard into my flesh.

This is the end, I wanted to sob. I will not be able to escape it.

Suddenly, its grip loosened. I saw Hyacinth behind me, his teeth bared, his jaws having sliced straight through the branch. "Hurry," he growled.

A surge of strength rushed through me. I swung the sword as hard as I could, and the last of the thorns before me parted. I stumbled out of the branches and fell to my knees against solid ground.

When I lifted my head, I was staring at the entrance to the castle, its front gates wide-open. Bundles of dead, dried grasses were tied to every iron bar. It reminded me of the billowing grass in the valley of the arrow, and I shivered at the memory of the wind's whispers.

The sword in my hand had turned dull, its surface slowly eaten away by the poison of the thorns. I dropped it, watching the blade vanish into nothing, leaving only the hilt. Then I struggled to my feet and went on without looking back.

I walked across a barren courtyard where great processions must once have marched through. Dark drapes hung across every castle window. Old flags of a once-great kingdom now hung in tatters from the castle ramparts. When my eyes lingered long enough on their faded embroidery, I could make out the hint of a sun, great golden waves of thread radiating out from a central circle. It was such a familiar symbol. I frowned at it, trying to place where I might have seen it before.

As I went, the hairs on the back of my neck rose. No one walked these grounds, and not a sound came from anywhere except my own feet against the stone, but still, I could sense the gaze on me, coming from some hidden place.

At last, I reached the tower entrance. The stairs that curved upward, the same from the château, were wet, and water pooled in the dip of each step's worn stone. I peeked at the shadows to make sure no one stood there, but I could only see to where the stairs disappeared into the darkness.

I began to climb.

The windows were smaller than ever, their dark drapes billowing, and the little light they let in was not enough for me to see the steps in front of me. I kept my hand pressed against the curve of the wall. Against the stone hung tapestries of the kingdom's royal family. They were worn with age and weathered by water and wind, but I could still make out the face of the king in his youth, smiling and confident, with his young queen at his side. The same sun symbol from the flags shone behind them, and in their arms were cradled two infant children.

I paused on the steps to linger on the likeness of the queen. Her dress was white and gold, trimmed with lace, with a sweeping skirt that pooled like water near her feet. The gown, the curve of her cheekbones, the arch of her neck . . . everything about her looked so familiar.

The stairs seemed to grow taller and narrower as I went, so that sometimes I had to pause and feel for the top of the next step before I could continue. My feet made no sound against the wet stone. Occasionally I heard a tiny splash as I stepped into the puddles formed by the water.

Something glided past one of the windows. I thought I heard the whisper of the wind as it went. Behind me came a sound. I thought it might be Hyacinth, but when I turned to look down, all I saw were the billowing drapes of a lower window, as if something had slithered inside. Again, my skin prickled with the sensation of another presence.

Footsteps, slow and laborious, came from somewhere far below me.

"Hyacinth?" I whispered into the dark. No one answered.

Panic started to rise in my throat again. I continued my climb, as fast as I could without losing my footing against the slippery stone. Behind me, the footsteps followed.

The stairs grew narrower still. I was nearing the top. The night flower pricked me in my pocket as I went. I patted it to reassure myself it was still there, and did not look back.

Then, abruptly, I saw a figure sitting on the top curve of the stairs.

It looked like a person, but I could not be sure. It sat hunched against the wall, veiled in black, and its face stayed hidden inside its hooded cloak. I thought I could hear it humming.

If you see someone on the stairs, do not look at them.

I quickly turned my eyes down to the steps. My heart began to pound. Slowly, I started to make my way up again, pressing myself tightly against the opposite side of the stairs as I neared the figure. Behind me, the footsteps continued from the darkness swallowing the stairs.

The seated figure drew near. I could only see it as a blurry shadow from the corner of my eye. Everything in me screamed to look at it, but I forced myself not to as I quickened my pace. I wondered if I should run past the figure and risk provoking it, or creep slowly by and risk being within its grasp. I steadied myself against the wall. I had no choice. I had to keep going.

I could hear its humming distinctly now. It sang a strange tune, a song that changed from common time to notes that came in thirds, lighthearted notes mixed in with sharp, off-key bridges. The music reminded me of what I'd heard in the château, on the day I'd refused Hyacinth's request.

I edged close to the figure, and then I was directly across from it, and my nightgown brushed silently against the ends of its robe. Goose bumps peppered my skin.

Careful, I told myself through my terror. If I tripped, I might fall down into the waiting grasp of the unseen creature following me. *Hyacinth, help me. Where are you?*

I slowly passed the seated figure. It did not move. The humming grew slightly fainter. The tower's ceiling was close to me now—I was nearly at the top.

Then, the seated figure spoke. Its voice came out as a whisper that wrapped around me.

"Nannerl."

At my name, I instinctively turned. The figure was looking straight at me, one of its bony hands outstretched from its robes. My eyes unwittingly settled on its face.

Under the shadow of its hood, the face had nothing but a mouth filled with teeth. "Will you play something for me?" it whispered.

If they ask you a question, do not answer.

Then it lurched forward, clawing its way up the stairs toward me. At its feet came another, each one stirred to life by the one before it. They were the same creatures that had glided around the tower and outside our home during my illness.

I broke my careful walk and ran. My feet slipped, and I fell hard against the wet stones. I gritted my teeth and scrambled up the stairs on my hands and knees. Behind me came the clatter of bone scraping against stone. The creatures were following in my wake.

Above me, the door to the top of the tower came into view, a heavy, rusted chain hanging on its knob.

My hands clawed at the closed door. One of the creatures on the stairs called out to me again. *Nannerl.* Its words hung, haunted and rasping, in the air. *Won't you give me the flower?*

Through my panic, I remembered Hyacinth's warning about the night flower.

Do not give it to anyone.

I took the night flower out of my pocket and began to crush it in my hands. Its thorns cut at my skin. I bit my lip hard until I could taste blood in my mouth, but I did not stop. The flower crumbled into ash, the petals hard and brittle, and the thorns turned into powder. The creatures crawled closer on the stairs, their voices turning into a cacophony of snarls. All I could see were their teeth.

I took the powder in my hands and rubbed it against the door's chain.

Nothing happened at first. Then I saw the lock start to melt, the rusted metal turning into thick globs of liquid. It pooled at my feet in a bronze puddle. I pushed against the door as hard as I could.

The nearest creature reached out now and grabbed for me. I felt its bones close around my foot. A scream burst from my throat. I kicked out at it, forcing it to loosen its hold.

"Hyacinth!" I cried, and pressed both of my hands against the rotting wood of the door and gave it another mighty heave.

It swung open. I fell into a room with a floor layered in straw.

A worn clavier sat in one corner of the room. The scarlet sky peeked through a tiny window. And in front of me, curled in a ball in the center of the room, stirred a young girl who looked very much like myself, her hair in the same loose, dark waves as mine, her eyes the color of a midnight lake. Even her dress, a simple thing of white and blue, reminded me of the dress I'd worn when I first played for Herr Schachtner, on a day so long ago.

She sat up to look at me in horror.

"You have slain the river guardian," she whispered at me. "You have cut through the thorns my father erected."

The river guardian? But the thorns were not there because of the late king. Were they? I opened my mouth to tell her this, but no words came out.

I turned around at a sound behind me, sure it was the creatures on the stairs. But it was Hyacinth, his white skin still glistening wet from the river, his eyes narrow and pulsing as if freed of an ancient thirst.

The girl's eyes skipped to him. She shrank away. "You helped him across," she whispered at me.

And only then, as she met my stare, did the truth flash through my mind as surely as if she had sent the thought to me.

The familiarity of the sun symbol on the flags and the tapestries of the royal family. I recognized it because it had been emblazoned on the shield in the ogre's house.

The queen's high cheekbones had been the same cheekbones of the faery trapped in the grotto. The queen's white-and-gold dress had been the same white gown clinging tattered against the faery's slender figure, draping down to where her feet were molded into the grotto floor. Even her magic, what Hyacinth had called her terrible power of fire, was a gift from the Sun, who had cherished her.

The Queen of the Night was not a wicked witch, but the queen herself. The ogre in the clearing had not been an ogre at all, but the king's champion, who had failed to find the queen and her son.

And Hyacinth . . . I thought of the river monster that guarded against him, the bundles of dead grasses tied all along the castle's gates. They were the same grasses Hyacinth couldn't touch in the clearing with the arrow, the same that were poisonous to him. The grass was protection for the castle, meant to keep him out.

Hyacinth was never the princeling of the kingdom, the queen's missing son. He was the faery creature that had stolen the boy, the monster that the kingdom had tried to keep out.

I let out a cry. My arms came up to shield the girl. But Hyacinth leapt past me. And as I looked on, he lunged at the princess and devoured her.

Letters from a Midnight Wood

I WOKE WITH A START.

The morning had not yet ripened, and shadows still lingered behind the bedroom door and windowsill. My hands were outstretched before me, reaching blindly out to where I thought Hyacinth stood. My lips were parted in a silent scream and my eyes were still wide at the sight of his bloodstained teeth.

When my dream world at last gave way to the real one, I realized that Hyacinth was nothing more than my bedpost. I looked quickly to where Woferl slept, certain that I had stirred him, but he did not move, and his breathing stayed even.

A deep cold had settled into my bones, and I was shaking so hard that I could barely press my hands together. Something terrible had happened in the Kingdom of Back. Even as I fought to remember it, I felt the horror of the vision fading away, the sharp edges softening. The princess in the tower had my face, formed by my imagination. Had I even seen any of it? Hyacinth was *my* guardian, and surely that meant he could not have betrayed me.

But something seemed different about the haze in my mind this morning, like a hand had reached into my thoughts and stirred them, turning the clear waters murky. Like someone else had curled inside. I closed my eyes and let myself reach for the final moments of my dream. *The queen. The champion. The princess.*

Hyacinth was not the princeling of the Kingdom of Back. He had instead destroyed the kingdom, and I had been the one who'd helped him.

"Are you feeling well, Nannerl?"

I jumped at Woferl's voice. When I looked at him, his eyes were staring, unblinking, back at me. "I did not mean to wake you," I answered.

"I had a strange dream," Woferl said.

A thread of fear coiled through me and tightened. "What happened in it?" I asked.

"I was in a city. It was burning to the ground; the fire nipped at my skin and the smoke blinded my eyes."

"A city? Lille?"

"No." His voice was flat. "A city with no name."

It was Hyacinth's doing, this dream of his. I could feel his presence in the spaces between my brother's words, teeth sinking into the air. I waited for Woferl to speak again. When he just rolled away and closed his eyes, I turned on my side and stared at the strengthening light peeking in from the window.

Hyacinth was not the kingdom's princeling.

It meant that his wish, to reclaim his birthright and his throne, to reunite with his sister, was also a lie. What was his true wish, then? He had bargained with me . . . to what end? I thought of the hunger in his eyes at the top of the tower, all he had done and all he'd had me do.

He had wanted to devour the princess at the top of the castle.

My gaze returned to the fragile, curled form of my brother, his chest rising and falling in a gentle rhythm. A thought began to take shape. Had Hyacinth not once told me that the young prince was never found? That the Queen of the Night never knew what happened to her child? Hyacinth was not the kingdom's princeling, but someone was. And if Hyacinth had wanted all along to devour the princess, perhaps he now hungered for the princeling too.

Perhaps it was the reason Woferl pricked his finger on the night flower. The reason for his illnesses. The reason for his strange dreams, the faraway look in his eyes. Most of all, perhaps it was the reason for Hyacinth's promise to fulfill my wish. The air around me felt too thin now. I shifted, dizzy from the truth.

Woferl was the young princeling of the Kingdom of Back.

And perhaps, perhaps, everything Hyacinth had done was in order to find a way to claim Woferl's soul in the same way he had claimed the princess.

You can be remembered, if he is forgotten. So let me take him away, Fräulein. I heard the words whispered as clearly as if he were standing beside me.

Deep in a corner of my mind, Hyacinth blinked in the dark, stirred, and smiled.

As we left Lille for Amsterdam, then Rotterdam, then the Austrian Netherlands, strange things started to happen.

Snow fell during one of our concerts on a sunny afternoon. News came from London that an unusual plague had broken out in England's countryside. At the same time, we began to hear reports of vicious attacks across France, of man-eating wolf dogs prowling the mountain paths near Périgord.

"Herr von Grimm said the Beast of Gévaudan has a tail as long as I am tall," Woferl said, knees on his chair as we ate a supper of lentil soup and spaetzle. He stretched his arms out. "And twice the rows of teeth of any wolf."

Mama scolded him to sit down properly, while Papa chuckled. "And what makes you believe everything Herr von Grimm has to say?" he asked.

Woferl brightened, hungry to coax more smiles out of our father. "Well, he said I knew more at my age than most kapellmeisters in Europe." He glanced at me. "He said Nannerl had the finest execution on the harpsichord. Isn't that all truth?"

I looked up at my brother's praise. His eyes darted to me for an instant before flickering away. He was curious about my mood lately, my quiet spells and faraway expressions. This was his way of reaching out to me.

I gave him a careful, practiced expression of gratitude. "You are very kind, Woferl," I said to him. "Thank you."

Woferl's joy dampened at my response. He knew it was the kind of polite answer I gave to the nobility we played for, whenever I wanted to leave a good impression. He stared at me, searching for the truth beneath my trained response, but I just looked away from him and back to my plate. Perhaps he thought I was still angry with him because of my music. And perhaps I was. But I could not look at him without remembering what had happened in the kingdom, and what Hyacinth might want with him.

Papa sensed none of this odd tension between us. He laughed genuinely. Few things pleased him more than a reminder of courts impressed by our performances, and Herr von Grimm had indeed said those words when we'd played for the Prince of Conti in Paris during an afternoon tea.

Mama paused to meet our father's eye. "It might not be a bad plan to avoid the mountain paths," she said meaningfully to him.

Papa waved a nonchalant hand as he stirred his soup. "Nothing more than tales exaggerated by panicked witnesses, no doubt. There have been no reports from around Paris."

"Louis XV himself has put a bounty on any wolf corpse brought in to him," Mama said. "If the king fears this beast, then perhaps we should as well."

"Beast." Papa said the word through a twisting mouth, his distaste for the imaginary souring his good mood. "There is no such thing as a beast."

I ate quietly. The conversation swirled around me like the waters of a murky lake, and my family smeared into distortion. None of us had said another word about my music published as a birthday present for the Prince of Orange. It was possible that my father had already forgotten all about it, that he had been paid his coin and promptly tucked my music away in some dusty corner of his mind.

And yet, I could feel the weight of this betrayal hanging over the dinner table like a storm. Everyone knew. Sometimes I waited for my father's punishment to come, for him to finally confront me one day about my compositions and toss them into the fire, like I'd always feared.

I would have preferred that over this silence, this dismissal of what I'd written.

The thought sent such a chill through my bones that I shivered in the warm room, trying to stop my lips from snarling into a grimace.

I knew very well who was killing the people of Périgord and Gévaudan. I'd seen his form in my dreams last night, prowling through tall grasses. It was not a wolf dog, but a faery creature with a splitting grin and yellow eyes, hungry for more flesh now that I had finally helped him get a taste.

What I did not know was what I now wished. A part of me needed to return to the Kingdom of Back, to set right what I had done wrong. The Queen of the Night had tried to warn me, yet I had not believed her. The king's champion had called out for me to come back, and yet I had thought him an ogre and fled. The river guardian had tried to keep me out. And yet, I had helped a monster. I had to fix what I'd done.

But a part of me still yearned for my wish, feared that I had lost it forever. Could I be remembered, without Hyacinth's help? Was I now doomed to be forgotten, if I did not continue along with Hyacinth's demands? *I want what is mine,* I'd told him. I still did.

And a part of myself that frightened me—a whisper in the shadows, a figure waiting in the woods—wanted to see my brother walk into the air. He would turn lighter and lighter until you could barely make out his shape. And when you finally blinked, he would be gone.

Weeks later, we finally returned to Salzburg.

I leaned out of our carriage to admire the Getreidegasse as we passed through it, even though Sebastian and Mama told me to sit properly. The touch of the air, the smells that came with late autumn, the old wrought-iron signs that hung over the storefronts—it was all still there, in exactly the same spots they'd been when we'd first left years ago. For a moment, I forgot all about Hyacinth and my music and let myself indulge in the returning familiarity of this place. My heart hung on a hook, raw with anticipation, as we drew close to the row where our flat would be.

Here was home. Here, also, might be a letter from Johann, written and addressed to me. I tried to conjure up his hopeful face in my mind, the way we'd talked and laughed in my old dream. What might he say in a letter? Was he still traveling through Europe, visiting universities? Did he have plans to come to Austria? It didn't matter to me. All I knew was that, if his letters had arrived, I needed to get to them before my parents did.

Beside me, Woferl sensed my tenseness and turned his face up to study mine. In the light, I saw the first hints of his adolescent cheekbones. How quickly he had turned twelve. How swiftly I had turned sixteen. We did not have many years left together now. I looked nervously away from him and back to the street. The feel of his eyes on me seeped through my back.

Papa hopped out of the carriage when it'd just barely come to a stop. Down by the entrance to our building stood Herr Hagenauer, our landlord, and he beamed as Papa came up to him to close his hands in a hearty shake. There was a hasty conversation about the rent, about giving us more time to pay for the months we'd been gone. I waited until Woferl slid off his seat to follow our father before I reached out to touch my mother's arm.

"Mama, please," I whispered, my gaze darting to where Papa and Herr Hagenauer were chatting loudly. She glanced back at me. "Can you get our mail and see if there is any for me?"

Her brows lifted in surprise. "Just for you, Nannerl?" She knew to whisper it.

I flushed hot and hoped no one else could see it. "Yes, Mama," I murmured.

She frowned. "And from whom?"

"His name is Johann." I swallowed, suddenly unsure whether Mama would keep a secret like this for me. "He attended one of our concerts and said he wanted to write with his best wishes."

My words trailed off under my mother's stern gaze. "A boy," she murmured. "And does your father know who Johann is?"

"He did not like him very much. Please, Mama," I whispered, turning my eyes down. "He is writing from Frankfurt."

I didn't know what she saw in my face to make her take pity on me. Perhaps my expression triggered for her memories of long ago, of an age when she was not yet married. Whatever the reason, she sighed, shook her head, and stepped off the carriage, holding her hand out to take Sebastian's outstretched one.

"I will see to it," she said over her shoulder to me.

And sure enough, by the time we reached our flat with our luggage stacked around the door, and Papa had stepped out in a hurry to the archbishop's court, Mama found me alone in the bedchamber and handed me three brown envelopes, written in a curling script.

I glanced up at her, relieved, but she did not speak. Instead, she squeezed my shoulder once, then left the room and quietly closed the door. Outside, the muffled sounds of Woferl playing the clavier wafted to me. He would be preoccupied for a while yet, so much had he missed his instrument. My attention shifted back to the envelopes in my hand. I sat with my back to the door, so that

I could stop anyone who might want to come into the room, then fluffed my skirts out around me and pushed a finger underneath the first envelope's flap. The wax seal broke with a single pop.

The handwriting on the letter inside matched the script on the envelope itself—curling and beautiful, the writing of a cultured boy—and I found myself smiling as I read it.

To my Fräulein Mozart,

Do you know, when I returned home to Frankfurt, the very first thing I did was sketch what I remembered of you? I am sketching a great deal. I'm afraid my art is not as miraculous as yours, but I am doing it all the same, drawing just as you may be composing.

My father has decided to send me to law school here in Frankfurt. I'd wanted to find a university farther away, but staying in Germany will not be so bad, and I can hope to receive letters from you more frequently.

I think of you often. Sometimes I imagine I will catch you standing outside our local bakery shop, or out in the square, just like I'd seen you that day in London. But then I suppose I am just a simple young man, with optimistic thoughts. Please tell me if you'll come to perform in Frankfurt again. I will wait for you.

Until we meet again, I will be your hopeful

Johann

I was glad that no one was here to see the blush on my face, but I touched my cheeks and did not feel ashamed of it. I folded the letter and reached for the second. Outside, Woferl finished playing one menuett and began another, one I'd never heard before. Perhaps he was making it up as he went. I didn't dwell on it as I eagerly began to read Johann's next letter.

To my Fräulein Mozart,

You may not know it, but word has reached Frankfurt that you and your family have taken the Dutch completely by surprise, and that they cannot believe their good fortune. I heard of this in passing on the street. Think, Nannerl, that you have as a young woman already earned such popularity as to be mentioned by strangers in passing! I am more astonished and impressed by you than anyone I've ever met.

I am writing a pœm. I have discovered that my writing skills are quite a bit stronger than my painting. I am relieved that you have never seen my art. I should be embarrassed.

If you are ever in Frankfurt, as you know, I will always be at leisure to see you.

Johann

I leaned my head back against the door and closed my eyes. Johann could not know that anything he heard on the streets about me was always solely in reference to my brother. Still, the brightness leaking from his words warmed me. The dream I'd once had of us sitting under a night sky now came back to me, as fully formed as if it had really happened.

I let myself savor it until I heard Woferl finish his second menuett and begin playing a third, a melancholy piece in a minor key. Then I broke the wax seal on the third letter and began to read.

I flung it away in fright. A soft cry escaped from my lips.

My Darling Fräulein,

You have helped me. A bargain is a bargain. Come to me in Vienna, and I shall take you to the ball.

There was no signature, but there did not need to be. This was not a letter from Frankfurt. This came from a forest under moonlight. The wildness in Hyacinth's voice was here in his words, in every jagged, hurried line. Even though I had never seen his handwriting before, I recognized it.

Outside, Woferl's menuett lifted in a crescendo. The notes tumbled after one another, beads on a glass counter.

Somewhere in my mind echoed a laugh, a sound of the night.

I took the letters and put them carefully back in their envelopes. Then I hid them in my trunk, underneath a pile of clothes. Later, I would burn them all.

Hyacinth was calling me back.

The Ghost on the Parchment

That evening, Papa burst home from the archbishop's court in a flurry. His mouth was pulled in a tight line.

Woferl, Mama, and I looked up in surprise from the dining table, but it was Mama who recognized with a single glance everything in our father's expression, for she darted up from her chair to rush to his side. She took his hat before Sebastian could, then touched his shoulder to comfort him before hanging up his coat.

Papa eyed the dishes on the table with a withering gaze. "Fish again?" he muttered. He leaned a hand against the back of his chair and shook his head, over and over, unsatisfied with something. His eyes scanned our home, searching for something he couldn't quite place.

"What is it, Leopold?" my mother finally asked.

"This flat is so cramped," he complained, waving a hand in

annoyance at the foyer. “I hadn’t realized how small it was until we returned. Look, Anna, at how we can barely fit all our luggage in here.”

“There’s plenty of room, Papa,” Woferl said. “I don’t mind it.”

“Of course you don’t,” Papa replied. “You are still a small boy.” It was rare to hear him short with my brother, and I leaned in, curious and intimidated. His eyes jerked to me and held my gaze. “Nannerl, though, is turning into a young lady. And here you two are, still sharing a room.”

Fear slithered cold into the marrow of my bones. I could hear the unspoken words behind it. *The older they are, the less magnificent they seem.* I would not be a miracle child for much longer.

Mama leaned over to him and put her hand on top of his. “We can certainly give Nannerl her own room,” she said, still searching for the root of Papa’s mood.

Woferl looked at our mother in shock. “Why?” he asked.

Mama frowned at his interruption. “Woferl. You are twelve. You cannot continue to stay in the same room as your sister—it’s not proper.”

My brother glanced at me, expecting me to protest. When I didn’t, he tightened his lips. I could see the fear in his face at sleeping by himself, left alone to his nightmares and midnight visitors. Perhaps he was still running from Hyacinth in his dreams. I thought of the beasts that had been prowling the French countryside and imagined the faery boy’s sharp teeth digging into my brother’s flesh.

“Yes.” Papa nodded at my mother’s words. “It’s decided, then. We’ll have to arrange accommodations for Sebastian in the next building. Nannerl can take his old chambers.”

Woferl and I exchanged a glance. What would haunt him at night without me there?

He looked down in silence at his dinner. Before my illness and what cracked us apart, he might have protested loudly. Now he twirled his fork against his dish and broke his fish into pieces. Somewhere deep in my thoughts, a figure watched him curiously with a tilted head.

Mama watched Papa closely, picking up something else under his temper. "There has been some news," she finally said, "that has been unfair to you."

At that, Papa's shoulders sagged. "It is absurd," he answered after a while. "I received no letters, no warning at all."

"What has the archbishop done?"

"They have stopped my salary, Anna, due to my extended absence. Now that I am back, they have lowered my pay another fifty gulden."

Mama stiffened at the revelation. "Fifty gulden," she breathed. "And no reason for it?"

"Only that we have been away," Papa replied, "which he knew about in advance. He will not agree to us leaving again."

"And Herr Hagenauer?"

He rubbed the crease between his brows, as if it might come out if he did it hard enough, that it might solve his problems. "He has agreed to give us another month to catch up on our rent. No more."

A lowered salary. Our unpaid rent. Papa's complaints about my age. Soon my parents would need to start talking about the matter of my dowry, too, another expense to weigh down the family. I could look into my future and see my path laid out clearly before me. My father would approve of a man who I could be matched with. He would ask for my hand in marriage. I would marry, and like my mother, I would bear children to carry on my husband's name, leave my family behind for his, and look on as Woferl headed off into the glittering world of operas and concerts and noblemen eager to commission him for his music.

The thought of my predestined future made me light-headed. I could not imagine life changing beyond what it currently was—could not picture a time when I wouldn't be riding beside my brother in a carriage and playing before a court.

I thought of my father up late at his desk, his sleeves pushed up to his elbows. Would he linger there tonight, long after we'd all gone to bed?

"Perhaps you should have a talk with him," Mama was saying. "The archbishop can be a reasonable man."

"He is still skeptical of Woferl's talents. He says that we have no reason to be running around the courts of Europe with what he deems a—a"—even the mere thought made Papa's cheeks redden in anger—"a traveling circus."

"What does he want?"

"Proof." Papa's face darkened even more. "As if he has not already heard Woferl perform, and Nannerl accompany. As if he has not already witnessed their miracles for himself! And he calls himself a man of God!"

"Leopold," Mama said sharply.

Papa knew he had spoken too much, and his voice hushed immediately, his eyes darting once to the window, as if the archbishop could hear his insult all the way from the court. He sighed and ran a hand through his hair. "I have agreed to his request."

"What kind of request?"

Papa looked to my brother. He seemed almost apologetic. "The archbishop wants to commission an oratorio from our Woferl."

Mama studied her husband's face carefully. "This is good news, isn't it?" she asked, knowing it wasn't.

"He wants it in a week," Papa said.

A week. "It's impossible," I whispered out loud before I could catch myself. Impossible, even for Woferl.

Across the table, Woferl stared intently at Papa, the circles dark under my brother's eyes, his face like a burning wick with the hope of pleasing our father. "I can do it," he said in the silence. "Please. I already have the most wonderful harmony in mind."

Mama didn't answer or disagree. We all knew that there was no use in it. Papa had not even mentioned a payment from the archbishop for the oratorio. It meant that the payment would be Papa being allowed to keep his salary.

As if he'd heard my thought, our father turned now to look at me. In his eyes, I saw the same light as on the day when he'd returned with the finished, bound volume of my music, ready to deliver it to the Princess of Orange.

I didn't know what came over me then. The spirit of Hyacinth stirring in my thoughts, perhaps, or the memory of what I had demanded from him. The recklessness of already having lost and knowing I could not lose more.

Anger that had been waiting in some corner of me, waiting for the right time to emerge.

I tilted my chin high, my eyes on my father's, and held his gaze like a challenge. He raised an eyebrow in surprise, but I did not back down. What could he do to me now, anyway? I'd been through the worst already. My life was charted before me, and there was little I could do to stop it. What difference did it make now for me to push back?

So I did not lower my stare. With it, I said, *You know what you want to ask of me. And if I help him, you will have to acknowledge my music, my true talent.*

You will have to admit what you did.

My father looked away first. Mama tried to console him by putting her hands on his shoulders and whispering something close to his ear. He would have none of it. "I am retiring to our room," he said. Before Mama could respond, he had brushed past her and hurried off in the direction of their bedchamber, his dinner forgotten.

Later that night, Papa called me into the music room and whispered to me by candlelight. "Nannerl," he said, his voice strangely subdued. "Woferl cannot finish such a work in eight days."

"I know," I replied, because it was true, and sat calmly with my hands in my lap, a shawl draped about my shoulders. I could tell my father wanted me to offer my help willingly, suggest that I work with my brother.

Instead, I was silent. My eyes stayed level with his, willing him to ask first. I had summoned the strength to challenge him earlier—I could not back down now.

Papa hesitated, his hands fidgeting restlessly. He was weary, the lines on his face pronounced tonight. He kept searching for the right words to say. I watched his eyes settle again and again on the window. Even though I knew he could not possibly be seeing Hyacinth, I still felt the hairs rise on the back of my neck, could sense the faery's presence in the room.

Finally, I said, "What does this have to do with me, Papa?"

For an instant, my father's eyes softened at me, and with their softness I felt myself lean instinctively closer to him, trying to remember this rare moment.

"You write in a style not unlike your brother's," he replied at last. His words were not stern, but reluctant, as if he was voicing a thought he had kept quiet for a long time.

The silence in the room weighed against us. I stayed frozen, unsure how to respond. His words echoed through me like a bell. This was it, his admission of what I'd done.

I wrote like my brother. I *wrote*. It was an acknowledgment of my volume of sonatas. Papa was telling me, without saying it directly, that he knew that music was mine. The quiver of candlelight trembled against my folded hands, disguising the shudder that coursed through my body.

"How do you know this?" I asked him quietly.

"Nannerl," he replied. His eyes fixed on mine. "You know how."

You know how. I looked around the music room, its shadows stretched and shaking from the candles. Any doubt I might have felt over what had happened now fell away. Here, at last, was his admission that he had indeed taken my music with intention, had put my brother's name on my work and published it.

"Why didn't you tell me?" I said.

"Would it have made any difference?" he mumbled. "Except to make you miserable sooner? Would you have fought me? It is the way it is."

He did not like coming to me like this, vulnerable in admitting to me the truth. I found myself thrilled by his discomfort. For once, I was not the one apologizing to my father, seeking his approval, trying to find a way to appease him. Now it was his turn.

I let him shift, his eyes first meeting mine and then darting away in frustration, trying to settle on anything else but my unspoken accusation.

"Why do you do all of this, Papa?" I asked. "Our lessons. Our tour. Sacrificing your own standing with the archbishop. What is the reason for it? I know it is about the money. But that cannot be the whole of it."

His posture was stiff and hunched, his fingers woven against each other. I waited patiently until he finally found his response.

"Do you know what I thought I would become, Nannerl?" When I shook my head, he said, "A missionary. My parents thought me a future priest, that my calling was a divine one." He was silent a moment. "For a long time, I thought that my entire purpose in life had been to become a missionary, and that I had failed it. Music and composition? I am good at it, but I am no lasting figure." He looked down at the creases in his hands. "Then I heard you and your brother at the clavier. I knew what God had put me on this earth to do. In a way, I have become a missionary. There is no greater purpose for me than to ensure that you are heard by as many as possible."

I studied his bowed head and realized how old he looked. In that moment, I felt sorry for him. I believed my father, but I did not think he understood himself as well as he thought. He wanted me to be heard, but not by name. He wanted me to be seen, but not for what I could create. And he thought himself a missionary, an ambassador of God, when what he really wanted was to validate himself.

The satisfaction I'd felt earlier at his admission and his vulnerability began to fade. I'd gotten what I wanted from him. Now, as I stared at his aging face in the candlelight, all I wanted to do was shake my head. Underneath his harsh exterior was just a pitiful man, mired in insecurity. I sighed. The thought of dragging this on suddenly brought me no joy.

"I'll help him," I said.

My father glanced up at me, surprised.

"I'll help him," I repeated. "It will be hard, but we can do it."

Papa opened his mouth, closed it, and searched my gaze. He did not smile. I waited, wondering if I might catch a glimpse of guilt, some semblance of an apology on his face.

But he had already admitted too much for one night. In the next instance, he leaned back and furrowed his brows. "Of course you will," he said. The authority had returned to his voice just as I had retreated to my meek position, the daughter at his command. "I want you and Woferl to do nothing else in these eight days, to go nowhere, until you have finished the oratorio. I will check on you both twice a day, at morning and at night, and your mother will bring you food. If Woferl tires, you will take his place."

"Yes, Papa."

"I cannot have the archbishop thinking that Woferl does not deserve the reputation he has earned across Europe. You understand this, of course, Nannerl."

I nodded. My father tightened his lips in approval, and then he rose from his chair with a single motion. I rose with him, watching him move out of the candlelight and back into the shadows of his bedchamber. When he disappeared, I turned away.

I had not bothered to ask if my name would appear next to Woferl's on the oratorio's title.

My father locked Woferl and I together in the music room the following morning, with nothing but fresh sheets of paper and quills and ink, and our clavier. We were not to leave until nighttime.

In a way, I was relieved. Hyacinth had always been able to find his way to us, but somehow, trapped in this room, I felt like even he might be unable to unlock a door that my father had secured himself. And even though Woferl and I still hung in an uneasy place, at least here we could speak only with music.

It was the original secret we shared between us, the ability to hear a world that others could not.

Woferl said little about why we were working together, but he seemed relieved too by my presence. He sat beside me at the clavier stand, his body turned unconsciously toward mine as he settled on a tune, the key, and the tempo. He sang part of the harmony to me, so that I would know what he wanted, and then he started to write out the first violin lines. Woferl had become faster at composing, if not for his growing experience then for the stress that Papa placed on him. He wrote in a nonstop fury, until he had completed three lines of continuous music, and then he paused. He looked at me and pointed out the phrases as he hummed.

I watched his thin hands at work. He was very pale this morning, and his dark lashes stood out against the white skin. "We'll start with arpeggios," he said to me. His hands floated across the paper. "Slow, and then grand, with other strings below it." He paused to scribble more notes below the ones he'd already written. "With an undercurrent of notes, to fill out the harmony."

"A flute," I said, after I watched him write out the second violins. "To carry your melody above the strings."

He nodded without looking at me. The music had already taken away his focus on anything else.

I went on, writing the lines for the flute and the horns. As I did, I noticed a shift in our styles, where his flowing melody met the abrupt sounds of my harmony. It was a subtle difference, so small that those unfamiliar with our work might never know.

Woferl would, though. He could distinguish what we wrote even when others could not.

I stared at the parchment. It was as if I were looking at a ghost of myself on the page. *I was here,* the harmony said.

When I started writing again, I did not change my style. I let it stay mine, the flutes and the horns. Every flourish, every trill and arpeggio. It was distinctly different from Woferl's work, but to me, it still matched the piece, made it whole. And perhaps no one would ever recognize my hand in this, no one would clap for me when it was performed—but my brother would see it, know it for what it was. So would my father.

Papa will tell me to fix it, I thought. This piece was not my own to do what I wanted.

But I left it anyway.

Woferl paused from his work on the strings to read over what I had written. I peered at him from the corner of my eye, wondering if he would tell me to change it too, if our father's voice would come out of his throat.

I knew he saw the shift in our styles. But a beat of time passed, and he said nothing.

Finally, he sighed. "Oh, Nannerl," he said.

It was not an exclamation of exhaustion or exasperation. Nor was it some desperate attempt for him to win me back, empty praise in the hopes of an affectionate response, or even some trick from Hyacinth with words laced in cunning. In his voice, I heard a yearning that reminded me of our younger days, when he would sit in the morning sun and lean his head against my shoulder, watching in wonder while I played. It was love for what I'd written. When I looked more closely at him, I could see tears at the corners of his eyes as he read my music over and over, as if playing it repeatedly in his head.

He didn't look at me, so he couldn't see the softness that came briefly over my face, the small smile that touched my lips.

He nodded at the measures, then bent his head again and continued on without a word. I felt the burden on my chest, there for so long since my illness, shift, turn lighter. His dark hair had grown into a longer tail tied at the nape of his neck. His feet still dangled a short distance above the floor, as they had when he was a child. As I stared at him, I felt a certain pity for this little creature, caught by a different limb in the same snare as me.

"Nannerl?" he said after a while.

I paused in my writing to look at him. "Yes?" I said.

He hesitated, then spoke again. "Thank you."

For helping me, was the part of his sentence that remained unspoken. In that moment, I thought he might address what happened with my sonatas. I halted in my work to look at him, my heart quickening, waiting for him to say it. Would he? The seconds dragged on. I realized that I was hoping he would, so that we could bring this ugly scar between us out into the open.

Woferl's shoulders seemed weighed down. He wrote a few more measures in silence before he spoke. "I saw Hyacinth last night, in my dream," he said softly.

I hadn't heard Woferl mention the kingdom in months. Even hearing his name on my brother's lips seemed to chill the air. "What did he want?" I asked.

"He runs after me," Woferl said. He looked pensive now, and

weary. "I cannot escape him. He lingers, now that I am alone."

The cold prickled my skin. Hyacinth was here, in our home. What was he telling my brother?

"If you're afraid," I said to him, "you can come to me. I won't tell anyone."

He nodded once, but his expression looked pale and unsure. There was more to Woferl's story, I could sense it—but he just kept writing, the light feverish in his eyes.

We wrote late into the night, until Woferl collapsed in exhaustion against the clavier stand. I helped clean his hands of ink stains, and then carried him to his room before retiring to my own. There, unable to sleep, hollow from the absence of my brother beside me, I lay awake and let my heart burn from what Hyacinth wanted with my brother.

Woferl was in danger. I could sense it now, the ice hanging in the air, waiting for him. Hyacinth was coming for him—somehow, someway. And I didn't know how to protect him. I turned to my side and stared at where moonlight painted a silver square against the floor. Would he climb through my brother's window in the night, while we slept? When would he do it? How?

The darkness in me, the *someone else* that I'd felt in my chest, stirred now. It painted for me a vision of Woferl whispering to Papa about where my compositions had been hidden. *Have you already forgotten?* the voice reminded me. *Why do you protect him?*

I tossed and turned, haunted by what the faery might do, until I finally heard my door creak quietly open to reveal Woferl stealing into my room. He hesitated by the door, not uttering a word.

How did he still look so small?

I stayed silent for a moment, unwilling to invite my brother inside. But then I pictured Woferl alone in his room, listening for Hyacinth to appear.

I waved him over. "Come here," I said.

He crawled into my bed and snuggled beside me, just like he used to. His small body trembled. I brushed my fingers through his hair and let the voice in me slowly fade. There Woferl stayed, listening to my humming, until he finally drifted into a dreamless sleep.

The Return to Vienna

We finished the archbishop's oratorio in nine days, a day later than he requested. The time passed so quickly, I couldn't remember the separation between one morning and the next. Everything blurred together into an endless string of feverish writing. We spoke little to each other, except to exchange ideas and notes about the composition. Every night, Woferl came to my room and huddled beside me in bed.

By the end of it, I could see the shadows clearly under my brother's eyes. My own cheeks were pale, my eyes even darker against my white skin.

Papa looked at the oratorio once with a hurried eye, made several changes, and then delivered it to the archbishop so that he could receive his payment. The archbishop approved, pleased enough with the work to forgive our brief lateness.

A marvelous feat, he told my father.

He had not believed Woferl could do it at all. For a man as powerful as the archbishop, this was just a game to him. But my father did not complain, because soon after, he received a modest sum for our work and his salary was reinstated. We paid our rent.

Woferl had signed the oratorio with his name. I could not bear to watch while he did it. Instead I stared at my father, until Papa had to turn away from my searing gaze, grumbling over the extra time it took us to finish it.

"Perhaps next time he will give the children more than eight days," Mama said at supper after Papa had told us about the archbishop's payment. "How can a man toy with his subjects so?"

"Perhaps next time the children will write a piece worthy of more," Papa replied.

"It was brilliant," Woferl suddenly said to our father before I could utter something in our defense. The whole of his small body tensed, leaning forward like a stag protecting his herd, and a fierce light appeared in his eyes. "If the archbishop cannot appreciate it, he is wholly incompetent."

Papa sucked in his breath at Woferl's words, but I smiled for the first time that morning.

"You and Woferl did very well," Mama said to me later that morning. We sat together in the music room for a moment's reprieve, for the sun had decided to come out on this late winter's day, and the room felt warm and lazy.

"I know," I said to her before turning away to stare out the window. "But it is never enough, is it? We could work ourselves nearly to death, and Papa would still hand us the quill and ink."

At that, Mama frowned. "Nannerl. Don't speak about your father that way. He loves you, and he loves your brother. He fears for your health and your brother's as much as his own. He just wants to ensure that our family—including you—is provided for."

I looked back at her. "Yes," I replied. "I know the lengths he'd go to in order to provide for us. So does Woferl."

There was a brief silence. "You are still angry with your brother," she said gently.

"No," I replied. "What is the use of such anger?"

Mama sighed. "Woferl is like your father. They are stubborn men, and as the women in their lives, we must learn to voice our opinions without letting them realize it. It is the way of things."

The way of things.

I looked down, unwilling to meet my mother's gaze. I wondered if, decades from now, I would find myself in the same position, comforting my own daughter. Would I repeat this advice to her?

"You are stubborn too, Nannerl, like your father," Mama went on. I could not help looking at her now, and when I did, she leaned forward and touched my cheek with her hand. "I know the little things you do to show your will." She was telling me something without saying it outright, and although I couldn't guess at exactly what she meant, I could sense the feeling of it hanging in the air.

Then she gave me a sad smile. "I know your compositions meant a great deal to you."

I had not prepared myself to hear her speak directly to me

about the bound volume of sonatas. Mama was our silent sentinel, always watching and sometimes disapproving, but she did not question our father's decisions for us. This was the closest she'd ever come to acknowledging my work.

For the first time, I thought about what Mama must have been like at my age. What dreams did my mother have as a young girl? Had she imagined this life with my father, moving always along the sidelines of our lives? When she looked at the night sky, did she ever think of some land far away, where the trees grew upside down and the paths ended along a white-sand shore? When did she become the mother that I now knew?

Suddenly, I feared that I would cry in front of her. I slid out of my chair, then knelt on the rug and put my head in her lap. She brushed my hair with soothing strokes, humming as she went. I savored the sound of her musical voice. Herr Schachtner was right. My mother had a wonderful ear.

We stayed this way for a long while, bathed in the light shining through the music room's windows.

Finally, the door to the music chamber opened and Papa came striding in. My mother and I looked up in unison, jointly shaken out of our quiet moment.

"The archbishop has given us his blessing," he said. "We are going to Vienna."

"Can we not wait until the following year? We've not been in Salzburg for long."

My mother's voice was hushed and hurried, tense tonight as

she spoke with Papa in the dining room after Woferl and I had gone to bed. I stayed near my door and listened, peeking through a crack at the sliver of my parents seated at the table.

"Next year Woferl may be several inches taller," Papa replied in his terse, gruff way.

"He's so small as it is. No one will question that he is a young prodigy, even if he grows a little."

"And what of Nannerl? She's a young woman now."

"Very young, still."

My father sighed. "We can barely afford to keep Sebastian as it is," he said, "and we must take the children before the courts while we can. I've already received an invitation from the empress."

I saw the corner of Mama's mouth twitch. "Is a celebration to happen in Vienna?" she said.

"The empress's daughter Maria Josepha is to be married to King Ferdinand IV of Naples. They will hold a huge feast and have days of celebrations—all of the royal courts and our patrons are to be there. Think of it, Anna!" Papa's eyes lit up. "We'll earn ten years' salary in a week."

My mother's voice lowered so that I could barely hear her. "It is not safe—"

Papa's voice cut into her words. "Archduchesses do not marry daily."

Their conversation ended there. I watched them sit in silence for a moment, their figures flickering in the candlelight. Finally, they rose and headed to their bedchamber. I watched them go until their door closed, then went back to my bed and crept underneath the covers.

When the flat at last became still, I sat awake in the dark and thought. In the adjacent room, I could hear Woferl stirring in his bed. Already, Papa had started arranging our trip, and before long, we would have our belongings packed once again into the carriage, be waving our farewells to Salzburg.

I shivered and pulled my blankets higher until they came up to my chin. It was not a coincidence, our trip to Vienna. I thought of the letter I had burned, the ink staining the paper until it blackened and disappeared against the coals.

Come to me in Vienna, and I shall take you to the ball.

What he would do there, I couldn't guess. How he wanted my brother, I didn't know. There were too many possibilities, and my mind whirled through each until I exhausted myself with fear. The part of myself I understood shrank away at the thoughts. The part of myself lost in the kingdom stirred and smiled.

All I knew for certain was this: we were headed to Vienna, just as Hyacinth had predicted. And he would be waiting for me there.

The Devil's Dance

It had been years since my first performance in Vienna before Emperor Francis I and Empress Maria Theresa. Now I barely recognized the city.

Banners hung from balconies in bright and festive colors, and fireworks lit the night sky. People streamed past our carriage with laughter and cheers. The air smelled of wine and of smoke from the fireworks, of bakeries busy putting out celebratory breads and cakes. Our driver shouted impatiently at the crowds that thronged before our carriage; as we lurched forward in increments, I kept my face turned to the commotion outside. People spilled into and out of the opera houses, dressed in their finest, and still others danced behind tall windows or simply out in the street.

Woferl pointed to the people. "They are like colorful birds," he said, and I thought of the opera we'd attended together so long ago, where I'd seen Hyacinth playing cards from a balcony seat.

My gaze swept the squares, searching for his sharp smile

in the throngs, listening for the off-key notes of the kingdom in between the music that filled the streets. But nothing seemed out of the ordinary yet.

We found lodgings that night on the second floor of a house in the Weihburggasse, at the courtesy of Herr Schmalecker, a goldsmith. He greeted my father with a wide grin when we stepped out of the carriage, then immediately started to help him bring our belongings inside. I stared at the house. It was finer than our own in Salzburg.

"It is splendid to see you, Herr Mozart!" he said to Papa. "What a time to stay in Vienna, don't you think? The city has been like this for several days already."

Papa smiled back at him. "You are most gracious, sir. We will not forget this kindness."

"No need to thank me, the pleasure is all mine. Do you think it such a burden for someone to host the Mozarts?" He laughed heartily, as if amused by his own joke, and my father laughed along with him. I smiled quietly next to my mother, while Woferl watched them move the luggage.

We dined with Herr Schmalecker's family that night. I spent my time moving my slices of baked chicken around with my fork, my thoughts clouded with visions of Hyacinth. Outside we could hear the sounds of merriment continuing late into the night, but in the living room it was quiet, except for Herr Schmalecker's booming voice.

"How long will you stay in Vienna this time, Leopold?" he asked my father. I glanced to Papa. He looked tired, although he kept a civil tongue.

"We'll stay until the marriage, and perhaps several weeks after."

"How splendid!" Herr Schmalecker laughed loudly. "I saw

the princess-bride in public a day ago. She stood on the palace balcony with the majesties. What a lovely one. She"—he paused to wave his fork at my father—"and the youngest one, that little Antonia, will make the best children, I tell you." I looked to Herr Schmalecker's side. His wife, a frail young creature with pale, dusty skin, sat eating her supper without a word to her husband. Two of his children played together with a bit of carrot underneath the table, and a third child slept at the table with her head tucked in her arms.

Papa did not tell Herr Schmalecker to speak of the princesses in more proper terms. If Woferl had said something similar, he would have surely sent him away to bed without his supper. I concentrated on the festive sounds outside and continued to pick apart my food.

The celebrations intensified as the days passed. On one occasion, Woferl and Mama and I accompanied our father to see the opera *Partenope*, and on another we attended a ball to toast our happiness for the princess-bride. I sat in the balcony and spent most of the time distracted, my eyes darting frequently to the seats around us. That slender figure. Those glowing eyes. I searched and searched for him until I was exhausted.

We went out daily, perhaps so that Papa could distract himself from wondering when the court would call for us to perform. Woferl practiced religiously on the clavier and violin when we stayed in our rooms. He continued to compose, this time starting on a new symphony that kept him up late into the night and sometimes early into the morning.

I continued to compose too, but I always waited to begin my work until the house had fallen silent, lest my new work end up again in Papa's hands. The noise from the festivities helped me to conceal my soft movements—my feet on the cold floor, the dipping of a quill into its inkwell, the faint scratching on paper. As I wrote, the composition I'd been developing grew louder, changing from its soft opening into something harsher, as if the noise from outside had agitated it. My hands shook now when I added to it, so that I had to stop at times to rest and steady myself.

The days passed by. Hyacinth did not appear. I slept poorly, always alert for some glimpse of his shadow moving through the house or his figure waiting in the city's alleys.

Then, finally, in the second week of our stay, he came to me.

In the first days of October we attended another opera, *Amore e Psiche*, a romance of sorts between the love god Eros and a mortal beauty. We watched the princess Psyche hunger to see her lover's face, only to be punished for her desire with death.

Papa leaned over and used Psyche's mistake as a chance to warn me. "Do you see, Nannerl?" he said. "This is the danger of desire."

He meant the danger of desire for Psyche, not for the god Eros, who had been the one who wanted her all to himself.

I stayed quiet while the young actress on the stage pressed a hand against her forehead and sank to the stage floor, her dress spilling all around her. My jaw tightened at my father's words. It was not fair, I thought, for a god to tempt a maiden and then condemn her for her temptation.

I do not know if it was my thoughts, my silent disapproval, that conjured him. Perhaps it was the tightness that coiled in my chest at Papa's reaction. As the opera entered its third act, a man in a dark suit stepped into our box. I looked instinctively at him, but my mother and father didn't seem to notice his presence at all, as if he were merely a shadow that stretched from the curtains. Beside me, Woferl shifted, but he did not turn his head.

The man leaned down toward me until his breath, cold as fog, tickled my skin. I did not need to look up at his face to know that I would see Hyacinth's familiar eyes.

"Fräulein," came his whisper. "Come with me." Then he disappeared, his form melting back into the silhouette of the curtains.

I trembled at his presence, at how no one else seemed capable of seeing what I'd seen. Down below, the goddess Venus handed Psyche a lamp, encouraging her to uncover the identity of her lover.

I rose from my seat without a sound. My parents did not stir. As I stepped out of our box and let the curtains fall behind me, I caught a glimpse of Woferl, turned halfway toward me in his seat. If he noticed my absence, he did not say anything.

Beyond the curtain, my slippers sank into the thickness of the rugs carpeting the marble hall. When I looked down, I realized that it was not carpet but moss, deep blue in the dim light, grown so thick that my feet nearly disappeared in it. The hall had become a path, and as I went, I began to recognize the gnarled trees in place of pillars, the deep pools of water their leaves formed.

The trees grew denser as I went, and the sounds of the opera faded behind me, until they sounded less like music and more like the call of crows that glided against the night. Up in the sky, the twin moons had started to overlap each other. Ahead of me, where the trees finally parted, the river that encircled the castle on the hill appeared, its dark waters churning steadily along.

The enormous fins of the river guardian no longer cut through the water. Instead, the wall of thorns that grew beyond the river had now twisted low, arching a gnarled bridge of sharp spikes across the water.

I hesitated at the sight of it, like standing before the gaping jaws of a great beast.

Fräulein.

Hyacinth's whisper beckoned to me on the other side of the thorns. I looked up, seeing where the castle's highest tower still loomed above the brambled wall. Then I moved one foot in front of the other, until my slippers scraped against the thorny floor of the bridge. Through the gaps in the bridge's floor, I could see the dark waters foaming, eager to take me back. Angry with me for stealing their guardian. I walked faster.

I crossed to the other side and in through the thorns, until I finally had stepped out of it and into the clearing before the castle.

Great tables had been laid out along the sides of the castle's courtyard, great golden apples and red pomegranates on porcelain plates. Vines curled around the table legs. There were no candles, reminding me again of Hyacinth's fear of fire. Instead, thousands of lights flickered across the courtyard, the wayward paths of the faeries that always followed Hyacinth, giving the entire space an eerie blue glow. They giggled at my presence. Several flocked near me, cooing and tugging on my hair, their voices tiny and jealous, their nips vicious. I swatted them away, but they would only return, incensed and determined.

"Leave her."

At Hyacinth's voice, the lights immediately scattered, twinkling their protest as they swarmed across the rest of the courtyard. I looked up to see him approaching me.

He smiled at me. Tonight, he glittered with a sheen of silver, wearing thousands of skeleton leaves carefully sewn together into a splendid coat. His hair was pulled away from his face, flattering his high cheekbones. His eyes glowed in the night. He would be beautiful, except I remembered the way he had looked the last time I'd seen him, pupils slitted with hunger, right before he lunged at the princess.

Courage, I told myself, and reminded myself instead of what my father had done with my music.

"How lovely you are tonight," he said, lifting a hand to touch my chin. He took my hand in his and gestured to the courtyard. "Dance with me. I have something to ask you."

I could feel the scrape of his claws against my palm. A vision flashed before my eyes of them covered with the princess's blood. But instead of cringing away, I followed him to the center of the courtyard and rested my hand gently against his shoulder. A sharp tug against my locks made me wince, and I recoiled from the faeries that now darted around my face, all of them eager to bite me.

"Away with you," Hyacinth snapped at them. They scattered again, protesting, flitting about his face and planting affectionate kisses on his cheeks. Then they lingered around us, forming a sullen blue ring as Hyacinth pulled me into a dance.

I followed his lead. The memory of his sugar-sweet kiss came back to me now. I could feel the cold press of his hand against the small of my back. If Johann were here, would he dance with me too? Would his hand be warm against my skin?

"You're quiet tonight," Hyacinth said to me in a low voice.

"Why did you bring me here?" I said to him.

He smiled, amused. "Is the Fräulein angry with me, I wonder?"

"You don't belong here, in this castle."

"I should. The queen had banned me from her court, distasteful woman that she was."

"You killed the princess in the tower."

"In some ways, she was already dead, wasn't she? Are you truly alive if you spend your entire life locked in a tower, hidden away for so long that you wouldn't even know to flee if the door opened for you?"

His words rang deep in my chest, as true and clear as the music of him that had first called to me all those years ago. His irises were gold, hypnotizing me. *Are you alive, Nannerl?* they seemed to say to me. *Don't you want to be?*

My lips tightened. Here we were, playing his games again. But today I was tired of them. "Tell me what you want with me," I said.

Hyacinth smiled and spun me again. The world turned in a dizzy circle, his face at its center. "Don't you remember what we'd agreed to in the very beginning?"

Make them remember me. How long ago that wish of mine seemed. How much had happened since then.

Hyacinth pulled me close, his hand cool against the small of my waist. "A bargain is a bargain. You have helped me, and so I shall help you. There is only one thing left for us to do."

"What is it?"

He drew close to whisper in my ear. His coat of skeleton leaves brushed roughly against me. "Bring your brother here, to the castle."

Woferl. Fingers of ice trailed their way down my spine. "What do you want with him?"

"Leave him here."

Leave him here.

"When you bring him and then return to your world, he will not come with you. Let me keep him here with me. He will bring his music to the kingdom, and you will bring yours to the world beyond. It will be a perfect trade."

Something in Hyacinth's voice had turned very dark, a growl trembling beneath his soft words. "You want to keep him in the kingdom forever?"

Hyacinth's eyes glowed. "Woferl was never meant to stay long in your world, after all," he said. "He is the princeling of this kingdom. You know that. From the moment he was born, you knew the fragility on his face and the paper of his limbs. One illness after another will continue to ravage him, until he is nothing more. That is his curse. He has always been suspended between one world and another. It's time. Give him to me, Nannerl, and you shall finally have what you've always wanted."

The beat of my heart crashed against my ears. This, at last, was the merging of our wishes. Hyacinth wanted my brother, the princeling of the Kingdom of Back—and if I helped hand him over, I would receive what I'd asked for.

Without my brother, I would be the only one my father had. What other name could appear on a volume of music? They would have no choice but to remember me.

I shrank away from him in horror. He waited, humored, as I stood a few feet apart from him, trembling from the suggestion, unable to look away from his golden eyes. The world around me blurred. A lightness pervaded my mind.

"I can sense the pull in your heart, Fräulein," he said, taking a step toward me again.

I pictured my brother sitting in the highest tower, looking down at the dark river. I pictured Hyacinth's eyes trained on him.

"What will you do with him, once he's here?" I whispered. My voice sounded like it came from somewhere outside of me. Hyacinth's eyes pulsed in rhythm with my heartbeat.

"Oh, Nannerl." Hyacinth sighed. He kissed me gently on the cheek. "The question is always about him, isn't it? What will *you* do, once he's here?"

My lips parted and nothing came out.

"Write a composition for me, Fräulein," Hyacinth whispered. "The song of your heart. When you play it, I will call for you. Bring your brother with you then. I'll be waiting for you both in this castle courtyard, underneath the aligned twin moons. Head nowhere else. Bring no light with you. And we shall finish what we started."

I closed my eyes as he spun me again, my head dizzy. His words surrounded me until I could barely think. He would be waiting for us. *Head nowhere else. Bring no light with you.*

And then, suddenly, a flash of clarity cut through the fog clouding my mind. I looked up at the sky to see the moons, half of each overlapped with the other.

On the night that the twin moons aligned in the sky, the trapped queen's magic would be at its strongest. She had told me that, the night we went to her grotto. It would be the time when her magic—her fire, her gift from the Sun, what Hyacinth feared most—was returned to her. It would be my chance to set things right, to release her from her underwater prison and restore her to the castle. Only she could stop Hyacinth, and only then would he leave us in peace.

Even now, the poison of Hyacinth's promise tugged at me, protesting. *Don't you want to be remembered? You have fought so hard to earn this.* I winced in the darkness, willing myself to steady.

The *someone else* in me bared her teeth and yearned to push me fully into Hyacinth's arms.

The part of me that kissed my brother's forehead, that pulled him protectively to my side when he was afraid at night, urged me back.

When I returned to the kingdom, would I do what I needed and free the queen? Or would I bring my brother with me and present him to Hyacinth? The two sides of me stirred, clashing, and in this moment, I could not tell which would win.

"I will do it," I found myself whispering. "I will meet you here."

Hyacinth did not respond. When I opened my eyes again, the courtyard was gone, along with the castle and the thorns and the moat. I was standing in the box again with my parents, Woferl beside me, and I was clapping, along with everyone, as the librettist down on the stage curtsied for her adoring audience.

Already, Hyacinth's hands on me felt like little more than the touch of a ghost. Woferl looked at me, his eyes curious and expectant. My heart hammered against my ribs. Perhaps Hyacinth had come to him too, coaxed him in the same way he coaxed me. I could feel the threads of his web tightening around us.

"Remember this lesson well, Nannerl," my father said, leaning over to me. "Think of all that Psyche suffers, for the sake of her love, and how noble her loyalty makes her."

I nodded but did not answer. Perhaps the fulfillment of her wishes was never worth what she had to sacrifice. Perhaps Psyche could have suffered for something other than love of a man. Perhaps, in another life, things could have been different for her.

The opera was a harbinger of things to come. That night, I dreamed over and over of Hyacinth in the tower, his teeth sinking into the young princess while I stood by. I felt his bloodstained hands touch my cheek. I called after Woferl as he walked ahead of me down the path through the woods, growing more and more distant until I could no longer see him.

And when I woke in a sweat the next morning, I heard the news. The princess-bride Maria Josepha had become stricken with what we all feared the most.

The smallpox.

The Harbinger of Death

"Rumor is that she caught it from the Emperor Joseph's late wife, at her funeral," Herr Schmalecker told us through a mouthful of eggs and ham slices as we sat for breakfast.

I looked at Mama. Her face was pale. At my side, Woferl picked at his food, his expression tired. I'd heard him toss and turn all last night, murmuring in his sleep.

"It is the will of God," Papa said. His mouth was pulled tight, and his head stayed bowed. "We will pray for her recovery."

"Recovery!" Herr Schmalecker chuckled. "Listen to this man. Still thinking about how you will make your ducats here, aren't you? Well, do not lose hope yet, Leopold. The emperor has not retracted his request to hear your children perform."

So we waited. I spent the night awake, shaking. Hyacinth was slowly setting the final act of his game. I knew it, could feel his hands at work, letting the claws of this epidemic creep ever steadily closer to us. In the middle of the night, when I could bear it no longer, I went to Woferl's room to make sure he was still there. He lay asleep, unaware of me as I crawled into his bed and cradled him in my arms until morning came.

Celebrations across the city were disrupted, canceled, shuffled around. We stayed at the house for longer stretches. My father spent much of his time listening to Herr Schmalecker's gossip and pacing the floors. Woferl buried himself in his writing. In the quiet hours after everyone had retired, and in the early morning before the birds had roused the city, I would work on my own composition for Hyacinth. Papa sat with us for endless hours as we played at the clavier. There was little else we could do now.

Days later, an announcement from the royal court confirmed all the rumors we'd heard. Papa wrote a hasty letter to our landlord Herr Hagenauer, to tell him that the smallpox rash had appeared on Maria Josepha and that our concerts would be delayed.

I prayed for the princess that night. She had been among the crowd during our very first performance in Vienna years ago, and as I prayed, I tried to remember what she looked like. Had I taken the time to smile at her? There had been so many archduchesses.

Papa prayed for her too, although he prayed first for Emperor Joseph to not cancel our royal performance in light of the dire circumstances.

Two days later we heard gossip about her gradual recovery, and for an instant Vienna returned to its festive state. She would pull through! A miracle from the heavens! The sounds of music and dance returned to the streets outside my window, and Papa brightened, began talking again about when we'd go to the court.

The happiness lasted until the next week, when the princess-bride took a turn for the worse.

Mama fretted quietly with our father by candlelight, when they thought Woferl and I had gone to bed. "Another archduchess has come down with it," Mama said. "So has the empress herself."

I tried to write that night, but my hands shook so badly that I finally had to stop. I closed my notebook, then wrapped it in a silk petticoat and pushed it far underneath my bed. In the silence, I thought I heard a sound. When I sat up and listened to it, I realized that it was Woferl weeping softly in his sleep, lost again in his dreams.

This was how we hovered for days and weeks, holding our breath along with the rest of Vienna, until the day finally came when the royal court issued a last announcement.

The princess-bride Maria Josepha had died. The empress followed her a day later.

The fanciful operas, plays, and fireworks that had lit Vienna for weeks suddenly came to a halt. Theaters shut their doors until further notice. Streets were stripped of colorful banners. In their place hung mourning notices, and instead of the sound of music, we heard wails in the streets, saw crowds gathered in the city plazas for masses in honor of their late empress and princess. Still others spread news of the smallpox appearing in corner houses and alleyways, rashes blistering the skin of their kin.

It was Hyacinth's whisper, the poison of him seeping through the city, searching for my brother. I could hear it in the air, the sharp pitch of it from the kingdom. My writing grew more urgent. There was not much time now for me to finish my composition for Hyacinth, for me to return to the kingdom before he came to claim Woferl.

"We must leave Vienna," Mama argued that evening with Papa. "There's nothing here for us now. I will not have Nannerl or Woferl catching the smallpox."

"Anna, be reasonable."

"Reasonable? The entire city has been thrown into a panic. What would you have us do? Surely you do not want to stay here."

"Well, we certainly can't leave. The emperor has not retracted his invitation, and we must wait for word from him. He may still wish to hear the children perform."

Mama made an angry noise as she threw up her arms, and I tensed at this rare display. "The emperor has not retracted his invitation because he's likely forgotten all about it. What man wants to hear a concert after the death of his wife and daughter? Meanwhile, we wait here like trapped mice." Her voice grew quieter as she reached for Papa's hand. "Listen to me, Leopold. Emperor Joseph will not begrudge us for leaving in haste. An epidemic will spread quickly in a city so overcrowded with revelers. How will you make our money if our children die? Many of the foreigners have already started to leave. You see their carriages lined up in the streets, more and more of them every day."

"No." Papa's voice was harsh with determination. "We will *stay* here for the time being. We will not go outside, unless we absolutely must. Let me think of a plan."

I sat on my bed in the darkness as their voices rose and fell, my eyes fixed on the bit of candlelight that crept underneath my door and into my room. The air was not cold, but I still trembled. I'd seen before what the smallpox could do to people, turn their skin red and angry, their eyes milky and blind. I thought of Sebastian, who waited for us in Salzburg. Then I thought of Johann and hoped that the epidemic would not spread to Germany.

A commotion in the hall woke me the next morning. I startled, still dazed with sleep, and realized that Mama was shouting at someone outside my door.

I opened it to see Mama opposite Herr Schmalecker, her face red with anger. Papa stood near her.

"Why did you not tell us of this?" Mama said to Herr Schmalecker. "You knew of it, for so long!"

"Calm yourself, Frau Mozart," he said. An embarrassed smile lingered on his face. "Augustine healed before you had even arrived—so I did not think of telling you."

"And what are we to do now?" Mama's voice became shrill. In it, I heard the fear of the mother who had lost so many children before Woferl and me. "Your two other boys have fallen ill. Soon we will all have the smallpox. This will be on your shoulders, Herr Schmalecker."

Over their arguing, I could hear the wails of Herr Schmalecker's stricken children coming from somewhere downstairs.

Papa looked at me. His eyes held a silent warning. "Nannerl," he said. "Go sit with Woferl in his room. I will come get you when I'm ready."

I nodded without a word and headed to my brother's door.

"What has happened?" Woferl asked me as soon as I stepped in. He sat unmoving on his bed, his head turned in the direction of Mama's voice. He looked startled to see me.

"Herr Schmalecker's youngest daughter had the smallpox

shortly before we arrived," I replied. "The others woke up feverish this morning."

Woferl searched my face with blank eyes. He looked distant this morning, his soul somewhere far away. I sat down on the corner of his bed and frowned at him. "What is it, Woferl?"

He shrugged. His vacant stare turned to the window, as it had for the past few days. "Hyacinth was in my room last night," he said. "He stood in the corner and watched me."

I tensed, my fingers closing tight on his blankets. He found us. "Why was he there?"

Woferl didn't answer. Perhaps he didn't know. Instead, he looked back down at the papers spread out on his bed, then pressed his hands to his ears. "I cannot concentrate," he said. "There is too much screaming."

I worked on my composition late into the night, urged on by the fear of Hyacinth watching my brother. *The song of your heart*, Hyacinth had asked of me. I flipped through the pages and listened to the music in my mind. It was a path that extended nowhere, long and winding, forever heading toward a place I might never see. I wrote and wrote until my eyes strained from the low light.

Outside, I could hear the sounds of horse hooves clattering against the cobblestones, the shouts of people as they carried their luggage to the carriages and prepared to leave Vienna. Still other voices were ones of terror, voices calling out for doctors to visit their homes, to see to family members that had fallen ill. I tried to shut out the sounds. They rang in my mind, tearing apart my thoughts.

Finally, when the moon rose high in the sky, I stood up gingerly and crept to my door. I did not know what I wanted to do. I simply did not want to stay in my room any longer.

I walked silently over to Woferl's door, then opened it and stepped inside. He had fallen asleep amidst the strewn papers of his composition, and his dark hair framed his face in wayward curls. His cheeks looked flushed. I closed the door behind me, then walked over to his bed and crawled in next to him. I hugged him to me. He stirred a little, then instinctively huddled closer to me and let out a sigh.

I tried to remember him as a tiny boy, when his fingers were still small and fresh and chubby, and his face was eager and innocent. I lay awake beside him, caught in my own emotions.

I had not stayed with Woferl for an hour when Papa suddenly burst into the room. I bolted upright, disoriented in my weariness.

"Papa?" I said.

His face was grave. He hurried over to the bed and began to wrap Woferl up in his blanket. My brother whimpered, then rubbed at his eyes even as Papa threw a coat over him. "Go back to your room, Nannerl," he said to me. "I will speak to you in the morning."

I watched Papa nudge my arm away from Woferl and pick him up. A sudden panic hit me. "Where are you going? Where are you taking Woferl?"

Papa ignored me, then stood with Woferl in his arms and promptly left the room. Through the open door, I saw Mama standing at the top of the stairs. Without waiting any longer, I swung my legs over the edge of the bed and rushed out to the hall. Papa had already started down the stairs. Woferl looked up at Mama and me with sleepy, startled eyes.

I put my hand on my mother's arm. "Mama, where are they going?"

"Hush, Nannerl," Mama said. Her face looked drawn, and full of fear. I looked quickly from her to Papa's back, and then to her again. "Your father is taking Woferl to a friend's home. He will be safer there."

"Safer?" I furrowed my brows. He was taking him away to a place where I could no longer watch over him. Hyacinth would find him and steal him in the night. The certainty of it clawed at me. "What about us?"

Mama looked at me. "We are to stay here," she answered.

I could not believe it. Instinctively, I broke away from her and started running down the stairs.

"Nannerl!"

I ignored Mama's calls. Papa and Woferl had headed out the front door by the time I reached the bottom of the stairs. I stumbled on one of the steps, then pulled myself upright and ran out toward the street. Herr Schmalecker and his wife stood in the living room and watched me go.

A coach was already waiting for Papa. I hurried to him before he could reach it, and with a strength borne from another world, I grabbed his arm in a tight grip. In that moment, I realized I wasn't angry with him for taking Woferl. I was angry because he was not taking me.

"Papa!"

He turned around to glare at me. "Go back inside," he snapped. "Do not stand out in the street in nothing but your nightdress."

"Why are you leaving us? Take us with you!"

"You cannot come," he said. He turned away from me and helped Woferl into the coach. "Stay here with your mother."

"Why?" I demanded.

"Woferl is in the gravest danger. You should know that,

Nannerl." Papa prepared to step into the coach. "His frail health cannot last in this house. A friend has agreed to let us stay with him, at least until the threat subsides. He lives near the edge of the city. He will only take two of us. The times are dangerous enough as they are."

"Why can we not leave Vienna?"

"You know very well why we cannot leave yet."

I realized that I had started to cry. When Papa turned away from me again and made to get into the coach, I grabbed him again and pulled him away with all my strength. "I'm frightened, Papa," I said, fighting to keep my voice steady. "How can you leave us behind and take only Woferl? What if we fall ill? What will happen then?"

Papa grabbed my shoulders and shook me once. "Your mother has come down with smallpox once before—she should not be harmed. You know how delicate your brother's health is. What will happen to this family if something were to happen to him? Have you ever thought of that?"

"And what if something were to happen to *me*? I can do everything that he can!" I had started to shout my words now. I no longer cared. "*I* can take care of our family! There are those in the audience who love me too, and who I can please. We are the same, Papa! Why do you not take *me* with you?"

Papa slapped me. I gasped, suddenly dizzy, and touched my cheek with my hand. "You are a selfish girl," he said. His eyes burned me. "Go back inside. I will not tell you again. Wait for me—I will come back for you and your mother." With that, he turned away one last time and stepped into the coach.

I watched as they pulled away. My hand stayed against my cheek. When I felt my mother touch my shoulder, I flinched and started to hurry back into the house. I ignored the looks that Herr Schmalecker and his wife gave me.

"Nannerl, darling!" Mama called out from behind me. I did not turn around.

Instead I ran up the stairs, then into my bedroom, and then to my bed, where I pulled my music notebook out from underneath my blankets. Hyacinth's smiling face appeared in my mind. *Leave him here,* his whisper reminded me. It was still something I could do. The side of me that believed this surged against me, dark and tempting. The light in me struggled against it.

I needed to return to the kingdom, to undo what wrong I'd done. But it was still not too late to let Hyacinth follow through with what he needed in order to fulfill my wish. It was not too late for me. I walked to the clavier, placed the notebook on the stand, and sat down. My wish came back to me now in a terrible wave. I saw my brother's flushed cheeks, his sleeping figure surrounded by music. I saw myself, walking down a path toward a place I could never reach.

I opened the notebook to the composition of my heart and began to play.

The Chosen Path

That night, I went to sleep in a haze of fear and grief. The music of my composition haunted my dreams. When I woke, I could still hear the measures I'd played so feverishly on the clavier, the notes hovering in the air.

How Hyacinth would come to me now, I couldn't say. What if he had found some way to trick me again? Perhaps all he needed from me was to hear my composition. Perhaps he didn't need me to bring Woferl to the kingdom.

Without my composition to work on, without Woferl at my side, all I could do was spend the day pacing. Awaiting word from Papa. Listening to the constant commotion in the streets. Letting my thoughts spiral deeper and deeper.

Mama and I did not attend church that Sunday. Finally, the day after, Papa came to visit us. I rushed to see him, anxious to ask about my brother, but I did not meet my father's eyes when I reached him. I simply curtsied, and then stood with my gaze pointed down.

"Woferl has developed a cough," he said to my mother. "It is nothing serious yet."

A cough. My hands trembled against my dress.

"How long do we stay in Vienna?" It was always Mama's first question.

"The emperor has not responded to my inquiries," Papa said. He looked defeated. "The archduchess is very ill. We will leave Vienna."

That was it, then.

We packed our things in a silent hurry, bid farewell to Herr Schmalecker and his wife, and headed into our waiting coach. When we went to the home of Papa's friend and helped Woferl into the seat beside me, I saw that my brother's eyes had turned so dark that they looked black.

I was in a city, Nannerl. His dream came back to me now, and I shivered at the truth of it. *It was burning to the ground. The fire nipped at my skin, and the smoke blinded my eyes.*

I took my brother's hand in mine and squeezed it tightly. "How is your cough?" I asked him as we headed on our way. Behind us, I could hear the city's cacophony of church bells and prayers and panic.

Woferl shrugged. Already, he looked suspended between here and somewhere else.

"It is just a cough," he replied.

We left Vienna and their royal family behind us, then arrived in Olmütz, a small city on the edge of the Morava River, on a day full of rain. I sat opposite our father, although neither of us looked at the other. Papa was not a man of many words, but today he seemed even quieter than usual, and his lips stayed locked in a tight line across his face. He kept his eyes turned toward the windows. Once, when I looked away and could see him only through the corner of my vision, I thought I saw him stare at me. When I turned my eyes back to him, he had returned to his silent study of the rolling terrain.

The room at our Olmütz inn did not help Papa's mood. When he smelled the dampness of it and saw the smoke that poured from its stove, he threw his hands up and cursed loudly. "God has punished my greed," he muttered.

Woferl's cough grew worse from the rain and smoke, so that he kept us awake throughout the night with his fits. I could not sleep, anyway, as the smoke forced my eyes to tear unabated.

I held my breath for much of the next day. There was no clavier here, not even separate rooms for us. I had nothing to do, nothing to distract myself. All I could think of was the music I had played, of when Hyacinth would come calling for me, and the smile in his raspy, haunting voice.

Woferl continued to cough. His black eyes watered without pause.

The next morning I awoke to the sound of our door slamming shut. My father had left.

"Where is Papa going?" I said to Mama as I sat up. Woferl was not sleeping at my side.

"Hurry and get dressed, Nannerl," she said to me. Behind her, Woferl swayed on his feet and shivered in his clothes. "We are moving to better rooms."

We switched to a room with less dampness and smoke, but

by now it was already too late. Woferl had trouble breathing properly that evening, and by the time Sunday came and we were to attend church, Woferl had become delirious with a high fever. Mama hovered over him, distraught and teary-eyed, and Papa told her he would ask the cathedral's dean about my brother's condition.

I already knew what would happen, although I did not say this to my parents. Hyacinth had found his way to my brother.

The dean, an old friend of my father's from Salzburg, sent the doctor Joseph Wolff to our inn straightaway, and confirmed that Woferl had smallpox. We moved again to the dean's house. There, under the surveillance of Herr Wolff and my family, we watched helplessly as Woferl's fever worsened and his eyes swelled shut with pain.

That night, I dreamed again of the clavier sitting on the dark sands of the kingdom's shores and of Woferl's milky, vacant eyes. I woke with tears streaking my face.

Woferl woke up crying one night, and as Mama rose from her slumber in a nearby chair and hurried to his side, he told her that he could not see. Even candlelight hurt his eyes so much that he kept them closed all the time. Red spots began to appear on his skin, slowly at first, and then more and more quickly, like a wildfire to an untouched forest. I could hardly recognize him through the smallpox rash. Whenever he burst into a fit of coughs, I thought it sounded like Hyacinth's laughter. I would look for him at night, but he did not appear to me.

I woke the next night, trembling. From my open door I could see candlelight still flickering in Woferl's sickroom. I rose then, wrapped my blanket around me, and made my way to him.

Mama slept quietly in the chair at the corner of the room,

while my father lay with his head in his arms at the writing desk. I saw an unfinished letter to Herr Hagenauer crushed beneath his elbow. I walked carefully, so that I would not wake them, and sat down beside Woferl's bed. Through the flickering candlelight and the windowpane I could see the hints of floating shapes, the cloaked figures that seemed to haunt us in a way that others could not notice, waiting patiently beside the glass. I turned to look at Woferl, who tossed and turned in his fitful sleep.

"Nannerl?" he whispered.

I blinked. Woferl suddenly turned his head in my direction, although his eyes—still swollen shut—could not see me. Instinctively, I reached for his hand and pressed it between mine. His skin was hot to the touch.

"I'm here, Woferl," I said.

He tried to smile, but the pain stopped him. "You came to see me," he said.

I swallowed. "Of course," I said. "You are my brother."

"Do you think I will get better? Is the smallpox very bad?"

The weakness in his voice cracked my heart. "It is not so bad," I lied. The shapes outside the window grew larger, so that I could see their bony arms and long, spindled fingers. "The smallpox will disappear in just a few days."

Woferl shook his head. He did not believe me. "I wish I could see you," he whispered. His hand slipped out of mine and reached up for my face. I let him touch my cheek and held it there for him, so that he could feel the coolness of my skin.

My music notebook called to me. I thought I could hear its notes coming from my room, fragments of my composition. A tingle ran through my body at the sound.

Hyacinth. He had come to call. The time was near.

“Nannerl,” Woferl said suddenly. He turned his face to me. “I’m sorry about your compositions.”

At that, I turned sharply back to him. “What?”

“I’m sorry,” he whispered.

I swallowed, afraid of what he might say next. “What do you mean?”

“The six sonatas that Papa took from you. He should not have done that.”

I was silent. My hands pulled Woferl’s away from my cheek so that he could not feel the tremble of my jaw. How long ago was it that our father betrayed me? I had tried to bury it away in my heart, didn’t think Woferl would ever speak aloud to me about it. Now the memory of it all came roaring back, stabbing so hard at me that I winced in pain.

From the corner of my eye, I saw a small movement. A tiny patch of mushrooms was growing on the dresser top, right beneath the light of the candle. They were a shiny black and dotted with scarlet.

Woferl struggled to get closer to me. “I didn’t tell him, you know,” he said. “I didn’t tell Papa about your music. I did not think that he would ever find them, but he did, for he was searching for a pair of cuff links he had lost. I could not stop him from going there.”

He spoke frantically, as if he knew he was fading away. I patted his hand, clucking to him softly so that he would not work himself into a frenzy. “I know,” I whispered. “It’s all right.”

“They are yours,” Woferl went on. “And they are better than anything I could have written.” He took a deep breath. “All I’ve ever wanted, Nannerl, was to be like you. It is still all I hope for. I need you to know. I need you to know.” He repeated it several times, urgent in his fever.

All I've ever wanted.

And suddenly I realized that, here, kept safe within the small chest of my brother, was my wish all along. I'd despaired so much of ever seeing it come true, had spent so much effort turning toward my father for validation, that I'd never taken the time to look in Woferl's direction for it.

It was my wish not to be forgotten, to have a place within hearts when I was gone. To be remembered by the world.

But it was my brother's wish to be like me. He was the one who handed me quill and ink. He was the one who remembered.

Tears blurred my vision. All around us, vines had begun creeping up the walls and around the bedposts, their leaves a glittering black, their flowers tiny and white. *Nannerl,* the whisper came, calling for me. The kingdom had finally come to claim my brother.

Woferl gave me a thoughtful expression. I hurriedly wiped away my tears. Although I knew he could not see them, he seemed to know I was crying.

"You did not look through the final volume," he said at last.

"No," I answered. "How could I? I saw your name printed on the cover."

"You did not see the final copy of *Die Schuldigkeit*, either. I remember you walked out of the room, complaining of the air."

I thought back to the oratorio we had written together. "I did not have the strength to see your signature on it."

"I did not sign either with my name, you know. I could not do it."

I continued to look at him, more surprised now than anything. Such a thought had never crossed my mind. "What did you sign them as, then?"

"I signed them *Mozart.*"

I leaned forward. "Just *Mozart*?"

"Yes. For both of us. We are both Mozart, are we not?"

Woferl paused and made a gesture with his hands, as if to write something down. I broke out of my thoughts long enough to see it, then rose and walked over to where Papa slumped on the writing desk. I carefully took the quill and inkwell, my hands brushing past ivy leaves and tendrils as I did, and then a sheet of paper. I returned to Woferl's side. With both hands, I helped him find the quill and dip it in the inkwell. He touched the paper, then pressed the quill down.

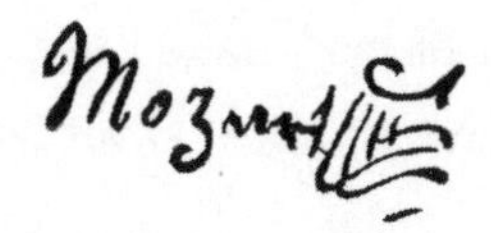

He smiled at me. I was too stunned to say anything in return. I simply leaned closer, then laid my head gently against his swollen cheek. His breathing became shallower, a hissing tide between lulls. I hummed for him. He tightened his grip on my hand.

The glow of blue fireflies had begun to flood the room, darting impatiently from one place to the next. This was the moment Hyacinth had been waiting for. I could hear him calling for me, the music of my composition seeping into the air and his whispers accompanying it.

Nannerl. My Fräulein. It is time.

Woferl was suspended between two worlds. The time had come to lead him out of this world forever and into the kingdom beyond.

When he had fallen asleep, I took the quill and ink and

placed them back on the writing desk. Then I left the sickroom, returned to my own room, and found the music notebook tucked underneath my bed. I cradled it in one arm. Black mushrooms dotted the floor, but disappeared wherever my feet landed.

I went to my brother's door. Then I passed it by and headed down to the main entrance.

Nothing stirred in the night except the kingdom itself, which had begun to grow faster, its dark grasses lining the steps of the stairs, its poisonous vines and leaves suffocating the buildings. I could hardly feel my bare feet against the rain-soaked pavement of the street.

The woods of the kingdom lined the side of the city, the path into them shrouded in black. I halted in my steps to gather my courage. My shadow wavered under silver light. I looked up to see the twin moons aligned at last with each other, forming a single bright disk in the sky.

Then I stepped onto the path and disappeared into the woods.

The Queen of the Night

THE PATH I TOOK WAS LIT BY NOTHING MORE than slivers of moonlight. The tortured trees of the kingdom sighed in the wind, leaning their bare branches and roots toward me as if to pull me to them. I went on, careful to avoid the dark water pooled near their bases. For a while, I couldn't be sure where I was heading. The path could have led down to the white shores, toward the hidden grotto where the trapped queen lived. Or to the castle, where I would meet Hyacinth.

I tried to turn in the direction that I thought would take me to the beach. My feet padded quietly down the winding path. My breaths came shallow and swift. What if Hyacinth kept me from going there? What if he appeared at the end of the path, waiting for me?

The piece I had composed played on the night air, a melancholy melody that drifted between the trees. There were no faeries lighting the path tonight, for all of them must have abandoned the woods to join Hyacinth at the castle. I was grateful for their absence. If one were here, it would surely tell Hyacinth the news of my presence. But he was distracted by the festivities he was throwing tonight, waiting for me to bring my brother to him.

At last, when I thought I could go no farther, the woods ended, and the path led out onto the shore of white sand. With a start, I realized that the kingdom had permitted me to take the path that my heart wanted to follow. And my heart had led me to the trapped queen.

The ocean was no longer the calm blue I remembered. Now it was so dark that I could no longer see the sand sifting at its bottom, and when I dipped a toe into the water, it no longer felt warm but as cold as the winter sea. I sucked my breath in sharply as I waded in, letting the icy water shock my skin. A short distance away rose the rocks beneath where the grotto lay.

I glanced back once at the woods behind me, half expecting to see Hyacinth waiting for me on the shore, his head tilted at me in expectation. But he was not there.

I turned back to the ocean, took in a deep breath, and dove.

At first, I could see nothing. The water swallowed me whole, pushing against me as I swam deeper, my arms searching for the rough surface of rock. I went on and on, until my lungs began to burn. Had it taken this long for Woferl and me to find the grotto's entrance when we last entered it? Had it only been a nice memory, the warm, sweet water and the glowing cavern?

What if it was no longer there? Perhaps I was too late, and the Queen of the Night had perished alone.

Just as I thought my lungs might burst, my hands scraped along rock that curved inward into a tunnel. I pushed myself frantically through the black water, reaching blindly, until I hit the end of the tunnel and felt it curve sharply upward. My legs kicked with the strength of my last breath.

I surfaced with a terrible gasp.

The cavern had grown darker since the last time I saw it. The blue flowers that had draped down from the cavern ceiling in sweet garlands, filling the air with their heady scent, had withered and died, leaving behind their shriveled shells. The night flowers that crawled along the walls, lighting the space with their blue glow, had turned scarlet as they died, their skeleton husks littering the cavern floor with an ominous red hue.

I swam toward dry ground. As I went, the silhouette of a figure hunched against the rock walls, her head in her hands, came into view.

Her shoulders shook as she cried. Her legs were still melted into the cavern floor, trapped eternally there. Her wings looked even more tattered and faded than I remembered, hanging limply against her back—but tonight, there was a golden glow about her, as if some remnant of magic were stirring in her blood.

The twin moons. Their alignment. I remembered that this would be the night when her power would be at its height, and then I recalled the Sun's love for the queen, how he had bestowed her with the magic of his fire.

She did not look up at my approach until I pulled myself out of the water. It must have been the sound of my dripping against the rock that shook her out of her reverie. Her face jerked up, and her dark gaze locked straight on to mine. There was no white in her eyes at all. Suddenly I remembered Hyacinth's old warning to me, that she was a witch who was not to be trusted, and I felt myself yearning, even now, to heed his advice.

Then her sobs quelled some as she took me in, tilting her head this way and that. At last, a glint of recognition appeared.

"You tricked me," she said. Her blue lips curled into a snarl as her voice echoed off the cavern walls, repeating the words over and over. *You tricked me, you tricked me.* "Hyacinth's little Fräulein."

I forced my hands to stop trembling and myself to move forward. "He tricked me too," I whispered. "He told me that you were the Queen of the Night, but not that you were once the queen of Back."

At my words, she froze. She eyed me suspiciously, as if not quite believing me, and for a moment I thought that perhaps she didn't remember her past at all.

Then she said, "How do you know this?"

I could barely force the answer from my lips. "Because Hyacinth entered the tallest tower of the castle and killed the princess confined there." There were tears in my eyes now. "Because I did not know any better, and helped lead him there."

The queen's suspicion changed to shock. In that shock, I suddenly saw not a faery, nor a creature, but a woman who'd once had a son and a daughter. Her dark eyes blinked, turned moist, filling up until fresh tears ran down her cheeks. It had been her daughter in the tower, and the realization made her crumple there, defeated.

I waited, frightened, for her to unleash her wrath on me. Instead, she looked up at me with a sad gaze and shook her head. "He tricked you," she said. "Just as he'd once tricked me."

"What do you mean?" I whispered.

"The girl in the tower was you, child," she said. "You still live, just as your brother does. But Hyacinth will take you both tonight, if you are not careful."

Both of us. I trembled, struggling to understand her.

If Woferl was the princeling of Back, then I was the princess. It was why I saw so much of myself in the girl trapped at the top of the tower, how I'd felt like I was looking into a mirror. Perhaps it was even why I seemed to *feel* the pain of Hyacinth's teeth sinking into her in that moment, why I woke with visions of blood staining my hands.

She was me, and I was her.

Hyacinth had devoured the part of my soul trapped in that castle. What he really wanted now was the rest of my heart. The entirety of me. And after I brought him my brother tonight, he would let the illness overcome me and take me with him too.

"Wicked souls always seek to trap us," the queen told me. Her voice was so lyrical, so sad in its sweetness, that I could feel the crack it made against my heart.

"What did he do to you?" I whispered.

"I was a young queen who loved her husband and was eager to rule her kingdom. Oh, I had so many ideas! The king would sit and listen to me for hours, writing down all I wanted to do for the villagers. Give food and homes to our poor." Her eyes shone for an instant with the past. A wistful smile played on her lips. "And then, in the woods, I encountered a young faery."

I could see it now, the queen's first encounter with Hyacinth, how she must have been as hypnotized by his charms as I once was.

"He cast a spell on me and led me farther and farther from home. When I tried to find my way back, I only stumbled upon the white sands of this shore." She looked away. "He imprisoned me in here, cursing my legs to forever be trapped as part of this cavern, until the day someone came to free me."

She turned her eyes up to me again. The glow around her pulsed with a life of its own. "Here I am. And here you are. Have you come to free me? Or are you his messenger again, to put me out of my misery?"

I stared back at her, remembering her fury and frustration the last time I'd seen her.

"Is it possible to find what you're looking for?" I finally asked her. "Is it possible to get what you want?"

"These are questions I cannot answer for you, child," she replied. "But we must still try."

My gaze shifted to the night flowers growing along the wall. There were only a few left now, dying because the spirit of the queen was dying as well. I walked closer to one and ran my finger delicately along its enormous black petals. It cast its scarlet light against my skin.

Fill the night flower with water, the queen had said to me when I last stood in this grotto. *Pour it on my feet. Free me!*

I closed my fist around the stem of the flower. I pulled hard. The stem cracked, the flower coming free into my hand. I walked to the grotto's pool and knelt over it, filling the flower with water. Then I returned to the queen and held it over her feet.

"Perhaps," I said, "we should have helped each other all along."

The Return of the Queen

From a distance, we must have looked like a timid pair, the queen and me. She walked behind me, her form small and fragile in a riding cloak. Beneath her hood, I could see nothing but the line of her lips. But there was a strength about her tonight. When she looked at the night sky, to where the twin moons hung aligned, her shoulders straightened and she tilted her head up as if to soak in the sight. The light of the moons was a reflection from the Sun, I realized, and even this small bit of heat seemed to feed her heart. I could feel the warmth emanating from her skin, see the yellow glow growing around her, highlighting her features underneath the cloak.

Her breaths quickened as the distance between us and the castle shortened. When the first tall spires began to peek through the trees, she paused in her tracks, as if she could no longer bring herself to move forward. I stopped to look at her.

She had not seen her kingdom since it had first fallen. Her

memory of this place was one steeped in beauty, filled with the love of her people and the affection of her king. Now it was emptied, the square no longer packed with smiling crowds or bustling merchants, the moat filled with dark water.

She stayed still for a long time, seemingly lost in thought. I wondered if she didn't have the strength to go any farther.

Then she took one step, and another. She came to my side and we walked together, our strides even. The glow around her strengthened the closer we drew to the castle.

As we reached the thorned bridge, the stems seemed to shrink away in fear from the heat that radiated from her. I gritted my teeth and continued to move forward. In my mind, I pictured my brother's blinded face, his weak gasps on his deathbed. The bridge trembled as the queen's bare feet walked across it. But she did not slow in her steps, and the thorns did not give way. They held together until we had reached the other side. Then the thorns, seemingly weakened from her magic, finally crumbled, falling into the churning waters below.

Hyacinth was already standing at the front gates of the castle, waiting for us.

His once-lithe body was now stripped of color, tall and sinewy like a creature of the forest, and his once-boyish cheekbones and delicate features had now grown so angular that he looked nothing like a human and every inch a faery. Perhaps this was what his appearance had always been, and I had simply never seen the real him.

His glowing eyes stayed fixed on me as we approached. He smiled as I stopped a few steps away from him. His gaze darted to the figure beside me, veiled behind the cloak. She stayed very still and did not move.

"My darling Fräulein," Hyacinth said to me. He drew closer. All around him tittered his ever-present faeries, their blue glow dancing from spot to spot. They whispered harsh, eager things at me. "You've done so well. You've brought him, as well as yourself."

As well as yourself. I stared into his lying eyes and saw the hunger there. The queen's warning echoed in my mind. He did not care if my wish was fulfilled. He would take me tonight, along with my brother, and neither of us would return to the world beyond.

I looked behind us. The path we'd come from had now closed entirely, the thorns cutting off the bridge and the moat.

"I'm here, as you asked," I said slowly.

Hyacinth's eyes darted again to the cloaked figure beside me. She stood so calmly. For the first time, I sensed in him a hint of doubt. His faeries flitted about, irritated and skeptical. Hyacinth lifted his face to the sky, closed his eyes, and took a delicate sniff. Then he looked at me again, and when he did, his pupils were narrowed into slits.

"Your brother?" he whispered to me.

I looked back at him as steadily as I had once looked at my father. I realized that I was not afraid now. When I didn't answer, Hyacinth swiveled his attention back to the cloaked figure and peered into the darkness that shrouded her face. His eyes then went to her hands, to the faint golden glow that came from her palms. When he peered more closely under her hood, he noticed the warm light against her features.

That was when the first hint of fear showed on his face.

"Who is this that you've brought with you?" he whispered to me.

I didn't move from my spot. I only looked to the figure at my side as she removed her hood.

"The queen," I replied, "the one who truly belongs here."

He took a step back. A stricken look came onto his face, replaced quickly by anger. In it, I saw a thousand realizations—who I'd brought before him, who had freed her, what she wanted.

The queen stared back at him with an unflinching expression. A small smile tilted up the edges of her lips. She was taller now, her bearing more regal. I wondered how I'd ever mistaken her for anything other than a queen.

"I thought we had a bargain," Hyacinth said to me. There was real terror in his eyes now. "Bring your brother to me, when the time has come, so that he may take his place in the kingdom. You betrayed me."

"My brother is on his deathbed," I replied, finding my strength, "because of you. If I'd brought him here today, you would keep him here eternally, so that he will disappear from my world. You would do the same with me."

"I am your guardian, Nannerl, not your demise."

I narrowed my eyes at him. "Everyone always thinks they are protecting me."

His mouth twisted into a grimace. His faeries flitted wildly, unsettled and angry. He did not like the look of understanding on my face. "Don't you want your brother gone? Isn't this what you've always wanted?"

Once, perhaps, when I didn't understand myself, I'd wanted it.

The queen stirred then, and Hyacinth backed uneasily away from her. She fixed her intense gaze on him and refused to let him look away. "The last time I saw you, you came to me with your glowing eyes and a charming smile on your face," she said. "You led me away from my children, and into a cavern where you imprisoned me."

Hyacinth growled, a low rumble that began in his chest and rose through his throat. "Stupid queen," he said, then glanced at me. "Stupid girl. All your life, you wanted nothing more than to stand tall next to your brother. Now you will be reduced to nothing but a brief mention in history. Perhaps not even that. And for what, my darling? Because you're afraid to harm your brother?"

I kept my face resolute. "Because I will not make a bargain with a liar. There are too many lies in my life."

His eyes slid anxiously to the queen again. Suddenly, with his persuasion taken away from him, he seemed weaker, his figure less menacing. The queen stood so tall that I couldn't even remember how she'd looked in the cave. Her skin began to glow with gold. Every line of her looked regal, unflinching and unafraid, finally ready to face the one who had brought her so much misery. The warmth from her wrapped around me in an embrace.

"You are in a castle where you don't belong," the queen said to Hyacinth. As she spoke, the castle stirred and sighed beneath its ivy-choked walls and soot-stained paths, as if remembering its mistress's voice. "Go back to the woods and torment us no more."

Hyacinth sneered at her, but already the castle was changing, revitalized by the magic of her warm presence, and as Hyacinth stood there, the thorns and ivy that had started to choke the courtyard walls began to crumple away. I heard the echo of laughter from long ago, the merry voices of villagers who had once strolled this place.

Hyacinth's smile reappeared. To my horror, his eyes were shifting . . . molding into something that looked surprisingly like my own eyes. "Little noble lady," he taunted. "So abruptly changed. But it is too late for you. You have made your choice, and you have decided to be forgotten."

This was a final lie. It was not too late yet.

Beside me, the queen lifted her glowing hands. Hyacinth shrank back in terror. His faeries darted away in a uniform wave.

"You're afraid of the light," I said to him. "Of warmth. Fire. Life."

"You will not do it," he said. His voice had turned into a whimper now as he looked between the queen and me. "You know I am your only chance to fulfill your wish. We have always helped each other, Fräulein. If you turn away from me now, there is no coming back."

"I've had enough of your temptations," I replied. "You are not the guardian of my destiny. I have already found my own way. You will not take my brother, and you will not steal me away to die."

"Everyone dies," Hyacinth said. He laughed, a high, nervous sound. "But not everyone, my darling, will be remembered."

I thought of what I'd written, the sonatas published under my brother's name. I thought of our oratorio, the measures of my own that I had kept. I thought of my brother's wide, admiring eyes, the way he would imitate my style, my composition, my music. I thought of his last words to me, his small voice, his hand in mine. It was my wish, in a form I could only now recognize.

All I've ever wanted was to be like you.

Perhaps I would never be remembered in the same way as my brother. Perhaps, in the world's eyes, I would never be what I wanted to be. Perhaps the only one who would ever hold me in his heart would be Woferl. But when I was gone, my work would survive, immortalized on paper, embedded in my brother's mind.

Locked away inside me, carried on through him. No one could take that piece of my soul away.

"What you offer me," I replied, "I have already achieved."

Hyacinth lunged toward me. The queen stepped forward, her arms outstretched, to protect me. The glow of her hands flashed a brilliant golden light, as bright as the Sun itself—and all at once, the entire castle seemed drenched in heat. Fire engulfed the dark grass near my feet, eating it away in great gulps. The queen lifted her arms to the sky, and the flames before us surged at her beckoning.

Hyacinth shrieked in anger and fear. Fire raced in a ring around me and swallowed the crooked black trees, the winding path, the vines and ivy and leaves, the clusters of mushrooms. It devoured the faeries in its path, the ivy staining the walls, the soot-charred stones. It devoured the ghosts of the past and the weight of the air. It fed on the dead silence of the castle, filling it instead with the roar of flames.

Hyacinth tried to run. He leapt over one column of fire, then another. For a moment, I thought that perhaps we would not be able to trap him at all, that he would end up escaping still into the woods, until the next time a poor fool crossed his path and he decided to use their lives for his pleasure.

Then the flames caught his arm. Hyacinth yelped, dancing in agonized fury amidst the flames and burning trees. His skin melted in the heat. His screams grew higher and higher. I watched as the flames ate away at his figure until he was no longer a tall, foreboding figure, not even the shy and mischievous boy I'd first seen so long ago, his eyes large with fear and his wide mouth twisted into a smile. He danced as he died, his body a column of fire raging in unison with everything around him.

Fräulein! he called to me as he went. *Help me!*

And even now, in spite of everything, I could feel the pull of his presence against my heart. But the queen and I watched in silence, until that pull weakened and weakened into nothing.

Then the fire engulfed him, and he at last turned to ash.

Before us was an empty castle, cleansed of its poison, drenched in light. The strange music that had always permeated the kingdom, the wind of Hyacinth's whispers, was gone now. In its place lingered something different. A sound as sweet as the earth, made not of magic but of something real and warm and alive. The music of a heart.

In the sky, the moons had begun to set. For the first time, I saw the beginning of a glow at the horizon, the first hour of dawn before sunrise. I stood transfixed by the pink streaking the sky.

The queen finally turned to me, her eyes steady again. She was no longer a cursed witch, but a human, her faded wings now transformed into her velvet cloak.

I didn't know what to say to her. What could I? I had let her stay trapped in her prison for so long. But when I couldn't speak, she did.

"Now I am free," she said. "And so will you be."

I didn't answer. I would return to my world, where Woferl would publish music and I would not. Where my future had already been laid out before me, a path that I could not hope to change.

The queen seemed to see my thoughts in my eyes, for she leaned forward and touched my chin. When she replied, I heard my mother's voice. "It is a long battle to fight," she said, "but you must still fight it. Speak for those less fortunate than yourself, who will need your help. Speak for the ones who will come after you, looking to you for guidance. Stay true, daughter. One day, you will see it all go up in flames."

She smiled at me, then turned back to her empty castle. Already, I knew she would transform it, change this broken place into something worthy again. Already, I knew I would never be able to return.

I turned my back and walked away. The thorns were gone, as was the moat. I followed the path until the streets of Olmütz returned and the cathedral reappeared before me. The fire left behind an abrupt silence. No traces of the kingdom remained. Only a few streaks of ash smeared against the street, already being washed away by a light drizzle.

I wrapped my arms around myself and began the journey back to our house.

The End of the Beginning

When spring arrived again in Salzburg, and the fear of the smallpox had long since faded, my father decided it was time to begin touring again.

I saw the carriage waiting on the Getreidegasse. For a moment, I stayed in the music room, seated on the bench of the clavier, tidying the white layers of petticoats that peeked through my blue silks. Down below, Mama looked on as the coachman helped Papa drag the last of his and Woferl's belongings into the carriage boot. They were headed to Italy, where my brother would play for the Hapsburgs and the Roman public.

The clavier, usually occupied in the mornings by Woferl, sat unopened and covered with a white cloth. I had not touched it in several weeks. Over the winter, I'd spent less of my time in this room and more time with Mama, reciting poetry with her and learning how to stitch a lace pattern.

Now I sat at the bench and ran a hand lightly across the instrument's covered surface. My hair hung loose about my shoulders, waves and waves of it, untouched and unruly. I smoothed it back as well as I could, then pushed it behind my shoulders with a few pins. It was not unlike the style I'd worn so long ago, on the bright autumn day when a court trumpeter had come to listen to me perform. I had been eight years old then.

I had turned eighteen in January. My years of performing before an audience were over.

Finally, when I felt ready, I rose from the bench. On the Getreidegasse, I saw my brother tilt his head up toward my window. He waved a hand at me. I smiled at him, then headed downstairs.

The air was warm today, the breeze ruffling the curls of my hair. I made my way to where my brother stood alone. When he heard my footsteps against the cobblestones, his eyes lit up and he ran at me, wrapping his arms around me in a tight embrace.

"Woferl," I said, laughing. "You are such a child, to run at me like that."

"I don't care," he said. "I will miss you. I'll write you letters, of course, and tell you everything that I see. You will feel as if you are right beside me."

I smiled at him. He had been growing steadily all winter, his limbs turning thin and awkward. Pockmarks lingered on his face from the smallpox, forever prominent, but through them I could still see the face of a young boy, at once too naïve and too mature for his age. "I will look forward to them every day," I said. I touched his cheek. "Tell me everything, Woferl. Even what you eat for breakfast."

He laughed. Behind him, Papa and Mama conversed in low

voices with the Hagenauers. They were financing part of this trip, and I could tell in Papa's gestures that he was thanking them for their continued generosity. Again, our rent was delayed. It was our endless state of being, teetering on the balance scale of the world, hoping always for better tidings.

"You will be safe here, with Mama?" Woferl asked. He stepped closer to me so that the others would not hear him.

I had told him, after he'd begun his recovery from the smallpox last autumn, what had happened to the kingdom on that night in Olmütz. That the kingdom was consumed by fire, that it was gone and had been rebuilt, and that we shouldn't talk about it anymore. He had taken it all in stride, as if the end of my imagination of it was the end of his as well. Since then, I had not been visited in my dreams. Neither, I think, had he, although he did not speak of it. There were no more visions of edelweiss growing on sheet music, or silhouettes of faery creatures waiting in our music room. There was no more magic permeating our lives, aside from the magic of the real world. Of music, his and mine, real and true.

"We will be safe, I assure you," I told him.

Woferl looked down. "Promise me you will write me too, and tell me everything. Send me your compositions. I hope you continue to write them down. I swear to you that I will not let them end up in our father's hands."

"I will send what I can." I opened my arms to Woferl and hugged him tightly.

Woferl's voice sounded muffled against my dress. "I've never been without you," he murmured.

I held him to me for a long time, savoring his embrace, and said nothing.

When Woferl finally released me and climbed into the carriage, I walked over to stand with Mama and said my goodbyes to my father. He patted my cheek and touched my nose with the tip of his finger.

"Be good, Marianne," he said to me. "Take care of your mother."

I nodded. He had stopped calling me Nannerl as soon as I'd turned eighteen. "Have a safe trip, Papa."

He smiled at me. Something sad lingered in his eyes.

For a moment, I wondered if he regretted leaving me behind, that he had also regretted what he'd done in Vienna, that forces outside of his powers made him act as he did. I thought for an instant he could see something in me, and he wished he could have created more with it.

Then it was gone, as always, and he leaned in to kiss my forehead. "I will write to you and your mother," he said.

I stayed at the music room's window long after their carriage had vanished down the Getreidegasse. I sat until the sun had shifted the shadows in the room and my mother called for me to join her. Only then did I rise, smooth my skirts, and leave.

Before I did, I stared out the window one more time and remembered the Kingdom of Back as I had first known it, with its upside-down trees and white sand beach, the little path and the wayward signpost. I remembered that first blustery day in autumn, ten years ago, when it had appeared in my dreams. I thought I could see it again now, a ghostly image imprinted over the Getreidegasse's wrought-iron signs and balconies, the faded castle rising up behind the buildings like a forgotten cloud.

It was the temple of my youth, the representation of so much that I had hoped for. Perhaps it had always existed and would always exist, ready for the next little girl to make a wish.

I did not imagine Hyacinth in the kingdom. I had long ago forgotten what he looked like.

Later that evening, I put away my old music notebook and my broken pendant, storing them in a place where I would not look every day.

Twenty-Three Years Later
Sankt Gilgen, Austria
1792

In February, as I rest in Sankt Gilgen with my husband and children, I receive a familiar guest from Salzburg who is coming to speak about Woferl's childhood. He arrives on a sunny, cold afternoon, right as I am braiding my daughter Jeanette's hair.

I have been expecting my guest. When my husband greets him at the door, he walks in with his usual air of merriness, shaking his hand before turning to me. He is slower now, his bones more brittle. Still, though, he is energetic in the way that he brushes leaves from the velvet of his justaucorps, and turns to smile at me.

I smile back, help Jeanette off my lap, and curtsy to him. "It is good to see you, Herr Schachtner," I say. "Thank you for coming. I hope you've been well."

He looks at me. How much has changed since that first

blustery morning when he heard me play. I am married now, mother to three young children. As for Herr Schachtner himself, he has become an old man, bent from the world.

"Thank you, Frau Berchtold," Herr Schachtner says. He bows to me. "How have you been keeping?"

"Well enough," I say. "Better than before." My words lodge in my throat for a moment before they come free. "It is slowly getting easier to accept Woferl's absence."

He gives me a sad smile and shakes his head. "Ah, I'm glad to hear it." We stay silent for an awkward moment, the consequence of many years apart and the lack of my father's presence. Papa would have known what to say.

Then Herr Schachtner clears his throat and reaches for a chair. "Let's begin, then," he says. "What is it that Herr Schlichtegroll needs to know?"

"He wishes to compile a biography of Woferl," I reply, "and has requested some information from his early life. I would like to have another's voice added to my own, so I thought of you. I'm sure you may remember some things about Woferl that I may have forgotten."

Herr Schachtner nods. Some of his early energy disappears as he begins to think of my brother. "Very well," he murmurs. He has brought with him a stack of papers, old letters and concert announcements, and he starts to sift through them. I bring over a stack of my own, and together we sit to pore over each one.

"Did you have a chance to speak to him before he died?" he asks me after a while, after we'd begun to compile a small list of anecdotes.

I look at him. "No," I say. "I spoke to him once, several years ago, but I did not know of his illness last winter until he had already passed." I pause there, suddenly uncomfortable with a topic that I've already needed to discuss on several occasions. I do not like to remember it. Sometimes I still wonder, on nights when the others have fallen asleep, what had ultimately caused my brother's early death. Woferl had been in the middle of a composition shortly before he fell ill. I never tried to ask his wife what the composition was. I was too afraid of recognizing in it some familiar, ethereal sound.

Perhaps Woferl had always been the boy suspended between worlds, never meant to stay here for long.

"There are still masses, you know," Herr Schachtner says. "All Salzburg mourns for him. I've heard of gatherings held in Vienna and Prague as well, attended by hundreds."

I picture Vienna, a city once plagued with smallpox, now in silent mourning for Woferl. I wonder how grand his mass was, or if it was simple like that of his funeral. I wonder if Marie Antoinette, the little archduchess to whom Woferl had once proposed, would have attended his mass if the French had not imprisoned her in the Tuileries Palace.

Herr Schachtner and I trade stories, some that we both know, some that I have to remind him gently of. I recall how Woferl had picked out thirds on the clavier with me, and his little frown when one of the keys seemed out of tune. Herr Schachtner remembers his fervent composing, even at a young age, and the tears that would spring to his eyes whenever he was forced to pause. I bring out my old music notebook, now yellowing with age, and point out pages where Woferl had composed menuetts or where my father had written notes. When Herr Schachtner asks me about the page torn in my notebook, I simply shrug and tell him I cannot be sure what had happened.

"You and Woferl were so close," Herr Schachtner remarks, when I become carried away in telling one of his childhood stories. A smile emerges on the edges of his mouth. "You were quite the pair, weren't you? You played for the kings of Europe, those who have changed our countries and written our histories."

The memory returns of our jostling carriage rides, the stories my brother and I would make up to entertain each other. I smile too, cherishing the warmth of this nostalgia. "Yes," I reply gently. "I suppose we were."

Herr Schachtner returns to his stack of papers, pulls out the next one, and holds it out to me. "Sebastian, your old servant, had this in his possessions. I found it and thought you might know more about it than I will."

I stare at the paper, momentarily unable to speak. It is the old map that Woferl and I had once asked Sebastian to draw for us, a map of the Kingdom of Back. Some of his sketching has faded away now, and the castle on the hill is smudged and ruined. I look at the little moat Sebastian had drawn, the upside-down trees and the white sand beach. I hear in my mind the crunch of leaves beneath our feet, the splash of water as we swim in the kingdom's ocean. I remember the dark, damp stairs in the castle tower, the scarlet sky and the children and the winding, crooked path.

I do not try to remember the faery's name.

"It was a childhood memory," I say after a while. "We called it the Kingdom of Back."

"The Kingdom of Back?" Herr Schachtner laughs a little. "How did such a name come about?"

Woferl had whispered it to me one afternoon, a long time ago. But to Herr Schachtner, I say something different. The kingdom, and all its secrets, were meant only for my brother and me. "I can no longer be certain," I say. "We used it to pass the time we spent in the carriage and on our journeys."

Herr Schachtner studies my face, as if he knows that there is more I want to say about it. I choose my words carefully, changing the kingdom into something that the rest of the world can understand. "We thought of ourselves as the rulers of this place," I say. "I suppose it was where we could escape to, with our joys and sorrows, and let them out to play." I look at Herr Schachtner. "Just a simple childhood game."

Herr Schachtner nods, satisfied with my story, and moves on to the next paper.

We sit together late into the afternoon. When it finally comes time for him to leave, he promises to visit me again and bring gifts for the children.

"I will let you know how Herr Schlichtegroll does with his writings," I say. "I hope he will portray Woferl as a great man."

Herr Schachtner bows to me. Then he seems to remember something and pauses halfway out the door to face me again. His hand digs into the pocket of his jacket. "I'm sorry," he mutters. "I'd almost forgotten. I have something for you."

I wait patiently.

The Herr pulls out a tiny package for me, wrapped in white silk and tied with a simple ribbon. "His widow, Constanze, told me that she found this among Woferl's possessions shortly after he died. She said that he meant this for you, as he had a little note on it with your name. She asked me to give it to you."

I turn the package over in my hands. Sure enough, a tiny scrap of paper is attached to its bottom. *Für Nannerl*, it says. I look at Herr Schachtner, who holds out his hands to me.

"I've no idea what it is," he says. "But I'm sure he would have liked you to receive it." He bows once again, tipping his hat to me. "Farewell, Marianne. History will remember the Mozart name."

I thank him, curtsy, and then watch his coach leave.

When he is gone, and I am still alone, I return to my seat and open the package. It's very light, as if it holds only air, and for a moment I think that when I open it, the silk will simply fall away to reveal nothing at all, a final bit of mischief from Woferl.

But when I unfold the silk, I find in my lap a seashell painted bright blue, a shell shaped like a near-perfect circle, with flecks of white showing through the paint inside its grooves. Like grains of sand.

"Mama," says a small, sweet voice beside me. I look down to meet the wide eyes of my little Jeannette, who has tilted up onto her toes to see what I am holding. "What is it?"

I smile at her, then lower my hand to show her the shell in my palm. "It is a gift from your late uncle," I tell her.

She studies it, turning her head this way and that at the curious object. Her hair is like mine, dark and wavy, held back in a simple state with pins. "Where did it come from?" she asks.

I am prepared to tell her something brief and careless, so that she does not ask again. Perhaps it is not worthwhile to mention the dreams and fears of our youth, of all that I had experienced. Perhaps it is unwise to trouble such a small girl with the pains of my past.

Then, from some distant place, a memory stirs. It is a whisper in the air, the voice of a mother.

Speak for the ones who will come after you, looking to you for guidance.

I fold the shell back into the silk wrap, re-tie the ribbon around it, and put it in the pocket of my petticoat. Then I pick Jeannette up to sit in my lap. My arms wrap securely around her. She snuggles against me.

"I am going to tell you a story you already know," I say to her. "But listen carefully, because within it is one you have never heard before."

Author's Note

The Kingdom of Back is actually a story I first wrote twelve years ago and have been finessing ever since. I grew up playing piano; Mozart's music always impressed me because it was easy to learn but incredibly difficult to master. And to think that he wrote so much, so young! How was this possible? I found myself constantly drawn to movies, articles, and books about him—but in what I read or watched, there was never any mention of him having a sister. The only hint that she existed lay in occasional paintings I came across online depicting Mozart as a boy or young man, playing the violin while a young woman accompanied him on the clavier. Who was she, and why did she appear so frequently at his side?

It wasn't until I read *Mozart: A Life* by Maynard Solomon (a wondrously detailed book I highly recommend) that I learned Wolfgang had a sister—and not just a sister, but one who both played the clavier with extraordinary skill and composed as competently as her brother. Nannerl, as she was affectionately known, was five years Wolfgang's senior and every bit a child prodigy. Before their father, Leopold, began teaching Wolfgang how to play the clavier, he taught Nannerl, marveling at how quickly she learned. In 1764, Leopold wrote in a letter, "My little girl plays the most difficult works which we have . . . with incredible precision and so excellently . . . although she is only 12 years old, [she] is one of the most skillful players in Europe."

A twelve-year-old girl who was one of Europe's most skillful players. I couldn't believe I'd never heard of Nannerl. Her brother was so celebrated! Yet here was his sister, his equal in talent, almost completely forgotten by history.

I learned that Nannerl and Woferl were incredibly close as children, often performing together as their father toured them throughout Europe. Woferl idolized Nannerl his entire life, as is evident in his letters, and most likely was inspired to play music because of her. It was during my reading of this time in their lives that a tiny detail caught my eye. With nothing to do during the long months they spent traveling in carriages, Nannerl and Woferl invented for themselves a magical place they called the Kingdom of Back. It became their way of passing the time during their often years-long tours, and they became so absorbed with it that they asked their manservant, Sebastian, to draw a map of the kingdom for them.

A world of fantasy and magic, invented entirely by the Mozart children. It was too interesting a premise for me to pass up, and I immediately knew I wanted to write a story around it. As the book evolved, it became a broader tale about Nannerl herself, what dreams and wishes she might have had, and what her compositions might have meant to her. How must it have felt to love something that the world refused to let her pursue? I am a writer, and telling stories is as much a part of me as my heart—I cannot fathom the agony of being barred from writing simply because of my gender. The thought of Nannerl living during a time when she not only couldn't share her compositions but also had to watch her brother take the world by storm . . . it turned her story into a personal one for me.

While there is no conclusive evidence that Nannerl ever composed under her brother's name, there are claims that her handwriting appears in the music notebook that belonged to her and was also used by Wolfgang during his lessons. In fact, a 2015 *Telegraph* article by Jonathan Pearlman reports that an Australian professor may have identified Nannerl's musical handwriting in pieces that her brother used to practice piano. What's more, we know that she composed her own music. In letters exchanged between the siblings, Woferl would enthusiastically ask her to send him her compositions. Could she have lent her hand to some of his work? We may never know for sure, but I'd like to think it is possible.

Tragically, none of her work has survived . . . under her own name, at least.

In the end, Nannerl lived to be seventy-eight years old. Her brother Wolfgang Amadeus Mozart died when he was only thirty-five, but in a way attained immortality through his work.

What legacy could Nannerl have left if she'd been given the kind of attention and access that her brother enjoyed? What beautiful creations were lost to us forever because Nannerl was a woman? How many other countless talents have been silenced by history, whether for their gender, race, religion, sexual orientation, or socioeconomic circumstances?

I wrote this book for the Nannerls of today and tomorrow, in the hopes that when they are ready to share their brilliance with the world, the world is ready to give them the attention and honor that they deserve.

Acknowledgments

I began writing *The Kingdom of Back* as a naïve, shaky twenty-three-year-old, still fresh out of college and finding my voice. I submitted the story to Kristin Nelson in all the wrong ways an aspiring writer could submit it—unfinished, just one hundred pages in, poorly formatted. Kristin saw something in the manuscript and was kind enough to encourage me to send the rest of the story when I finished it. I did, she took me on as her client, and that was the beginning of our partnership. Twelve years later, we have finally come full circle. This was the book that started us on our way, Kristin, and I remain forever grateful to you for being the first to believe in me and this story. *The Kingdom of Back* was always dedicated to you.

To the incredible, inimitable Jen Besser, who gave this book a wonderful home, to the absolutely brilliant Kate Meltzer, who helped shape this book into something a hundred times better than its original form, and to Anne Heausler, for saving my butt with every single copyedit: I cannot thank you all enough.

Immense gratitude to JJ, who read *The Kingdom of Back* over a decade ago. Thank you so much for your support and understanding of my strange, quiet story. Huge thanks to Tahereh Mafi, incredible writer and wonderful friend, who took the time to read the rough draft and give me invaluable feedback. Your wisdom is priceless.

Deepest thanks to Jen Klonsky and the fantastic Putnam/ Penguin team for giving *The Kingdom of Back* so much love and care, from its breathtaking cover and interiors to sharing it with the world. Nannerl's story is so special to me, and it means everything to me to know it is in your good hands. Much love to you all.

To the librarians, booksellers, teachers, and book champions around the world who work tirelessly to put books in the hands of readers: thank you so, so much for your support. Writers can't do what they do without you. I'm forever indebted.

Finally, to all the Nannerls out there. Whatever barriers are in your path, it is my deepest hope that you can shatter them, because we desperately need your talent.

Keep going. Don't give in. Light up this world.

MARIE LU is the #1 *New York Times* bestselling author of the Young Elites, Legend, and Warcross series. She graduated from the University of Southern California and jumped into the video game industry as an artist. Now a full-time writer, she spends her spare time reading, drawing, playing games, and getting stuck in traffic. She lives in Los Angeles with her illustrator-author husband, Primo Gallanosa, and their family.

The Kingdom of Back
The Starfishers' Realm
The Fields of Edelweiss
The Upside~Down Forest
The Witch's Grotto